Introduction to Cisco Router Configuration

Laura Chappell, Editor

CISCO SYSTEMS

CISCO PRESS

MACMILLAN
TECHNICAL
PUBLISHING
U·S·A

Macmillan Technical Publishing
201 West 103rd Street
Indianapolis, Indiana 46290 USA

Introduction to Cisco Router Configuration

Laura Chappell, Editor

Copyright © 1999 Cisco Systems, Inc.

Cisco Press logo is a trademark of Cisco Systems, Inc.

Published by:
Macmillan Technical Publishing
201 West 103rd Street
Indianapolis, IN 46290 USA

Printed in the United States of America   5 6 7 8 9 0

Library of Congress Cataloging-in-Publication Number 98-85495

ISBN: 1-57870-076-0

Warning and Disclaimer

This book is designed to provide information about Cisco router configuration. Every effort has been made to make this book as complete and as accurate as possible, but no warranty or fitness is implied.

The information is provided on an "as is" basis. The author, Macmillan Technical Publishing, and Cisco Systems, Inc. shall have neither liability nor responsibility to any person or entity with respect to any loss or damages arising from the information contained in this book or from the use of the discs or programs that may accompany it.

The opinions expressed in this book belong to the author and are not necessarily those of Cisco Systems, Inc.

Feedback Information

At Cisco Press, our goal is to create in-depth technical books of the highest quality and value. Each book is crafted with care and precision, undergoing rigorous development that involves the unique expertise of members from the professional technical community.

Readers' feedback is a natural continuation of this process. If you have any comments regarding how we could improve the quality of this book, or otherwise alter it to better suit your needs, you can contact us at cisco-press@mcp.com. Please make sure to include the book title and ISBN in your message.

We greatly appreciate your assistance.

Associate Publisher	Jim LeValley
Executive Editors	Julie Fairweather
	John Kane
Cisco Systems Program Manager	H. Kim Lew
Managing Editor	Caroline Roop
Acquisitions Editor	Brett Bartow
Development Editor	Laurie McGuire
	Kezia Endsley
Coordinating Editor	Amy Lewis
Project Editor	Tim Tate
Technical Editor(s)	Merilee Ford
	Doug MacBeth
	Karen Bagwell
Cover Designer	Karen Ruggles
Production Team and Book Design	Argosy
Indexer	Kevin Fulcher

Trademark Acknowledgments

Acknowledgments

Special thanks to Carol Lee for her tremendous commitment to content organization and development of this title. Thanks also to Jill Poulsen, of ImagiTech, Inc., for her assistance with the coordination of this project on behalf of Ms. Chappell. Thanks to Merilee Ford and Doug Macbeth for their time and effort on technical review of this material.

This book is the product of many contributors within the Cisco education department including, but not limited to, Cisco course developers, course editors, and instructors. We would like to acknowledge the efforts of training developers Elizabeth Goga, Bob Martinez, Ilona Serrao, and Diane Teare.

We also are grateful for the efforts of Macmillan Technical Publishing in developing this title and bringing it to press. Specifically, we would like to thank Brett Bartow, Julie Fairweather, Amy Lewis, John Kane, Laurie McGuire, and Kezia Endsley.

About the Technical Editor

Merilee Ford has held various technical training and technical support positions in the internetworking industry for eight years. Since joining Cisco Systems, she has authored or co-authored three multimedia CD-ROM titles. Currently, she is the course developer for the Advanced Cisco Router Configuration course.

Contents

Contents

Foreword

In April 1998, Cisco Systems, Inc. announced a new professional development initiative called the Cisco Career Certifications. These certifications address the growing worldwide demand for more (and better) trained computer networking experts. Building upon our highly successful Cisco Certified Internetwork Expert (CCIE) program—the industry's most respected networking certification vehicle—Cisco Career Certifications enable you to be certified at various technical proficiency levels.

With *Introduction to Cisco Router Configuration*, Cisco Press presents Cisco's most popular instructor-led certification preparation course as a single-volume book. *Introduction to Cisco Router Configuration* is not intended to replace the instructor-led course of the same name. Instead, it supplements and reinforces topics presented in the course.

Cisco and Cisco Press together present this material in a text-based format in order to provide another learning vehicle for our customers and the broader user community in general. Although a publication cannot replace the instructor-led environment, we must acknowledge that not everyone responds in the same way to the same delivery mechanism. It is our intent that presenting this material via a Cisco Press publication will enhance the transfer of knowledge to our audience of networking professionals.

This is the first of many course supplements planned for Cisco Press. Cisco will present existing and future courses through these coursebooks to help achieve Cisco Worldwide Training's principal objectives: to educate Cisco's community of networking professionals and to enable that community to build and maintain reliable, scalable networks. The Cisco Career Certifications and classes that define these certifications are directed at meeting these objectives through a disciplined approach to progressive certification. The books Cisco creates in partnership with Cisco Press will meet the same standards for content quality demanded of our courses and certifications.

It is our intent that you will find this and subsequent Cisco Press certification and training publications of value as you build your networking knowledge base.

Thomas M. Kelly

Director, Worldwide Training
Cisco Systems, Inc.
August 1998

Introduction

As today's internetworks grow and expand to support multiple sites, protocols, and operating systems, the interconnecting devices are the critical elements along the data path. Understanding these devices and how to configure them and integrate them into efficient, reliable network designs is essential to anyone supporting network communications. Cisco Systems, the premier designer and provider of internetworking devices, is committed to supporting network administrators, designers, and builders in the use of its products.

The content, organization, and goals of this book are based on Cisco's highly successful "Introduction to Cisco Router Configuration" course. As such, the book provides a comprehensive introduction to internetworking LANs and WANs using Cisco routers. Technical background and functionality specifications for the most popular internetworking protocols today, including TCP/IP, Novell IPX, and AppleTalk networks, are covered. In addition, the book surveys wide-area networking (WAN) techniques. Throughout, important general principles are balanced with configuration specifics for Cisco routers.

Many configuration examples are included to demonstrate management and troubleshooting techniques for internetworking communications. If you are using this book as a study aid in preparing for one of Cisco's certification exams, you will find the end-of-chapter tests useful. The tests are designed to help you evaluate your understanding of the concepts contained in the chapter and your ability to apply the configuration techniques available for Cisco routers. Chapters also contain sidebars in the form of Tips, Cautions, and Key Concepts to help emphasize critical details.

A follow-up title, *Advanced Cisco Router Configurations* (Cisco Press), provides more advanced details on traffic management and router configurations.

WHO SHOULD READ THIS BOOK

This book contains a broad range of technical details on routing models, processes, and design; it can be used as a general reference for anyone designing, implementing, or supporting an internetwork with TCP/IP, IPX/SPX, AppleTalk, SNA, DECnet and Banyan VINES protocols. If you anticipate taking one or more of the Cisco certification exams, particularly the Cisco Certified Network Associate (CCNA) exam, this book is a logical starting point.

Even if you're not using Cisco routers, this book can increase your understanding of the underlying technologies affecting network communications and security.

PART 1: INTRODUCTION TO INTERNETWORKING

Part 1 provides the foundation of knowledge required to build and configure a multi-protocol network. It examines the various layers of functionality and introduces the startup sequences and configuration options for Cisco router products.

Chapter 1, "The Internetworking Model," introduces concepts that enable us to move from local to global internetworks. The chapter provides an introduction to the communication processes seen in local, national, and international/global LANs and WANs. You'll learn how the data is built, packaged for end-to-end transport, addressed for internetwork routing, and addressed for local transit.

Chapter 2, "Applications and Upper Layers," focuses on the connection-oriented and connectionless communications defined by the transport layer of the OSI model. It also examines higher layer functions such as text and data formatting and conversion; image conversion; and sound and video conversion. Flow control and congestion avoidance mechanisms and are also covered.

Chapter 3, "Physical and Data Link Layers" focuses on the functionality supported by internetworking routers. You'll learn the difference between the Media Access Control (MAC) and Logical Link Control (LLC) sublayers of the data link layer. You'll learn the basic functionality and specifications defined for Ethernet/802.3, Token Ring/802.5, and FDDI networks. This chapter also introduces various WAN technologies including SDLC, HDLC, LAPB, Frame Relay, PPP, X.25, and ISDN communications.

Chapter 4, "Network Layer and Path Determination," focuses on the layer that defines router functionality and compares routing technologies available for TCP/IP, IPX/SPX, and AppleTalk networks. The chapter describes routing problems such as routing loops and the count-to-infinity problem, as well as the available solutions, such as split horizon, poison reverse, hold-down timers, and triggered updates. Link state, distance vector, and hybrid routing protocols are introduced and compared.

Chapter 5, "Basic Router Operations," delves into the Cisco-specific procedures required to start up and configure a router using a console port, auxiliary port, virtual terminals, or TFTP server. This chapter surveys the methods a Cisco router uses to obtain its routing configurations, including RAM/DRAM, NVRAM, Flash, and ROM memory. The process of changing router modes from user EXEC to privileged EXEC mode is also described. The chapter concludes with coverage of how to view the router startup, interface, and protocol status.

Chapter 6, "Configuring a Router," examines the process of loading configuration files and changing router modes. This chapter examines password configurations as well as the steps used to configure an interface, shut down an interface, and verify configuration changes. Finally, you'll look at how to manage the configuration environment through backup images and setup modes.

Chapter 7, "Discovering and Accessing Other Cisco Routers," focuses on Cisco Discovery Protocol (CDP) and its ability to discover other Cisco routers. You'll look at how to use CDP on a local or neighboring router.

PART 2: NETWORKING PROTOCOL SUITES

Part 2 details the most popular internetworking protocols: TCP/IP, Novell IPX, and AppleTalk. In this section, you'll examine the addressing system, service discovery, and routing techniques used by each of these protocol suites.

Chapter 8, "TCP/IP Overview," defines the elements in the TCP/IP stack with particular emphasis on the network and transport layer protocols, Internet Protocol (IP), User Datagram Protocol (UDP), and Transmission Control Protocol (TCP). Related elements of the TCP/IP suites, such as Address Resolution Protocol (ARP) and Internet Control Message Protocol (ICMP), are also discussed since routers typically support these elements.

Chapter 9, "IP Addressing," lays the groundwork for IP addresses that use standard class-based default masks and various subnet masking techniques. Examples deal with how to plan a Class B or Class C internetwork considering future network expansion and the current limitations of a class-based addressing scheme. This chapter also focuses on general and directed broadcasts as defined by the IP address format used. Finally, the chapter illustrates how to use simple and extended ping techniques to test communications between TCP/IP devices.

Chapter 10, "IP Routing Configuration," explains how IP routers learn of network destinations and assign a distance to each network. The chapter introduces and compares the RIP and IGRP routing protocols, and provides configuration examples of each. General elements of interior and exterior routing protocols are also compared in this chapter.

Chapter 11, "Configuring Novell IPX," introduces the IPX network routing techniques and the 10-byte addressing system used on NetWare networks. Service Advertising Protocol (SAP), Get Nearest Server (GNS), and encapsulation methods are examined as they relate to a router's functionality in this environment. Finally, you'll look at how to configure path splitting and path costs and validate router configurations for NetWare networks.

Chapter 12, "Configuring AppleTalk," examines the AppleTalk protocol stack and features, including nonextended and extended networks. The AppleTalk addressing process, service discovery, and network printing are covered.

Chapter 13, "Basic Traffic Management with Access Lists," defines the purpose of traffic filtering and management on LANs and WANs. The chapter explains both standard and extended access lists and provides examples of TCP/IP, Novell IPX, and AppleTalk access lists to control network traffic.

PART 3: WIDE-AREA NETWORKING

Part 3 deals with WAN communications, including serial connections, X.25, and Frame Relay networking. You'll examine the addressing system, link establishment, and routing techniques used by each of these protocol suites.

Chapter 14, "Introduction to WAN Connections," surveys the types of serial communications used today and the elements used for call establishment, maintenance, and authentication. This chapter provides substantial detail on Point-to-Point Protocol (PPP) link establishment, authentication, and configuration verification.

Chapter 15, "Configuring X.25," covers the protocol stack, logical elements, addressing, encapsulation, and circuit types of X.25. Complete configuration details, including X.25 packet sizes and window parameters, are also covered in this chapter.

Chapter 16, "Configuring Frame Relay," focuses on the terminology and operation of point-to-point and multipoint frame relay configurations. The chapter looks at star, full-mesh, and partial-mesh topologies and discusses reachability issues for frame relay communications.

VERSION INFORMATION

This book is based on the Cisco "Introduction to Cisco Router Configuration" course which covers IOS v11.3. Although some references are made to earlier versions of IOS, the examples shown throughout this course are based on IOS v11.3. For more information on Cisco router configuration options and commands, refer to the Cisco documentation maintained online at www.cisco.com.

PART 1

Introduction to Internetworking

The Internetworking Model

Network managers today face tremendous challenges from evolving network demands and capabilities. This chapter opens with a brief history of how networks have evolved and the resulting service improvements to network users. Some constants remain even in the face of rapid network evolution, however, including design principles and the model that networking technologies follow to enable complex devices from diverse vendors to communicate in a network. The second half of this chapter covers these topics, including a review of the International Organization for Standardization/Open Systems Interconnect (ISO/OSI) reference model.

THE EVOLUTION OF NETWORKS

The evolution of networks is in large measure an evolution of functions and capabilities. Each new phase in the evolution of networking incorporates and expands on the functionality (such as communications methods and access speeds) of the previous phase, beginning in the 1960s and continuing to the present day.

1960s and 1970s: Centralized Processing

In the 1960s and early 1970s, computer communication typically was organized in the form of dumb terminals connected to a host (mainframe). The processing power and much of the memory resided in the host as opposed to the terminals (hence the term "dumb" terminals). This centralized computing environment required low-speed access lines that the terminals used to communicate with the centralized host. This networking technology enabled users to access shared centralized data and printer resources.

IBM computers with Systems Network Architecture (SNA) networks and non-IBM computers with X.25 public data networks are typical examples of this type of environment. Figure 1–1 illustrates a simple host-based communication environment.

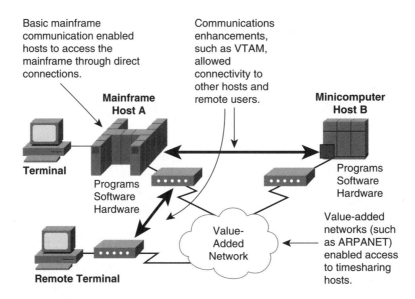

On a single computer, accessing resources, running programs, and copying files are relatively straightforward tasks. The computer must identify the requesting user and the desired destination device or program and then coordinate access between them. The single computer in this scenario is the master of all resources and thus can easily manage and coordinate them.

In a network—even one consisting of only two computers—coordinating resources becomes much more complex. Transferring information requires addressing, error detection, error correction, synchronization, and transmission coordination, among other things.

1970s and 1980s: Networks

The introduction of PCs revolutionized traditional communication and computer networks. Initially, PCs were standalone devices that put processing capabilities and ample memory on each user's desktop. As businesses realized the flexibility and power of these devices, their use increased.

However, some network resources—such as printers and hard disks for memory-intensive applications—were not cost-efficient for every desktop. LANs (local-area networks) evolved primarily to enable sharing of such expensive resources. As such, LANs permitted the combination of the best characteristics of standalone PCs and centralized computing.

The strategic importance of interconnected networks was quickly realized. Organizations began to move toward linking previously isolated LANs, as shown in Figure 1–2. Interconnected networks provided the basis for enterprise-wide applications such as e-mail and file transfer. These applications in turn increased overall productivity and competitiveness.

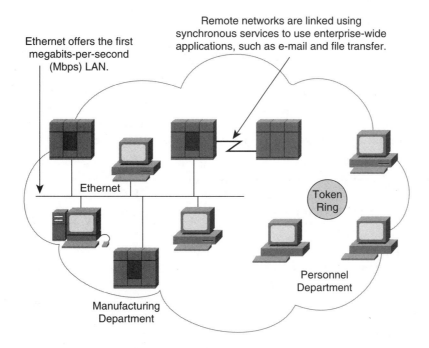

Ethernet offers the first megabits-per-second (Mbps) LAN.

Remote networks are linked using synchronous services to use enterprise-wide applications, such as e-mail and file transfer.

Ethernet

Token Ring

Manufacturing Department

Personnel Department

Figure 1–2
LANs and WANs enabled enterprise communications between multiple departments.

In addition to PCs and LANs, minicomputers and shared WANs (wide-area networks) evolved in the 1970s and 1980s. Minicomputers, often located away from the central data center, facilitated the emergence of distributed data processing, enabling the actual processing of information to occur outside the minicomputer on a terminal that supported a processor and memory. The Digital Equipment Corporation VAX systems and DECnet networking are typical of this era.

In general, however, applications from different computing environments (such as the mainframe environment and LAN environment) remained separate and independent from each other. Different communication protocols were developed to support communications between the various environments. For example, the mainframe environment used SNA (Systems Network Architecture) as a communication method while many LANs used Novell's IPX/SPX (Internetwork Packet Exchange/Sequenced Packet Exchange) and TCP/IP (Transmission Control Protocol/Internet Protocol) as communications methods.

Both IPX/SPX and TCP/IP were designed to allow connections between multiple networks through routers. This facilitated the growth of LANs within departments and companies.

1980s and 1990s: Internetworks

Internetworks tie LANs and WANs, computer systems, software, and a variety of different devices together to form the corporate communication infrastructure. For example, Figure 1–3 depicts a network that consists of mainframe, minicomputer, and PC-based devices attached through a variety of media and interconnected through private and public (Internet) WAN links. This internetwork moves information anywhere within a corporation and to external partners and customers. By serving as the organization's information highway, the internetwork has become a key strategic asset and a competitive advantage.

Routers are a key element in these internetworks because they allow (or deny) the communications between LANs and WANs. Understanding how routers function enables you to properly configure routers and select the routing protocol(s) most appropriate for an internetwork.

Today's internetworks combine a variety of devices, media types, and transmission methods. For many businesses, today's networks are an ad hoc mixture of old and new technologies. For example, older IBM networks might operate virtually in parallel with the newer LAN interconnected networks, electronic commerce, and messaging systems. Local networks, public data networks, leased lines, and high-speed mainframe channels have been added to internetworks in a "just in time" approach, often with little regard for network design, management, and overall efficiency. As applications migrated from central hosts to distributed servers, they caused changes in traffic patterns.

The approach to computer communication in most organizations is changing rapidly in response to new technologies, evolving business requirements, and the need for "instant" knowledge transfer. To meet these requirements, the internetwork, whatever form it takes, must be flexible, scalable, and adaptable to suit any organizational level (branch, regional, headquarters). It also must be thoughtfully designed to reflect the

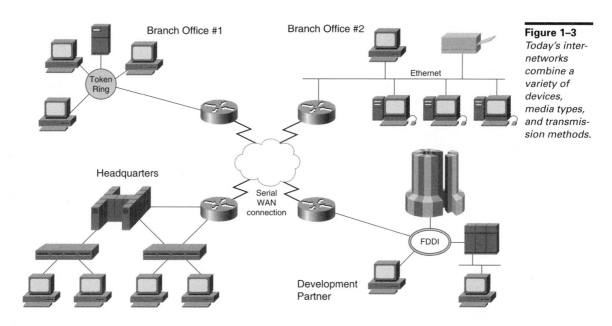

Figure 1–3
*Today's inter-
networks
combine a
variety of
devices,
media types,
and transmis-
sion methods.*

expected network traffic patterns. Network engineers and administrators must know
and understand how data packets are routed through a network to ensure that the most
efficient interconnection system is put in place to handle the demands of today's rapidly
growing networks.

**Internetworking removes barriers associated with the physical network con-
nections, hardware platforms, or software.**   **Key
                                                                  Concept**

1990s: Global Internetworking

The biggest pressure on networks in the immediate future is the globalization of business, and the support of applications required to conduct business internally and with
customers and clients around the world. It is not unusual now to find that a company
requires over 100 applications to function in a global internetwork.

Studies show that networks increasingly require more bandwidth to support these added
applications and internetwork connections. Networks will need to meet these demands
as well as provide low delay, bandwidth on demand, and other new services. New
devices will take their place alongside the router as additional network tools. Current

and future networks will have more functions distributed and must provide for the integration of voice, data, and video.

Consider Figure 1–4, for example. This global internetwork supports a variety of devices and applications that have varying bandwidth and speed needs. In the case of video streaming, audio transfer, and large graphics file transfers, for example, the data path must provide low delay, reliable end-to-end communications offered through ATM switching technology. The minicomputers and portable computers, which typically require support for bursty traffic, can be routed globally through serial connections.

Figure 1–4
Global inter-networks must support a variety of traffic types.

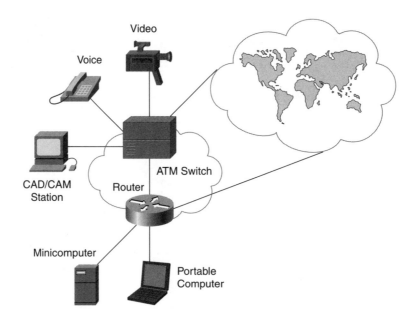

The following are characteristics of global networks:

- Increasing use of graphics and imaging
- Larger files
- Larger programs
- Client/server computing
- Bursty network traffic

Global internetworking will provide an environment for emerging applications that will require even greater amounts of bandwidth. Many of these applications are driven by the evolution of multimedia requirements that have a high-definition image, full-motion video, or a digitized audio component.

NETWORK TYPES AND DEVICES

Today's global internetworks can be categorized in three distinct types:

>Local-area networks (LANs)
>Wide-area networks (WANs)
>Enterprise networks

Each network type uses a different set of internetworking devices. Although this book is primarily concerned with routers, you need to be familiar with the other devices that each internetwork type uses, and you need to understand how those devices relate to routers and how their traffic may or may not be routed through an internetwork.

Local-Area Networks (LANs)

Local-area networks (LANs) are designed to operate within a limited geographic area and allow multiple users to simultaneously access high-bandwidth media. Typically, LANs connect physically adjacent devices and are controlled privately by local administration.

The major characteristics of LANs are:

- The network operates within a building or floor of a building. As increasingly powerful LAN desktop devices run more powerful applications, the trend is to reduce the size of individual LANs and connect smaller LANs together using routers.

- LANs provide multiple connected desktop devices (usually PCs) with access to high-bandwidth media.

- An enterprise purchases the media and connections used in the LAN; the enterprise can privately control the LAN as it chooses.

- Local services are usually available; LANs rarely shut down or restrict access to connected workstations.

Figure 1–5 shows a sample LAN that consists of two connected networks (network A and network B). This LAN connects physically adjacent devices on the media.

Figure 1–5
*LANs typi-
cally are used
to connect pri-
vate network
resources.*

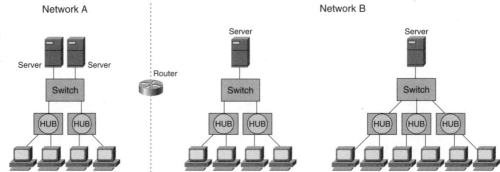

LAN devices include:

- Switches that connect LAN segments and devices and help filter traffic. (Bridges that used to provide connectivity between segments have been largely replaced by switches that can connect to segments as well as directly to the desktops.)

- Hubs that concentrate LAN connection and allow use of twisted-pair copper media

- Workgroup concentrators that deliver 100-Mbps service over fiber or copper cabling

- Ethernet and Token Ring switches that offer full-duplex, dedicated bandwidth to segments or desktops

- Routers that offer many services, including internetworking and broadcast control

- Asynchronous Transfer Mode (ATM) switches that provide high-speed cell switching

Wide-Area Networks (WANs)

WANs are designed to operate between a large mixture of telecommunications carriers and typically allow access over serial interfaces operating at lower speeds. WANs can be designed to provide part-time (dial-on-demand) or full-time connectivity over wide, even global, areas.

The major characteristics of WANs are:

- The network operates beyond the local LAN's geographic scope. It uses the services of carriers such as Regional Bell Operating Companies (RBOCs), Sprint, and MCI.

- WANs use serial connections of various types to access bandwidth over wide-area geographies.

- An enterprise pays the carrier or service provider for connections used in the WAN; the enterprise can choose which services it uses. Carriers are usually regulated by tariffs.

- WANs rarely shut down, but because the enterprise must pay for services used, it might restrict access to connected workstations. Not all WAN services are available in all locations.

Figure 1–6 shows a sample WAN that connects LANs located in different cities. This WAN connects physically remote devices through an ISDN network. For example, the WAN permits users in the remote office to access the servers FS-A, FS-B, and FS-C at the corporate headquarters.

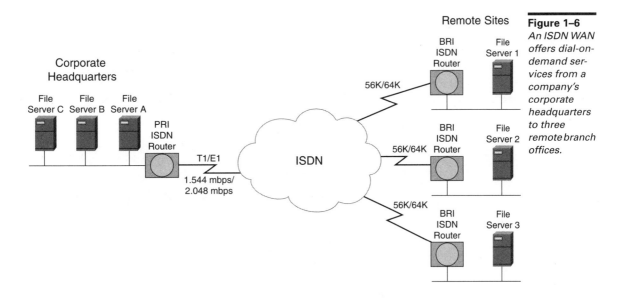

Figure 1–6
An ISDN WAN offers dial-on-demand services from a company's corporate headquarters to three remote branch offices.

WAN devices include:

- Routers that offer many services, including internetworking and WAN interface controls

- Switches that connect to WAN bandwidth for X.25, Frame Relay, and voice, data, and video communication. These WAN switches can share bandwidth among allocated service priorities, recover from outages, and provide network design and management systems

- Modems that interface voice-grade services; channel service units/digital service units (CSU/DSU) that interface T1/E1 services; Terminal Adapters/Network Termination 1 (TA/NT1) that interface Integrated Services Digital Network (ISDN) services

- Access servers that concentrate analog (or modem) dialin and dial-out user communication and provide other services, such as protocol translation between Telnet and X.25 protocol assembler (PAD)

- Multiplexers that share a WAN facility among several demand channels

- ATM switches that provide high-speed cell switching

Enterprise Networks

The enterprise is a corporation, agency, service provider, or other organization that ties together its data, communication, computing, and storage resources. An enterprise network usually contains a hybrid of both private and public network elements. Any or all of the LAN and WAN devices described so far can be found on the enterprise network.

Developments on the enterprise network include:

- LANs interconnected to provide client/server applications integrated with the traditional legacy applications from mainframe data centers

- End-user needs for higher bandwidth on the LANs, which can be consolidated at a switch and delivered on dedicated media

- Integration of formerly separate networks so the nonbursty traffic from voice and video applications coexists on a single network

- Relaying technologies for WAN service, with very rapid growth in Frame Relay and more gradual growth of cell relay (for example, ATM)

Cisco was the first company to offer a set of products to accommodate an entire enterprise network—that is, a set of products that work from the desktop all the way to the central office switch of the telecommunications carriers. The company's products have always supported the LAN aspects of the enterprise. With its acquisition of StrataCom, Cisco has added the missing pieces that work inside the WAN cloud.

NETWORK DESIGN GOALS

Regardless of whether it is local-area, wide-area, or enterprise-level, a network is just a collection of hardware and software. As mentioned earlier, the global networks of the present and future must be designed to meet the unique needs of the organizations they support. The role of a network manager is to create and refine that master design. In doing so, the manager must satisfy four major design goals:

- *Connectivity*—The internetwork must serve those in the organization who depend on it. Regardless of the range of media attachments, transmission speeds, and other technical details, the network design connects previously separate resources.

- *Reliable performance*—The organization becomes increasingly dependent on its internetworking tools, including the operator interface, the ability to distribute network software updates, utilities to log and monitor performance, redundant and backup operations, and the functions to secure access to resources. Building reliability into the network is critical to ensuring that the organization can operate competitively.

- *Management control*—An internetwork provides crucial functions; it also expends critical resources. Administrators continually ask how they can improve management controls through tasks such as performance measurement and analysis, resource usage, trouble-ticketing, utilization requirements, and security reporting. After the network is designed and operational, troubleshooting tasks follow.

- *Scalability*—Various pressures put on networks mean that flexibility is an important design goal. For example, expansion and consolidation of networks require overcoming physical or geographic boundaries. Also, as enterprises seek ways to provide new services and products to a network-accessible, global economy, they will require new or different network applications. Networks must be designed to be scalable—that is, to anticipate future demands and to evolve in a smooth, cost-efficient manner.

THE LAYERED MODEL

For a complex, multivendor internetwork to operate, its devices must be able to communicate with each other. The networking industry uses a model—the OSI model—that provides guidelines for that communication. This section explains the concept of data encapsulation and demonstrates how data is encapsulated as it travels down the layers of the OSI model.

Routing functions occur at the network layer of the OSI model, so the network layer is particularly important in this book. However, all the layers of the model are overviewed here (and covered more extensively in later chapters) because you need to have a general understanding of the hierarchy of processes that define network operation.

Why Use a Layered Model?

Most communication environments separate the communication functions from application processing. This separation of networking functions is called *layering*. For the OSI model, shown in Figure 1–7, seven numbered layers indicate distinct functions.

Figure 1–7
The OSI model includes seven layers that define network functionality.

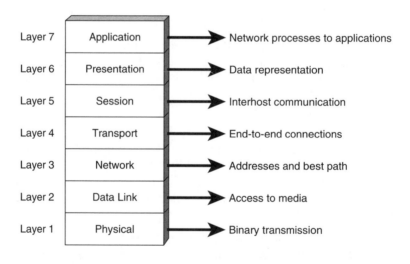

Layer 7	Application	→ Network processes to applications
Layer 6	Presentation	→ Data representation
Layer 5	Session	→ Interhost communication
Layer 4	Transport	→ End-to-end connections
Layer 3	Network	→ Addresses and best path
Layer 2	Data Link	→ Access to media
Layer 1	Physical	→ Binary transmission

As another example, the distinct functions of the Transmission Control Protocol/Internet Protocol (TCP/IP) fit into five named layers defined by the Department of Defense

(DoD) five-layer model. Regardless of the number of layers, there are several reasons for dividing network functions:

- To divide the interrelated aspects of network operation into less complex elements

- To define standard interfaces for "plug-and-play" compatibility and multivendor integration

- To enable engineers to specialize design and development efforts on modular functions enabling new applications and services to be deployed without redesigning each lower layer

- To promote symmetry in the different internetwork modular functions so they interoperate

- To prevent changes in one area from affecting other areas so each area can evolve more quickly

- To divide the complexity of internetworking into discrete, more easily learned operation subsets

---

A layered model provides a framework for, but does not define an inter-networking application or protocol. That is, applications and protocols do not conform directly to the OSI reference model, but they do conform to the standards developed from the OSI reference model principles.   **Key Concept**

Vendors use the definitions of the layered functions in the OSI model as guidelines in designing their network products. In examining each of the functions, the following sections use NFS (Network File System) as a sample application to apply to some of the model layers. NFS provides a distributed approach to UNIX file system access.

Application Layer

The application layer provides network services to user applications. For example, the NFS user interface can be mapped to this layer of the model.

Presentation Layer

The presentation layer provides data representation and code formatting. It ensures that the data that arrives from the network can be used by the application, and it ensures that information sent by the application can be transmitted on the network. Examples of these representations include ASCII, EBCDIC, JPG, TIFF, and encryption. External Data Representation (XDR) is SUN's presentation layer protocol for NFS.

Session Layer

The session layer establishes, maintains, and manages sessions between applications. For example, the tasks of starting, ending, interrupting, resuming, and abandoning a session are functions defined by the session layer. Remote Procedure Call (RPC) is at the session layer for the NFS stack.

Transport Layer

The transport layer segments and reassembles data into a data stream. It defines reliable and unreliable end-to-end data transmission, such as connectionless UDP (User Datagram Protocol) and connection-oriented TCP (Transmission Control Protocol).

Network Layer

The network layer determines the best way to move data from one place to another. It manages device addressing and tracks the location of devices on the network. The router operates at this layer. IP (Internet Protocol) is an example of a protocol that provides the functionality defined at this layer.

Data Link Layer

The data link layer provides physical transmission across the medium. It handles error notification, network topology, and flow control. Ethernet, Token Ring, and FDDI are media access methods that offer the functionality defined by this layer of the model.

Physical Layer

The physical layer provides the electrical, mechanical, procedural, and functional means for activating and maintaining the physical link between systems. Wires, signals, taps, and repeaters are examples of elements that interface to the physical media.

Peer-to-Peer Communications

Each layer of a transmitting system uses its own protocol to communicate with its peer layer in the receiving system. Each layer's protocol exchanges information, called protocol data units (PDUs), between peer layers. A given layer can use a more specific name for its PDU.

For example, in TCP/IP, the transport layer of TCP communicates with the peer TCP function using segments, as shown in Figure 1–8.

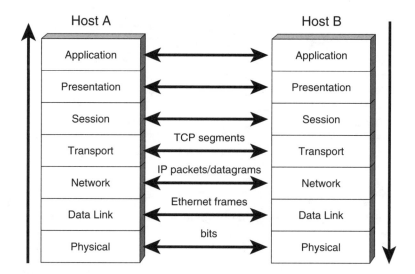

Figure 1–8
TCP/IP defines the transport layer protocol data unit as TCP segments.

This peer-layer protocol communication is achieved by using the services of the layers below the communicating layer. The layer below any given layer provides its services to that layer. Each lower-layer service takes upper-layer information as part of the lower-layer PDUs it exchanges with its layer peer.

Thus, the TCP segments become part of the network-layer packets (also called datagrams) exchanged between IP peers. In turn, the IP packets must become part of the data-link frames exchanged between directly connected devices. Ultimately, these frames must become bits as the data is finally transmitted by the physical-layer protocol using hardware.

Data Encapsulation and Headers

Each layer of the OSI model depends on the service function of the layer below it. To provide service, the lower layer uses encapsulation to put the PDU from the upper layer into its data field; then the lower layer can add whatever headers and trailers it will use to perform its function. Figure 1–9 shows the headers added by each layer. If you have the chance to use an analyzer on your network, you can see the headers embedded in the packet.

Figure 1–9
Headers are appended to the front of data as it is passed down through the functional layers.

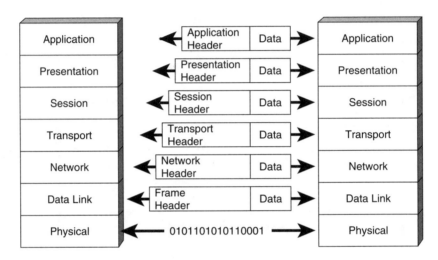

The data link layer in turn provides a service to the network layer. It encapsulates the network-layer information in a frame. The frame header contains information required to complete the data-link functions. For example, the frame header contains physical addresses.

The physical layer also provides a service to the data link layer. This service includes encoding the data-link frame into a pattern of ones and zeros for transmission on the medium (usually a wire).

As internetworks perform services for users, the flow and packaging of the information changes. Beginning at the transport level, five encapsulation steps occur:

1. Build the data.
2. Package data for end-to-end transport.
3. Append network address in header.
4. Append local address in data-link header.
5. Convert to bits for transmission.

Examine each of these steps to see how the data is affected as it is prepared for transmission. As an example, suppose that a Netscape client is browsing a Web server, as shown in Figure 1–10. This Web browsing operation requires the use of HTTP (Hypertext Transport Protocol) between the hosts.

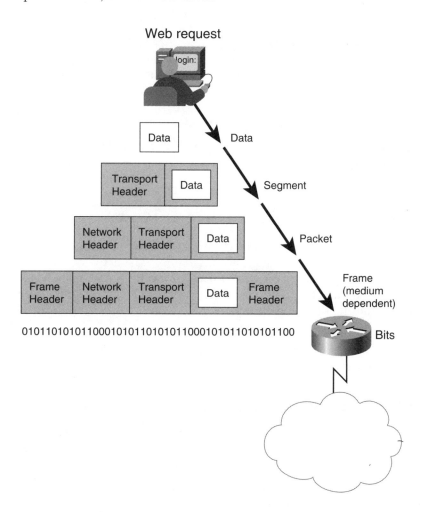

Figure 1–10
A user is browsing a Web server with a Netscape client. At each level of the OSI model, the appropriate information is encapsulated in a header to ensure the HTTP request gets to its destination.

Step 1: Build the Data

A user makes a request to open a specific page by sending the URL (Uniform Resource Locator) to the Web-serving host daemon process. The request, including the URL, is converted to data that can traverse the internetwork.

Step 2: Package Data for End-to-End Transport

The data is packaged for the transport subsystem. A transport-layer header is appended to the beginning of the data. In this example, the header is a TCP header, and it indicates that the data is directed to an HTTP server process.

Step 3: Append Network Address in Header

The data is put into a *packet* or *datagram* so the transport function can direct it over the internetwork. The packet includes a network header with source and destination logical addresses (for example, IP addresses). These addresses help network devices send the packets across the network along a chosen path.

In the example shown in Figure 1–10, the network layer header would be an IP header containing the Netscape user's IP address (the source address) and the HTTP server's IP address (the destination address).

Step 4: Append Local Address in Data-Link Header

Each network device must put the packet into a *frame* so the device can communicate over the local interface to another specific interface on the network. The frame allows connection to the next directly connected network device on the link. The frame type must match the data-link type. For example, if your data is sent on an Ethernet network using the Ethernet II frame type, the network packet is placed inside an Ethernet II frame.

In the example shown in Figure 1–10, the HTTP request will be addressed to the local router in an Ethernet II frame.

Step 5: Convert to Bits for Transmission

The frame is then converted into a pattern of ones and zeros for transmission on the medium (usually a wire). Some clocking function enables the devices to distinguish between these "1" and "0" bits as they traverse the medium.

The medium on the physical internetwork can vary along the path used. For example, the HTTP request can originate on a LAN, cross a campus backbone, go out a low-speed WAN link, and use a higher-speed WAN link until it reaches its destination on another remote LAN.

SUMMARY

Now that you have a basic understanding of the way protocols at different layers of the OSI reference model interact, the next chapter delves more deeply into the upper four layers of the model. In particular, Chapter 2, "Applications and Upper Layers," focuses on the transport layer, because routers are sometimes involved in transport layer functions.

Keep in mind the design criteria introduced in this chapter: connectivity, reliability, management control, and scalability. These principles of design inform the protocols, processes, and functions of networking you will read about in subsequent chapters. Also keep in mind the pressures that globalization and evolving functions put on the task of contemporary network design.

Chapter One Test
The Internetworking Model

Estimated Time: 15 minutes

Complete all the exercises to test your knowledge of the materials contained in this chapter. Answers are listed in Appendix A, "Chapter Test Answer Key."

Question 1.1

Define the three types of networks that make up today's global internetworks:

_____

_____

_____

Question 1.2

What are the seven layers of the OSI model?

Layer 7: _____

Layer 6: _____

Layer 5: _____

Layer 4: _____

Layer 3: _____

Layer 2: _____

Layer 1: _____

Question 1.3

How can data be defined when it is at the network layer? (Choose two.)

[a] segments

[b] packets

[c] bits

[d] datagrams

[e] frames

Question 1.4

Which of the following statements accurately describes the functions defined by the network layer of the OSI model?

[a] defines data representation and code formatting

[b] sends/receives binary information using device interfaces

[c] defines network addressing and determines the best path through an internetwork

[d] synchronizes communications between applications on different hosts

Question 1.5

Which line correctly represents the steps of converting data to bits for transmission?

[a] data, segments, frames, datagrams, bits

[b] data, frames, segments, bits

[c] data, packets, frames, datagrams, bits

[d] data, segments, packets, frames, bits

Applications and Upper Layers

This chapter discusses the upper four layers of the OSI reference model—application, presentation, session, and transport—with particular emphasis on the transport layer. Although these layers are not directly related to routing, understanding their interaction with and relationship to one another will help you understand important basic facets of network behavior. Understanding the transport layer is of particular importance because routers may be involved in transport layer functions at times.

APPLICATION LAYER

The application layer (Layer 7) of the OSI model supports the communicating component of an application. The term *application* in this sense does not refer to computer applications—such as word processing, presentation graphics, spreadsheets, databases—but rather to network applications. Examples of network applications include:

- File transfer
- Electronic mail
- Remote access
- Client/Server processes
- Information location
- Network management

Computer applications do not have knowledge of an underlying network and, therefore, cannot use the network directly. Rather, a computer application, such as a word processing program, can incorporate a network application, such as a file transfer component, that allows a document to be transferred electronically over telecommunication facilities. This file transfer component qualifies the word processor as an application in the OSI context and belongs in Layer 7 of the OSI reference model.

Many of the network applications offer services for enterprise communication. However, networking needs in the 1990s and beyond often extend beyond the enterprise. Information exchanges and commerce between enterprises increasingly involve internetworking applications such as those shown in Figure 2–1.

Figure 2–1
When computer applications use both network and internetwork application components, they become internetwork enterprise applications.

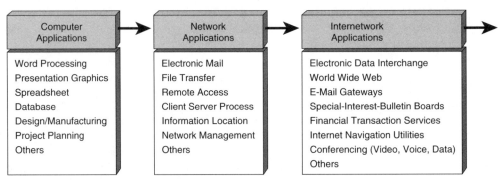

Internetwork application components include:

- Electronic data interchange (EDI) offers specialized standards and processes to improve the flow of orders, shipments, inventories, and accounting information between businesses.

- The World Wide Web links thousands of servers using a variety of formats including text, graphics, video, and sound. Browsers such as Internet Explorer and Netscape Navigator simplify access and viewing.

- The e-mail gateways might use the X.400 standard or Simple Mail Transfer Protocol (SMTP) to pass messages between different e-mail applications.

- Thousands of special-interest bulletin boards connect people who can chat with each other, post messages, and share public-domain software.

- Transaction services aimed at the financial community obtain and sell information including investment, market, commodity, currency, and credit data to subscribers.

- Special-purpose applications such as Gopher, Fetch, and Wide Area Information Server (WAIS) help navigate the way to resources on the Internet.

- People located in different regions use conferencing applications to communicate via live and prefilmed video, voice, data, and fax exchange.

PRESENTATION LAYER

The presentation layer provides code formatting and conversion services. Code formatting ensures that applications have meaningful information to process. If necessary, the presentation layer translates between multiple data representation formats for text, data, audio, video, and graphics, as shown in Figure 2–2.

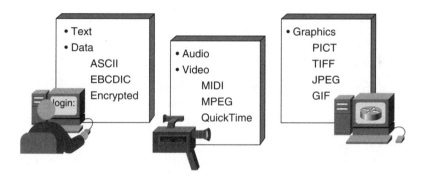

Figure 2–2
The presentation layer handles conversion and formatting for text, graphics, video, and audio elements.

Text and Data Formatting and Conversion

The presentation layer deals not only with the format and representation of actual user data, but also with data structure used by programs. Therefore, the presentation layer negotiates data transfer syntax for the application layer.

For example, the presentation layer is responsible for syntax conversion between systems that have differing text and data character representations, such as EBCDIC and ASCII. Another example of a presentation environment is Hypertext Markup Language (HTML), which describes how multimedia used on the Web should appear when viewed by a browser such as Internet Explorer and Netscape Navigator.

Presentation-layer functions also include data encryption. Processes and codes convert data so that the data can be transmitted with its information content protected from unauthorized receivers. Other routines compress text or convert graphics images into bitstreams for transmission across a network.

Graphics Formatting and Conversion

Graphics formats include PICT, a picture format used to transfer QuickDraw graphics between Macintosh or PowerPC programs; Tagged Image File Format (TIFF), a standard graphics format for high-resolution, bitmapped images; and JPEG, a picture format standard defined by the Joint Photographic Experts Group.

Audio and Video Formatting and Conversion

For audio and video, presentation layer standards include Musical Instrument Digital Interface (MIDI) for digitized music. Acceptance is growing for the Motion Picture Experts Group's (MPEG) standard for compression and coding of motion video for CDs, digital storage, and bit rates up to 1.5 Mbps. QuickTime handles audio and video for Macintosh or PowerPC programs.

SESSION LAYER

The session layer establishes, manages, and terminates communication sessions between applications. Essentially, the session layer coordinates service requests and responses that occur when applications communicate between different hosts.

For example, the session layer might set the exchange to be full or half duplex, define and group formatted data, and offer some session recovery or checkpoint mechanism between the applications coordinated between the hosts.

Figure 2–3 depicts a communication between two different hosts. The session layer has set the exchange to be half duplex (single request, single reply) and includes a checkpoint (session number) to ensure the transactions are matched between requests and replies.

Figure 2–3
The session layer coordinates session requests and replies.

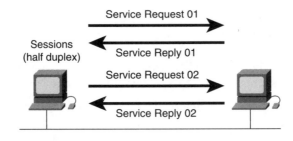

The following are examples of session-layer protocols and interfaces:

- *Structured Query Language (SQL)*—Database language developed by IBM to give users an easier way to specify their information needs on local and remote systems.

- *Remote Procedure Call (RPC)*—General redirection mechanism for distributed service environments. RPC procedures are built on clients and then executed on servers.

- *AppleTalk Session Protocol (ASP)*—Establishes and maintains sessions between an AppleTalk client and a server.

- *Digital Network Architecture Session Control Protocol (DNA SCP)*—DECnet session-layer protocol.

TRANSPORT LAYER

The transport layer defines end-to-end connectivity between host applications. Transport services include four basic functions:

- *Segment upper-layer applications*—Transport services can segment and reassemble several upper-layer applications onto the same transport-layer data stream.

- *Establish end-to-end operations*—This transport-layer data stream provides end-to-end transport services. It constitutes a logical connection between the endpoints of the internetwork: the originating or sender host and the destination or receiving host.

- *Send segments from one end host to another end host*—As the transport layer sends its segments, it can also ensure data integrity through the use of checksum calculations on the data and provide flow control mechanisms. Flow control avoids the problem of a host at one side of the connection overflowing the buffers in the host at the other side. Overflows can cause lost data.

- *Ensure data reliability (optional)*—Transport services also allow users to request reliable data transport between communicating end systems. Reliable transport guarantees that a stream of data sent from one machine will be delivered through a functioning data link to another machine without duplication of data or data loss. Data reliability may also ensure that data is received in the same order in which it was sent. A connection-oriented relationship between the communicating end systems is required for reliable transport. Connection-oriented sessions are discussed in more detail later in this chapter.

The following sections examine the transport-layer technologies available for controlling and optimizing communications.

Multiplexing

Multiplexing refers to the capability of multiple applications to share a transport connection. One reason for different layers in the OSI reference model is to accommodate multiplexing.

Transport functionality is accomplished segment by segment. Each segment is autonomous. Different applications can send successive segments on a first-come, first-served basis. These segments can be intended for the same destination host or many different destination hosts.

Software in the source machine must set the necessary port number for each software application before transmission, as illustrated in Figure 2–4. When sending a message, the source computer includes extra bits that encode the message type, originating program, and protocols used. Each software application that sends a data stream segment uses the same previously defined port number.

Figure 2–4
Transport layer functionality enables the multiplexing of application data within a data stream.

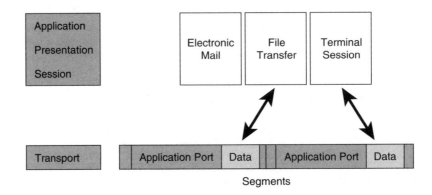

When the destination computer receives the data stream, it can separate the individual segments and reassemble each application's segments. This process allows the transport layer to pass the data up to its destination peer application.

TCP uses port numbers to multiplex from the transport layer to the application layer. These port numbers, listed in RFC 1700, have assigned ranges. Port numbers 1 to 1023 are called well-known port numbers and are reserved for particular protocols. For example, if the graphics file transfer is FTP, the initial application port value uses ˘1; the terminal session Telnet uses port 23. Table 2.1 shows defined port numbers for some of the commonly used protocols.

Number	Protocol
Port 20	File Transfer Protocol [Default Data]
Port 21	File Transfer Protocol [Control]
Port 23	Telnet
Port 25	Simple Mail Transfer Protocol
Port 53	Domain Name Server
Port 67	Bootstrap/DHCP Protocol Server
Port 68	Bootstrap/DHCP Protocol Client
Port 69	Trivial File Transfer Protocol
Port 70	Gopher
Port 80	World Wide Web HTTP
Port 119	Network News Transfer Protocol
Port 123	Network Time Protocol
Port 161	SNMP
Port 162	SNMP TRAP
Port 179	Border Gateway Protocol

Table 2–1
TCP-defined port numbers.

Port numbers in the range 1024 to 65,535 can be registered for convenience but are not assigned exclusively to one protocol; they can and often do have local significance.

Other protocol suites, such as NetWare's IPX/SPX protocol suite, have similar definitions (although they use the term *sockets*).

Connection-Oriented Sessions

To use the reliable transport services, two users of the transport layer must establish a connection-oriented session with each other. In essence, one machine places a call that must be accepted by the other.

For data transfer to begin, both the sending and receiving application programs inform their respective operating systems that a connection will be initiated. Protocol software modules in the two operating systems communicate by sending messages across the network to synchronize connection parameters, verify that the transfer is authorized, and acknowledge that both sides are ready. After all synchronization has occurred, a connection is said to be established, and the transfer of information begins. During transfer, the two machines continue to communicate with their protocol software to verify that data is received correctly.

Figure 2–5 depicts a typical connection between sending and receiving systems. The first "handshake" segment requests synchronization. The second and third segments acknowledge the initial synchronization request, as well as synchronize connection parameters in the opposite direction. The final handshake segment is an acknowledgment used to inform the destination that both sides agree that a connection has been established. After the connection has been established, data transfer begins.

Figure 2–5
In a connection-oriented session, the process of establishing a connection must be completed successfully before any data can be exchanged between peers.

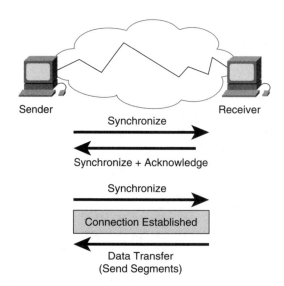

As an example, TCP/IP applications that require connection-oriented communications use the TCP (Transmission Control Protocol) connection establishment routine. This routine uses a "three-way handshake." The side that wants to establish a connection sends a Synchronize Sequence request packet. The other host replies with an Acknowledgment and a Synchronize Sequence request back. Finally, the originator sends an acknowledgment.

Flow Control and Congestion Avoidance

Once data transfer is in progress, congestion can arise for two different reasons. First, a high-speed computer might be able to generate traffic faster than a network can transfer it. Second, if many computers simultaneously need to send datagrams through a single gateway or to a single destination, that gateway or destination can experience congestion, even though no single source caused the problem.

When datagrams arrive too quickly for a host or gateway to process, they are stored in memory temporarily. If the datagrams are part of a small burst, this buffering solves the problem. If the traffic continues, the host or gateway eventually exhausts its memory and must discard additional datagrams that arrive. In this case, the host or gateway becomes the communications bottleneck.

Instead of allowing data to be lost, the transport function can, in a connection-oriented session, issue a "not ready" indicator to the sender, as illustrated in Figure 2–6.

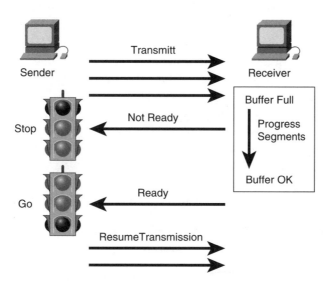

Figure 2–6
When the receiver's buffer space fills, it sends a message indicating that it cannot handle more data.

Acting like a red light, this indicator signals the sender to stop sending segment traffic to its peer. When the peer receiver can handle additional segments, the receiver sends a "ready" transport indicator, which is like a go signal. When it receives this indicator, the sender can resume segment transmission.

Flow Control with Windowing

In the most basic form of reliable connection-oriented data transfer, data segments must be delivered to the recipient in the same sequence that they were transmitted. The protocol in question fails if any data segments are lost, damaged, duplicated, or received in a different order. The basic solution is to have a recipient acknowledge the receipt of every data segment.

If the sender has to wait for an acknowledgment after sending each segment, throughput will be low. Because time is available after the sender finishes transmitting the data segment and before the sender finishes processing any received acknowledgment, the interval is used for transmitting more data. The number of data segments the sender is allowed to have outstanding—without yet receiving an acknowledgment—is known as the *window*.

Windowing is a method to control the amount of information transferred end-to-end. TCP/IP uses a Window field in the TCP header to indicate the buffer space available for incoming data. When the window size equals 0 (zero), the sender must stop sending until it receives a packet with a non-0 window size.

Figure 2–7 contrasts a window size of 1 with a window size of 3. With a window size of 1, the sender waits for an acknowledgment for every data segment transmitted. With a window size of 3, the sender can transmit three data segments before expecting an acknowledgment.

Windowing is an end-to-end facility between sender and receiver. In Figure 2–7, the sender and receiver are workstations on a small network. No router intervenes in the windowing function between them, which is fine because there is little or no chance that their acknowledgments and packets will intermix as they communicate.

Figure 2–7
Larger window sizes increase communication efficiency.

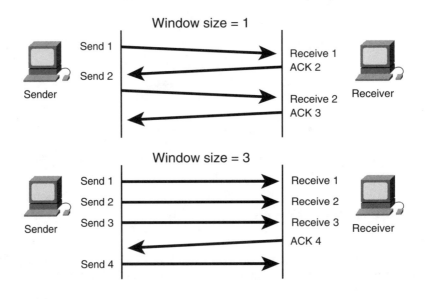

Positive Acknowledgment

Positive acknowledgment with retransmission is one technique that guarantees reliable delivery of data streams. Positive acknowledgment requires a recipient to communicate with the source, sending back an acknowledgment message when it receives data. The sender keeps a record of each segment it sends and waits for an acknowledgment before sending the next segment.

This system of waiting for an acknowledgment before sending more data is called an *expectational acknowledgment system*; it is used by TCP and SPX transport layer protocols.

The sender also starts a timer when it sends a segment, and it retransmits a segment if the timer expires before an acknowledgment arrives.

In Figure 2–8, the sender transmits segments 1, 2, and 3. The receiver acknowledges receipt of the segments by requesting segment number 4. The sender, upon receiving the acknowledgment, sends segments 4, 5, and 6. If segment number 5 does not arrive at the destination, the receiver acknowledges with a request to resend segment number 5. The sender resends segment number 5 and must receive an acknowledgment to continue with the transmission of segment number 7.

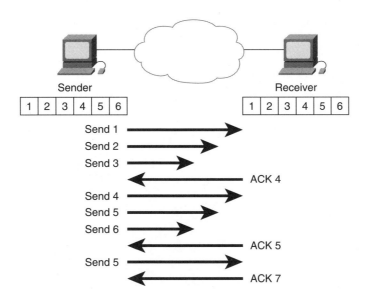

Figure 2–8
A sender must wait for acknowledg-ment of the transmitted data set before send-ing another set of data.

Various protocols handle retransmissions in different ways. Some protocols request resending only the missing segment. Others may request retransmission of the entire set of segments. As you might imagine, a protocol that requests only retransmission of missing segments is more efficient and effective.

Key Concept **Remember that flow control is achieved through windowing, while reliability is achieved through error detection and retransmission.**

SUMMARY

This chapter has focused on the upper layers of the OSI model with a particular emphasis on the transport layer. Routing functionality is defined at the network layer of the OSI model, which is discussed in more detail in Chapter 4, "Network Layer and Path Determination." However, a router may act at the transport layer as an endpoint for TCP communications. For example, a router may operate as a sender or receiver for windowing in synchronized, two-way traffic or for voice traffic over TCP.

The next chapter focuses on the physical and data link layers.

Chapter Two Test
Applications and Upper Layers

Estimated Time: 15 minutes

Complete all the exercises to test your knowledge of the materials contained in this chapter. Answers are listed in Appendix A, "Chapter Test Answer Key."

Question 2.1

Provide three examples of each application type:

Computer application:

_____

_____

_____

Network application:

_____

_____

_____

Internetwork application:

_____

_____

_____

Question 2.2

Match up the following transport layer elements with their functions:

A. flow control

B. windowing

C. retransmission

D. multiplexing

E. connection-establishment

F. positive acknowledgment

1. ____ guarantees receipt of data; ensures data integrity

2. ____ occurs when receipt timers expire or an error is detected

3. ____ often referred to as the "handshake" process

4. ____ enables many applications to use a single data stream

5. ____ uses buffering and congestion avoidance to prevent data from overflowing memory and possibly being lost before it can be processed

6. ____ controls the amount of data sent end-to-end by defining how many data segments may be sent before acknowledgment is required

Question 2.3

Host A has successfully established a connection-oriented session with host B. Host A transmits three data packets (data1, data2, and data3) to host B. One of these packets, data2, is involved in a collision during the transmit process.

How should the data transfer recover and resume?

_____

_____

_____

_____

_____

CHAPTER 3

Physical and Data Link Layers

This chapter focuses on the first two layers of the OSI model: the physical layer (Layer 1) and data link layer (Layer 2). It looks at the physical and data-link functionality of the three most commonly used LAN topologies: Ethernet, Token Ring, and FDDI. Finally, this chapter covers the basic WAN technologies available today.

BASIC DATA-LINK AND PHYSICAL LAYER FUNCTIONS

The data link layer provides data transport across a physical link. To do so, the data link layer handles the following operations:

- Physical addressing

- Network topology

- Line discipline

- Error notification

- Orderly access to the physical medium

- Flow control (optional)

The physical layer specifies the electrical, mechanical, procedural, and functional requirements for activating, maintaining, and deactivating the physical link between end systems. The physical layer specifies characteristics such as:

- Voltage levels

- Data rates

- Maximum transmission distances

- Physical connectors

These requirements and characteristics are codified into standards. For example, EIA/TIA-232 standardizes a physical connection to voice-grade access.

You can best understand physical and data link layers by considering WAN and LAN protocols separately. Therefore, for the rest of this chapter, the functions of these layers, and the standards that represent them, are discussed in distinct sections on LANs and WANs. As shown in Figure 3–1, certain layer standards are used with LAN links, and certain other layer standards are used by WAN links.

Figure 3–1
There are separate physical and data link elements for LANs and WANs.

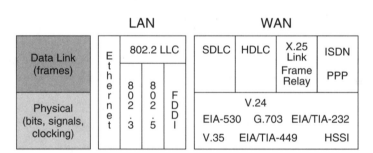

LAN STANDARDS AND STANDARDS ORGANIZATIONS

Today, much of the LAN standards work is performed by the Institute of Electrical and Electronic Engineers (IEEE). The IEEE 802 committee was formed in February 1980 ('80, 2nd month—hence, the name of the committee) to standardize LAN technology. The committee has the following subcommittees:

- 802.1 covers common issues concerning all LANs such as the spanning-tree protocol specified in IEEE 802.1D.

- 802.2 is responsible for the logical link control (LLC) sublayer.

- 802.3 is responsible for LANs based on the carrier sense multiple access collision detect (CSMA/CD) access methodology. Ethernet is an example of a CSMA/CD network.

- 802.4 is concerned with token bus networks. Token bus was developed by General Motors for computer-controlled manufacturing and is not commonly used today.

- 802.5 is responsible for Token Ring networks. The IBM Token Ring and IEEE802.5 standards are functionally equivalent.

One other type of LAN technology in common use today is based on the Fiber Distributed Data Interface (FDDI). The FDDI standard is the responsibility of the American National Standards Institute (ANSI).

TIA (Telecommunications Industry Association) is the formal organization responsible for the standards of telecommunications equipment that connects to the U.S. telecommunications network. TIA is closely aligned with the Electronic Industries Association (EIA), founded in 1944. The ITU (International Telecommunications Union), headquartered in Geneva, Switzerland, is an international organization within which governments and the private sector coordinate global telecom networks and services.

Table 3–1 lists Web address information for these standards organizations.

Organization	URL
ANSI	www.ansi.org
EIA	www.eia.org/eng/enghome.htm
IEEE	www.ieee.org
TIA	www.tiaonline.org
ITU	www.itu.ch

Table 3–1
Web contact information for computer and communications standards organizations.

LAN DATA LINK SUBLAYERS

LAN protocols occupy the bottom two layers of the OSI reference model: the physical layer and data link layer. The IEEE 802 committee subdivided the OSI data link layer into two sublayers: the logical link control (LLC) sublayer and the media access control (MAC) sublayer. Figure 3–2 illustrates the layers. This section covers each sublayer in sequence.

Figure 3–2
*LLC refers
upward to
higher-layer
software func-
tions, and
MAC refers
downward to
lower-layer
hardware
functions.*

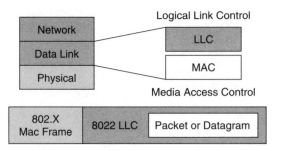

LLC Sublayer Functions

The LLC sublayer provides for environments that need connectionless or connection-oriented services at the data link layer. The LLC sublayer (specified by IEEE as 802.2) manages communication between devices over a single link of a network. It defines the fields that allow multiple higher-layer protocols to share use of the data link.

The LLC sublayer rests on top of the other 802 protocols to provide interface flexibility. Upper-layer protocols (for example, IP at Layer 3) can operate autonomously without regard for the specific type of LAN media. This independence occurs because, unlike the MAC sublayer, LLC is not limited to a specific 802 MAC protocol. Instead, the LLC sublayer can depend on lower layers to provide access to the media.

The LLC sublayer uses a set of fields, the Destination Service Access Point (DSAP) and Source Service Access Point (SSAP), to define a link to the upper OSI layers.

LLC sublayer options include support for connections between applications running on the LAN, flow control to the upper layer by means of ready/not ready codes, and sequence control bits.

MAC Sublayer Functions

The MAC sublayer provides access to the LAN medium in an orderly manner.

For multiple stations to share the same medium and identify each other, they must have unique addresses. The most important function of the MAC sublayer is defining a unique hardware or data-link address, called the MAC address, for each LAN interface.

On most LAN-interface cards, the MAC address is burned into ROM, hence the term *burned-in address (BIA).* When the network interface card initializes, this address is copied into RAM.

Before exploring details of MAC addresses, you need to have a bit of background on physical and logical addressing in networks.

Physical and Logical Addressing

Locating computer systems on an internetwork is an essential component of any network system. There are various addressing schemes used for this purpose, depending on the protocol family being used. In other words, AppleTalk addressing is different from TCP/IP addressing, which in turn is different from IPX addressing, and so on.

Two important types of addresses are *link-layer addresses* and *network-layer addresses*. Link-layer addresses (also called *physical* or *hardware addresses*) are typically unique for each network connection. In fact, for most LANs, link-layer addresses are resident in the interface circuitry.

Because most computer systems have one physical network connection, they have only a single link-layer address. Routers and other systems connected to multiple physical networks can have multiple link-layer addresses. As their name implies, link-layer addresses exist at Layer 2 of the OSI reference model; they are the addresses for which the MAC sublayer is responsible.

Network-layer addresses (also called *virtual* or *logical addresses*) exist at Layer 3 of the OSI reference model. Unlike link-layer addresses, which usually exist within a flat address space, network-layer addresses are usually hierarchical. In other words, they are like mail addresses, which describe a person's location by providing a country, a state, a Zip code, a city, a street, street address, and finally, a name. One good example of a flat address space is the U.S. Social Security numbering system, where each person has a single, unique social security number.

MAC Addresses

The MAC address is a 48-bit address expressed as 12 hexadecimal digits, as shown in Figure 3–3.

The first six hexadecimal digits of a MAC address (the first 3 bytes) contain a manufacturer identification (vendor code) also known as the Organizational Unique Identifier (OUI). To ensure vendor uniqueness, the IEEE administers OUIs. The last six hexadecimal digits are administered by each vendor and often represent the interface serial number.

Cisco's assigned vendor code is 0x00000c. The MAC address shown in Figure 3–3 is a device address for an interface manufactured by Cisco.

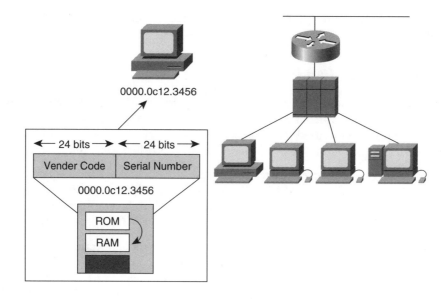

Examples of MAC addresses for different vendors include:

- Cisco: 00-00-0c-12-34-56
- Sun: 08-00-20-12-34-56
- Apple: 08-00-07-12-34-56

Finding the MAC Address

Before a frame is exchanged with a directly connected device, the sending device needs to "resolve" the logical address to the data-link, or MAC, address. *Address resolution* provides a mapping between the two different addresses: logical and data link. A commonly implemented address resolution protocol is ARP (RFC 826), used in TCP/IP networks.

Figure 3–4 illustrates two scenarios in which a sending device discovers the MAC address of the target device by broadcasting an address resolution request.

In the first scenario, host Y and host Z are on the same LAN. Host Y broadcasts a query onto the LAN indicating that host Y is seeking a data-link address for host Z. Because host Y has sent out a broadcast, all devices including host Z will process the request. However, because the request is only for host Z, only host Z will respond with its own MAC address. Host Y receives a reply from host Z and saves the data-link address for

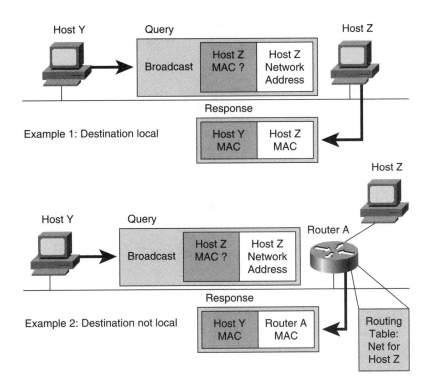

Figure 3–4
Devices can broadcast a request to locate a device; if a router answers, the destination device is not local.

host Z in local memory. The next time host Y needs to communicate with host Z, it recalls, from memory, the MAC address for host Z.

In the second scenario, host Y and host Z are on different LANs but can access each other through router A. When host Y broadcasts its query, router A recognizes the logical address as belonging to host Z on a different LAN. Because router A knows that it relays any packets for host Z, router A provides its own MAC address in reply to the query. Host Y receives the response and saves the MAC address of router A in memory. The next time host Y needs to communicate with host Z, host Y recalls the stored MAC address of router A.

Hardware addresses are used to get a packet from one local device to another local device; *logical addresses* are used to get a packet end-to-end through an internetwork.

It is important to understand how routers use hardware and logical addresses; these addresses define the data path along which a packet will be routed.

Key Concept

This discussion of address resolution is closely tied to TCP/IP. As you will see later, other protocols (such as IPX) have no need for this type of data-link participation in finding addresses. There are actually three different methods of obtaining a MAC address:

- A node asks for and receives an answer through Address Resolution Protocol (ARP). This is the method illustrated in Figure 3–4.

- A node is notified of another station's address with a hello packet.

- Addresses are assigned in a predictable way, as with DECnet.

COMMON LAN TECHNOLOGIES

This section extends the discussion of data link layer functions by exploring them in the context of specific LAN technologies. Physical layer functions are covered as well. The most commonly used LAN technologies are:

- Ethernet
- Token Ring
- FDDI

Ethernet and IEEE 802.3

Ethernet currently runs the largest number of LANs. Xerox developed Ethernet initially and was joined by the Digital Equipment Corporation (Digital) and Intel to define the Ethernet I specification in 1980. The same group subsequently released the Ethernet II specification in 1984. The Ethernet specification describes a Carrier Sense Multiple Access/Collision Detection (CSMA/CD) access method.

The IEEE 802.3 subcommittee adopted Ethernet as its model for its CSMA/CD LAN specification. As a result, Ethernet II and IEEE 802.3 are identical in the way they use the physical medium.

The two specifications differ in their descriptions of the data link layer. IEEE 802.3 splits the data link layer into two separate entities: the MAC sublayer and the LLC sublayer. The Ethernet II specification does not split them or offer LLC services. These differences do not prohibit manufacturers from developing network interface cards that support the common physical layer, MAC addressing, and software that recognizes the differences between the two logical link control layers.

Both Ethernet and 802.3 are now administered by the IEEE 802.3 committee.

The Ethernet/802.3 Physical Layer

The Ethernet and IEEE 802.3 standards define a bus-topology LAN that operates at a baseband signaling rate of 10 Mbps or 100 Mbps (Fast Ethernet) or 1000 Mbps (Gigabit Ethernet). (An earlier version of Ethernet that operated at 3 Mbps is now obsolete.) There are several defined wiring standards (some of which are illustrated in Figure 3–5):

- 10Base2 (known as thin Ethernet) allows network segments up to 185 meters on coaxial cable.

- 10Base5 (known as thick Ethernet) allows network segments up to 500 meters on coaxial cable.

- 10BaseT carries Ethernet frames on inexpensive twisted-pair wiring.

- 100BaseFX is a 100-Mbps implementation of Ethernet over fiber-optic cable.

- 100BaseT4 is a 100-Mbps implementation of Ethernet using four-pair Category 3, 4, or 5 cabling.

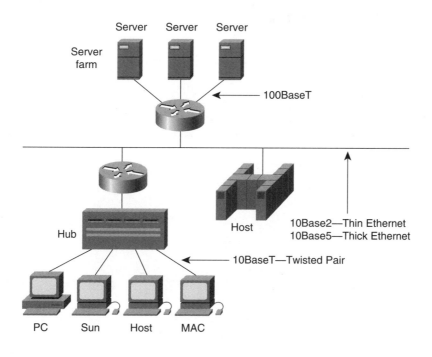

Figure 3–5
A network can use a combination of different Ethernet/802.3 access types.

- 100BaseTX is a 100-Mbps implementation of Ethernet over Category 5 and Type 1 cabling.

- 100VG-AnyLAN—The IEEE specification for 100-Mbps implementation of Ethernet and Token Ring over four-pair Category 3 UTP, two-or four-pair Category 5 UTP, STP, or fiber. The MAC layer is not compatible with the 802.3 MAC layer.

The 10Base5 and 10Base2 standards provide access for several stations on the same segment. Stations are attached to the segment by a cable that runs from an attachment unit interface (AUI) in the station to a transceiver that is directly attached to the Ethernet coaxial cable. In some interfaces, the AUI and the transceiver are co-located in the station itself, in which case no cable is required.

Because the 10BaseT standard provides access for a single station only, stations attached to an Ethernet LAN by 10BaseT are almost always connected to a hub. In a hub arrangement, the hub is analogous to an Ethernet segment.

The Ethernet/802.3 Interface

The Ethernet and 802.3 data links provide data transport across the physical link joining two devices. Each device on a network has one or more interfaces to the physical cabling media. Typical device interfaces, such as workstations and servers, are identified by their MAC addresses.

Figure 3–6 shows three devices directly attached to each other over an Ethernet LAN. The Apple Macintosh on the left and the Sun workstation in the middle show MAC addresses used by the data-link framing. The router on the right also uses MAC addresses for each of its LAN-side interfaces.

Figure 3–6
A Cisco router's data link to Ethernet/802.3 uses an interface named E plus a number (for example, E0).

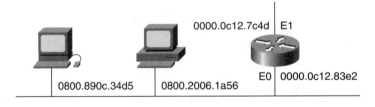

For indicating the 802.3 interface in the Cisco router configuration statements, you will use the Cisco IOS interface type abbreviation E followed by an interface number (for example, E0, as shown in the Figure 3–6).

Ethernet/802.3 Operation

In a CSMA/CD network using a linear bus design, one node's transmission traverses the entire network and is received and examined by every node. When the signal reaches the end of a segment, terminators absorb it to prevent it from going back onto the segment.

On a 10BaseT network, one node's transmission is repeated out all connected ports of a hub. Again, receiving stations examine these packets.

For example, in Figure 3–7, two Ethernet/802.3 networks are shown: a linear bus network and a 10BaseT network. In each case, Station A transmits a packet addressed to Station D. This packet is received by all stations. Station D recognizes its MAC address and processes the frame. Stations B and C do not recognize their addresses and discard the frame.

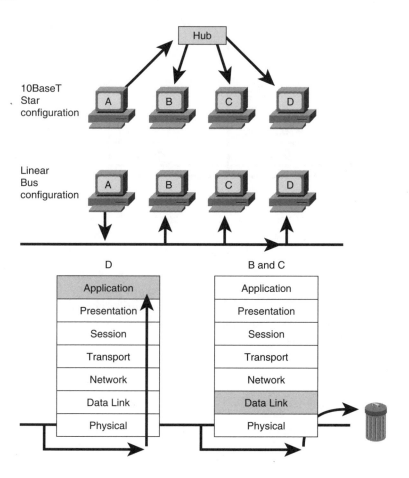

Figure 3–7
Comparing a linear bus to a 10BaseT star.

Ethernet/802.3 Broadcasts and Multicasts

The Ethernet/802.3 network definitions include methods for sending packets to all or a group of Ethernet devices using broadcasting and multicasting techniques. These technologies enable a device to transmit a single packet that is processed by many stations.

Broadcasting is a powerful tool that sends a single frame to all stations at the same time. Broadcasting uses a data-link destination address of all ones (FFFF.FFFF.FFFF in hexadecimal). As shown in Figure 3–8, if Station A transmits a frame with a destination address of all ones, Stations B, C, and D will all receive and pass the frame to their respective upper layers for further processing.

Figure 3–8
Broadcast packets go to all stations.

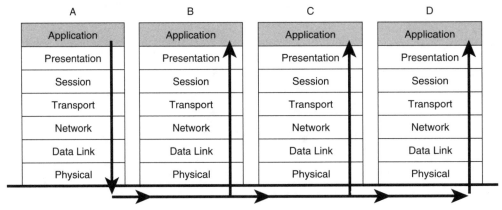

When improperly used, however, broadcasting can seriously impact the performance of stations by interrupting them unnecessarily. For this reason, broadcasts should be used only when the MAC address of the destination is unknown or when the destination is all stations.

A multicast address is a MAC address used to identify a group of destinations and is indicated by the first transmitted bit of the destination address being set to 1. For Ethernet, this bit appears as the low-order bit (for example, xxxx.xxx1) in the first byte of the destination MAC address.

Ethernet Frame Types

Figure 3–9 shows the two basic frame types: Ethernet and 802.3.

Both Ethernet and IEEE 802.3 frames begin with an alternating pattern of 1s and 0s called a *preamble*. The preamble tells receiving stations that a frame is coming.

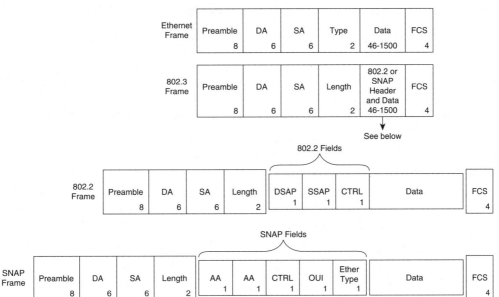

Figure 3–9
Several frame variations are possible on Ethernet/ 802.3 networks.

Immediately following the preamble in both Ethernet and IEEE 802.3 LANs are the destination and source physical address fields. Both the Ethernet and IEEE 802.3 addresses are six bytes long. Addresses are contained in hardware on the Ethernet and IEEE 802.3 network interface card (NIC). The first three bytes are specified by the Ethernet or IEEE 802.3 vendor. The source address is always a unicast (single node) address, while the destination address may be unicast, multicast (group), or broadcast (all nodes).

In Ethernet frames, the two-byte field following the source address is a type field. This field specifies the upper-layer protocol to receive the data after Ethernet processing is complete.

In IEEE 802.3 frames, the two-byte field following the source address is a length field, which indicates the number of bytes of data that follow this field and precede the frame check sequence (FCS) field.

In the case of Ethernet, the upper-layer protocol is identified in the type field. In the case of IEEE 802.3, the upper-layer protocol must be defined within the LLC portion of the frame.

If data in the frame is insufficient to fill the frame to its minimum 64-byte size, padding bytes are inserted to ensure at least a 64-byte frame. A packet that is smaller than 64 bytes is a *runt* packet. A packet that is larger than the maximum if 1518 bytes is a *giant*. Runts and giants are considered errors.

The actual data contained in the frame follows the type field in an Ethernet frame. The actual data contained in the frame follows the LLC or SNAP field in an IEEE 802.3 frame. After physical layer and link-layer processing is complete, this data is sent to an upper-layer protocol.

Following the data field is a four-byte FCS field containing a cyclical redundancy check (CRC) value. The CRC is created by the sending device and recalculated by the receiving device to check for damage that might have occurred to the frame in transit.

Following the length field is an 802.2 header for Logical Link Control (LLC). The LLC header consists of a Destination Service Access Point (DSAP), a Source Service Access Point (SSAP), and a control field.

The DSAP is a one-byte field that simply acts as a pointer to a memory buffer in the receiving station. This field tells the receiving NIC in which buffer to put this information. This functionality is crucial in situations where users are running multiple protocol stacks. The SSAP in the LLC header is analogous to the DSAP and specifies the service access point (SAP) of the sending process.

Although the original 802.3 specification worked well, the IEEE realized that some upper-layer protocols required an Ethernet type number to work properly. For example, TCP/IP uses the Ethernet type number to differentiate between ARP packets and normal IP data frames. To allow proprietary protocols in the 802.2 LLC frame, the IEEE defined the Subnetwork Access Protocol (SNAP) format. To specify that a frame is a SNAP frame, the DSAP and SSAP are both set to AA (hex).

The first three bytes of the SNAP header make up the vendor code, or OUI. For example, Apple's OUI is 00 00 F8. Following the vendor code is a two-byte field containing the Ethernet type for the frame. Here is where the backward compatibility with Version II Ethernet is implemented.

As with the 802.3 frame, a four-byte FCS field follows the data field and contains a CRC value.

Ethernet/802.3 Reliability

To understand how CSMA/CD provides an orderly transmission method, consider what typically occurs when a station transmits. When a station wants to transmit, it checks the network to determine if another station is currently transmitting. If the network is not being used, the station proceeds with the transmission. While sending, the station monitors the network to ensure that no other station is transmitting. Two stations might start transmitting at approximately the same time if they determine that the network is available. If two stations send at the same time, a collision occurs, as illustrated in Figure 3–10.

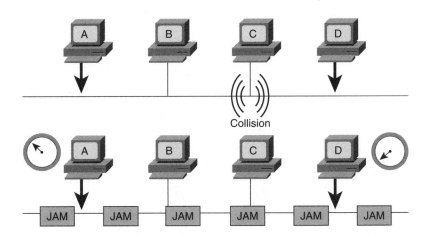

Figure 3–10
The jam signal ensures all nodes recognize a collision.

When a transmitting node recognizes a collision, it transmits a jam signal that causes the collision to last long enough for all other nodes to recognize it. All transmitting nodes then stop sending frames for a randomly selected time period, called the *backoff time*, before attempting to retransmit. If subsequent attempts also result in collisions, the node tries to retransmit up to 16 times before giving up.

If the two backoff times are sufficiently different, one station will succeed the next time it tries to transmit. The mean backoff time doubles with each consecutive collision up to the 10th retry, thereby reducing the chance of collision in subsequent transmits. From the 10th to the 16th retry, the stations do not increase the backoff time anymore but keep it constant.

High-Speed Ethernet Options

New applications can cause end users to experience delays and other problems such as insufficient bandwidth between end stations. In response to these problems, Ethernet networks have moved forward with the availability of 100-Mbps technologies, such as these:

- *100BaseFX*—A 100-Mbps implementation of Ethernet over fiber-optic cable. The MAC layer is compatible with the 802.3 MAC layer.

- *100BaseT4*—A 100-Mbps implementation of Ethernet using four-pair Category 3, 4, or 5 cabling. The MAC layer is compatible with the 802.3 MAC layer.

- *100BaseTX*—A 100-Mbps implementation of Ethernet over Category 5 and Type 1 cabling. The MAC layer is compatible with the 802.3 MAC layer.

- *100VG*—AnyLAN-The IEEE specification for 100-Mbps implementation of Ethernet and Token Ring over four-pair UTP. The MAC layer is not compatible with the 802.3 MAC layer.

Increasing Ethernet bandwidth to 100 Mbps solves part of the bandwidth problem. Backward compatibility can be an important consideration, however. The 100BaseFX, 100BaseT4 (four-pair Category 3, 4, or 5 cabling), and 100BaseTX implementations are compatible with the 802.3 MAC sublayer, but the 100VG-AnyLAN specification is not compatible with the other technologies.

Another part of the solution is reducing the contention for the Ethernet media. One method of reducing contention is built into the Ethernet standard, namely the CSMA/CD approach. Users of traditional Ethernet, a shared-media LAN, must submit to CSMA/CD so that no two users can simultaneously communicate over the shared LAN segment.

Switching also reduces contention for the media by creating multiple segments for desktop devices and high-end applications. Figure 3–11 shows a switch on the left that splits the Ethernet to reduce the number of users per shared segment. The switch makes multiple 10-Mbps or even 100-Mbps data pipes available. A limited number of users share a single 10-Mbps or 100-Mbps segment. These users work in a smaller collision domain with less contention from other nodes. For users with high bandwidth needs and servers, you can provide a single dedicated segment per user or server.

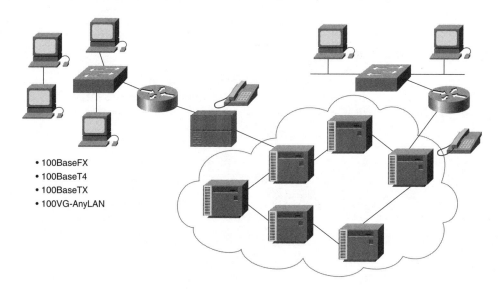

Figure 3–11
*High-speed
Ethernet
options and
switching
technologies
offer more
bandwidth to
users and
workgroups.*

- 100BaseFX
- 100BaseT4
- 100BaseTX
- 100VG-AnyLAN

Segmentation of Ethernet traffic can lead to a point where the switch dedicates a LAN segment to a single user. Figure 3–11 shows an Ethernet switch on the right that illustrates this situation. Two high-end workstations use their own Ethernet segment to receive a dedicated 10 Mbps to 100 Mbps for high-bandwidth applications, such as medical imaging systems. Servers typically get one or more dedicated 100-Mbps pipes. A final part of the solution to insufficient bandwidth is providing the network administrator with the tools needed to design, deploy, and manage a graceful transition to this complex switched internetworking environment.

Coinciding with the rapid growth of high-speed Ethernet options is the deployment of ATM, another high-speed technology.

Another high-speed technology is Gigabit Ethernet, or 1000BaseX. Gigabit Ethernet builds on top of the Ethernet protocol but increases speed tenfold over Fast Ethernet to 1000 Mbps, or 1 Gbps. This Media Access Control (MAC) and Physical Interface (PHY) standard, which was approved June 25, 1998, promises to be a dominant player in high-speed local-area network (LAN) backbones and server connectivity. Customers will be able to leverage their existing Ethernet knowledge base to manage and maintain Gigabit networks.

It has been decided that Gigabit Ethernet will look identical to Ethernet from the data link layer upward. However, to accommodate the increased speed from 100 Mbps Fast Ethernet to 1 Gbps, several changes need to be made to the physical interface. The challenges have been resolved by merging two technologies: IEEE 802.3 Ethernet and ANSI X3T11 FibreChannel. Figure 3–12 shows how key components from each technology have been combined to form Gigabit Ethernet.

The resulting standard takes advantage of the existing high-speed physical interface technology of FibreChannel while maintaining the IEEE 802.3 Ethernet frame format, backward compatibility for installed media, and use of full or half duplex (via carrier sense multiple access with collision detection or CSMA/CD). Leveraging two existing technologies helps minimize the complexity of the resulting technology, produces a stable technology, and shortens the development time.

With the approval of the 802.3z (Gigabit Ethernet) standard, Ethernet may gain a leading edge among LAN technologies in pushing bandwidth speed.

Token Ring and IEEE 802.5

Token Ring was developed by IBM and Texas Instruments in the 1970s. It is still IBM's primary LAN technology. The IEEE 802.5 specification is almost identical to IBM's Token Ring. A single Token Ring specification is now administered by the IEEE 802.5

Figure 3–12
*Gigabit Ether-
net protocol
stack.*

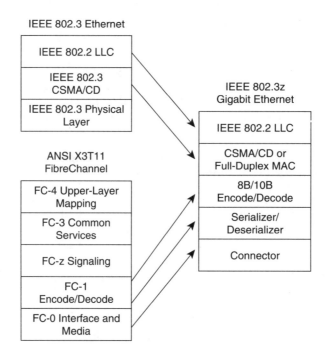

IEEE 802.3 Ethernet

IEEE 802.2 LLC

IEEE 802.3
CSMA/CD

IEEE 802.3 Physical
Layer

ANSI X3T11
FibreChannel

FC-4 Upper-Layer
Mapping

FC-3 Common
Services

FC-z Signaling

FC-1
Encode/Decode

FC-0 Interface and
Media

IEEE 802.3z
Gigabit Ethernet

IEEE 802.2 LLC

CSMA/CD or
Full-Duplex MAC

8B/10B
Encode/Decode

Serializer/
Deserializer

Connector

committee. The term *Token Ring* is generally used to refer to both IBM's Token Ring network and IEEE 802.5 networks.

Physical Layer: Token Ring/802.5

The logical topology of an 802.5 network is a ring in which each station receives signals from its nearest active upstream neighbor (NAUN) and repeats those signals to its downstream neighbor. Physically, however, 802.5 networks are laid out as stars, with each station connecting to a shared central hub called a *multistation access unit (MSAU)*. The logical and physical configurations are illustrated in Figure 3–13. Physically, the stations connect to the central hub through shielded or unshielded twisted-pair wire.

Typically, an MSAU connects up to eight Token Ring stations. For increased performance and port density, you can replace shared MSAUs and hubs with stackable Token Ring switches such as Cisco Catalyst 3920 24-port switch.

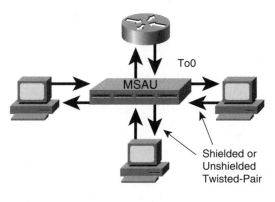

Figure 3–13
Token Ring networks are logically a ring but physically a star configuration connected to a central hub or switch.

Logical Topology

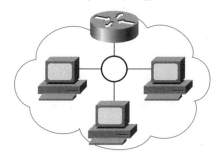

The Token Ring/802.5 Interface

The IEEE 802.5 Token Ring protocol parallels IEEE 802.3 by providing both MAC sublayer and physical-layer services in a single standard. Token Ring relies on the IEEE 802.2 LLC sublayer and upper-layer protocols for point-to-point services. Token Ring differs considerably from 802.3 in its use of the LAN medium, however.

All Token Ring stations use MAC addresses, including the router shown in Figure 3–13. For configuring the 802.5 interface on the router, you will use the Cisco IOS software interface type abbreviation for Token Ring (To), followed by an interface number (for example, To0, as shown in Figure 3–13). Token Ring networks can operate in either a 4 Mbps or 16 Mbps access speed.

Mixing speeds on a single ring destroys the integrity of the ring and prohibits proper operation.   **Key Concept**

Token Ring/802.5 Operation

Station access to a Token Ring is deterministic; a station can transmit only when it receives a token. This method of access is known as the *token passing method*. Although exceptions can be negotiated, stations are allowed to transmit only a single frame when they possess the token. Because no station can dominate the cable as it can in a contention-based access (CSMA/CD) network, administrators can quite accurately determine and plan network performance.

If a station receiving the token has no information to send, it simply passes the token to the next station. If a station possessing the token has information to transmit, it claims the token and then appends the information it wants to transmit and sends the information frame to the next station on the Token Ring, as shown in Figure 3–14.

Figure 3–14
Token Ring LANs continuously pass a token or an information frame.

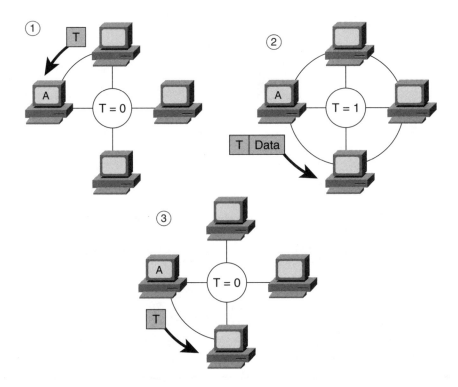

The information frame circles the ring until it reaches the destination station, where the frame is copied by the station and tagged as having been copied. The information frame continues around the ring until it returns to the station that originated it and is removed.

Unless early token release (ETR) is used on the Token Ring, only one frame can be circling the Token Ring at any one time; other stations wishing to transmit must wait until the current frame is removed from circulation and a token becomes available.

Early token release (ETR) is an optional feature that allows a station to insert the token onto the Token Ring immediately after transmitting an information frame. When early token release is in use, more than one frame can circle the Token Ring at a time.

Because frames proceed serially around the ring, and because a station must claim the token before transmitting, collisions do not occur in a Token Ring network.

Token Ring supports broadcasting and multicasting to enhance efficiency of one-to-many transmissions. Broadcasting may be used to locate a path to a destination. Multicasting is used to send packets to special Token Ring management addresses for ring integrity and error reporting purposes.

Token Ring/802.5 Media Control

Token Ring networks use a priority system that permits certain user-designated, high-priority stations to use the network more frequently than other stations can. Token Ring frames have two fields within the access control field that control priority: the priority field and the reservation field.

Figure 3–15 illustrates the bits in the access control field that are used to define the current priority and reservation priority.

Access Control Field

| P | P | P | T | M | R | R | R |

P   Priority bits
T   Token bit
M   Monitor bit
R   Reservation bits

Figure 3–15
The priority and reservation fields within the access control field in a Token Ring frame determine priority and reservation for sharing media.

Only stations with a priority equal to or higher than the priority of a token can claim that token. After the token is claimed and changed to an information frame, only stations with priority higher than the transmitting station can reserve the token for the next pass around the network. When the next token is generated, it includes the highest priority of the reserving station. Stations that raise a token's priority level must reinstate the previous lower priority level after their transmission is complete.

Token Ring/802.5 Active Monitor

One device on each Token Ring configures itself as the active monitor to provide clocking services and maintain the integrity of the token. In case the token is lost or destroyed, only the active monitor can purge the ring of all current data and transmit a new token for recovery purposes. The active monitor also ensures that frames do not circulate endlessly around the ring.

Figure 3–16 depicts the removal of an "old frame" from the ring. The station at left claims the token and transmits a frame on the Token Ring LAN with the monitor bit set to 0, indicating that this frame has not passed the active monitor. When the frame reaches the active monitor, the active monitor sets the frame's monitor bit to 1. Before the frame returns to the originating station, that station fails. Because the failed station is not able to remove its frame, the frame is allowed to start a second circuit of the Token Ring LAN. The active monitor detects a frame that it has seen before because the monitor bit is set to 1, removes it from the ring, and inserts a new token onto the ring.

Figure 3–16
The active monitor ensures token operation on the ring for media access.

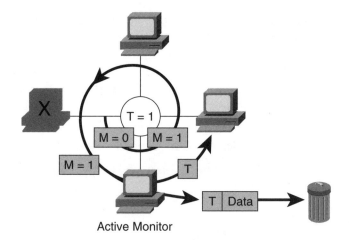

Active Monitor

Token Ring/802.5 Reliability

The IEEE 802.5 specification describes two bits in the frame status field: the A bit, which stands for address recognized (the receiving station sets this bit when it recognizes the incoming frame is addressed to the station's own address), and the C bit, which stands for copied (the destination Token Ring station copied the frame into its buffers).

These two bits are used to indicate the status of an outstanding frame, as shown in Figure 3–17. An originating station generates a frame with the A and C bits turned off (set to zero). Because the originating station always views the returning frame, it can exam-

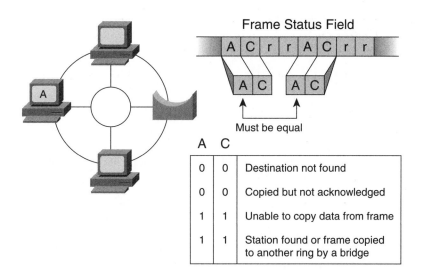

ine these two bits to determine whether they have been modified during their journey around the ring.

The A and C bits are duplicated in the Access Control field to provide error detection because this frame is not included in any error checking mechanism. Both sets of A and C values must be identical, or the frame is considered invalid.

FDDI

FDDI is an American National Standards Institute (ANSI) standard that defines a dual Token Ring LAN operating at 100 Mbps over a fiber-optic medium, as shown in Figure 3–18. The FDDI standards were published in 1987 in the ANSI X3T9.5 standards. Currently, FDDI is a popular campus and backbone LAN technology.

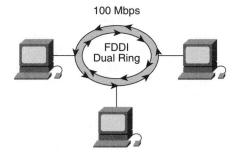

ANSI also has defined a Twisted-Pair Physical Medium Dependent standard. Based on this standard, Copper Distributed Data Interface (CDDI) provides operation of FDDI but with the more commonly used copper cabling.

Token Ring and FDDI share several characteristics, including token passing and a ring architecture.

Physical Layer: FDDI

FDDI standards describe the physical layer and MAC sublayer. The physical layer is further divided into two sublayers: the physical layer protocol (PHY) sublayer and the physical medium dependent (PMD) sublayer.

The PHY standard deals with data encoding, clocking, and symbols. PHY specifies a group-encoding method called 4B/5B. This encoding method translates upper-layer octets into pairs of five-bit symbols that perform the dual functions of passing data and maintaining clock synchronization between nodes.

Because FDDI specifies communication over fiber-optic cable, it is well suited for operations where nodes are separated by large distances or where networks must operate in electronically hostile environments such as factory floors.

FDDI specifies the following limits:

- 500 nodes per FDDI LAN

- 100-kilometer (km) maximum ring circumference

- 2-km maximum distance between FDDI nodes using multimode fiber media

FDDI can operate at high speeds that make it suitable for network applications requiring large bandwidth; for example, video and graphics applications.

FDDI uses a token-passing protocol that operates on dual counterrotating rings, as shown in Figure 3–19. Under normal operation, data flows on a primary ring, while the secondary ring is idle. Some stations, known as dual attachment stations (DASs), attach to both rings. Single attachment stations (SASs) have only a single physical medium dependent (PMD) connection to the primary ring by way of a dual-attached concentrator (DAC). Devices that are attached to two rings can still communicate in case of a single ring failure, because FDDI has the self-healing capability to enable one ring to loop back on the other to maintain integrity.

Mission-critical stations such as routers or mainframe hosts can use a technique called dual homing to provide additional fault-tolerance and help guarantee operation. With

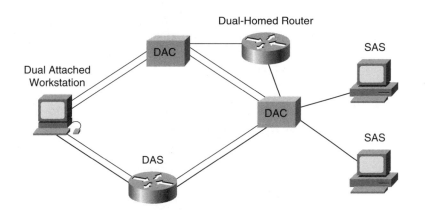

Figure 3–19
*Devices
attached to
FDDI use a
token passing
method; they
can be singly
or dually
attached.*

dual homing, a station is single-attached to two DACs, thereby providing an active primary link and a backup path to the FDDI LAN.

The FDDI Interface

FDDI is logically and physically a ring topology. Although it operates at higher speeds, FDDI is similar to Token Ring. The two network types share many features, such as token passing and predictable deterministic delays.

All FDDI LAN stations use MAC addresses, including the router shown on the right in Figure 3–19. The FDDI frame format uses five-bit symbols rather than eight-bit octets. Thus, the 48-bit MAC address for FDDI has 12 four-bit symbols.

To configure the FDDI interface on the router, use the Cisco IOS interface type abbreviation F followed by an interface number (for example, F0).

FDDI Dual-Ring Reliability

Access to the FDDI dual ring is determined by token possession. However, stations attach new tokens to the ends of their transmissions, and a downstream station is allowed to add its frame to the existing frame. Thus, at any given time, several information frames can be circling the ring.

All stations monitor the ring for invalid conditions such as a lost token, persistent data frames, or a break in the ring. If a node determines that no tokens have been received from its NAUN during a predetermined time period, it begins transmitting beacon frames to identify the failure and its suggested location.

If a station receives its own beacon from upstream, it assumes that the ring has been repaired. If beaconing continues beyond a certain time limit, DASs on both sides of the failure domain *loop* (or *wrap*) the primary ring to the secondary ring to maintain network integrity (as illustrated in Figure 3–20).

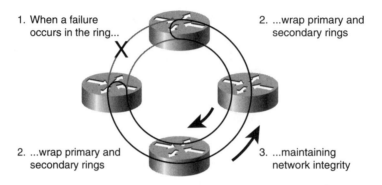

Figure 3–20
FDDI has the capability to detect faults and maintain integrity by wrapping the primary and secondary rings.

1. When a failure occurs in the ring...

2. ...wrap primary and secondary rings

2. ...wrap primary and secondary rings

3. ...maintaining network integrity

COMMON WAN TECHNOLOGIES

WAN physical-layer protocols describe how to provide electrical, mechanical, operational, and functional connections for wide-area networking services. These services are most often obtained from WAN service providers such as Regional Bell Operating Companies (RBOCs), alternate carriers, and Post, Telephone, and Telegraph (PTT) agencies.

WAN data-link protocols describe how frames are carried between systems on a single data link. They include protocols designed to operate over several different types of facilities, such as:

- Dedicated point-to-point facilities—For example, one office connected directly to another office through a WAN connection. One point connected to one other point.

- Multipoint facilities based on dedicated facilities—For example, a headquarters office connected to three branch offices (a multipoint connection) and the three offices connected to each other through the same type of multipoint connection.

- Multiaccess switched services—For example, a headquarters and three branch offices connecting into a WAN cloud, such as Frame Relay. Their communications are switched through the cloud, not necessarily taking the same path each time.

WAN standards are defined and managed by a number of recognized authorities, including the following agencies:

- International Telecommunication Union-Telecommunication Standardization Sector (ITU-T), formerly the Consultative Committee for International Telegraph and Telephone (CCITT)

- International Organization for Standardization (ISO)

- Internet Engineering Task Force (IETF)

- Electronic Industries Association (EIA)

WAN standards typically describe both physical layer and data link layer requirements.

Figure 3–21 identifies several popular WAN services used in internetworks today.

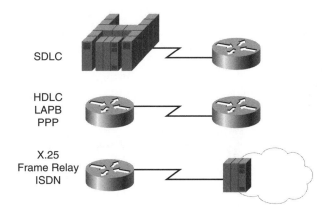

SDLC

HDLC
LAPB
PPP

X.25
Frame Relay
ISDN

Figure 3–21
There are three primary generations of WAN technology.

The Synchronous Data Link Control (SDLC) protocol was the original ancestor of most WAN framing, connecting remote devices with the central mainframe through point-to-point or point-to-multipoint connections.

The next generation of WAN technologies includes High-Level Data Link Control (HDLC) and, for X.25, Link Access Procedure Balanced (LAPB). For Internet WANs, Point-to-Point Protocol (PPP) connects peer devices.

Finally, switched or relayed services, such as X.25, Frame Relay, and Integrated Services Digital Network (ISDN), utilize a special device to interface a service provider's cloud.

Physical Layer: WAN

The WAN physical layer describes the interface between the data terminal equipment (DTE) and the data circuit-terminating equipment (DCE). Typically, the DCE is the service provider, and the DTE is the attached device. In this model, the services offered to the DTE are made available through a modem or channel service unit/data service unit (CSU/DSU), as shown in Figure 3–22.

Several physical-layer standards specify this interface:

- EIA/TIA-232
- EIA/TIA-449
- V.24
- V.35
- X.21
- G.703
- EIA-530
- High-Speed Serial Interface (HSSI)

Figure 3–22
Services are available to a DTE through a modem or CSU/DSU.

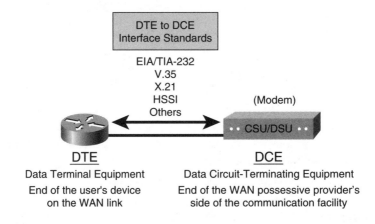

Note that EIA/TIA-232 and EIA/TIA-449 were known as recommended standards RS-232 and RS-449 before their acceptance as standards by the Electronic Industries Association (EIA) and Telecommunications Industry Association (TIA).

Data Link Layer: WAN Protocols

The common data-link encapsulations associated with synchronous serial lines are listed in Figure 3–23.

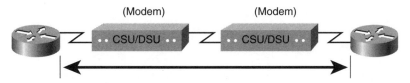

• SDLC-Synchronous Data Link Control
• HDLC-High-Level Data Link Control
• LAPB-Link Access Procedure, Balanced
• Frame Relay-Simplified version of HDLC framing
• PPP-Point-to-Point Protocl
• ISDN-Integrated Services Digital Network (data-link signaling)

Figure 3–23
Data-link encapsulations for synchronous lines.

Synchronous Data Link Control (SDLC)

SDLC is a bit-oriented protocol developed by IBM. SDLC defines a multipoint WAN environment that allows several stations to connect to a dedicated facility. SDLC defines a primary station and one or more secondary stations. Communication is always between the primary station and one of its secondary stations. Secondary stations cannot communicate with each other directly.

High-Level Data Link Control (HDLC)

HDLC is an ISO standard. HDLC might not be compatible between different vendors because of the way each vendor has chosen to implement it. HDLC supports both point-to-point and multipoint configurations.

Link Access Procedure, Balanced (LAPB)

LAPB is primarily used with X.25 but can also be used as a simple data-link transport. LAPB includes capabilities for detecting out-of-sequence or missing frames as well as for exchanging, retransmitting, and acknowledging frames.

Frame Relay

Frame Relay uses high-quality digital techniques in which the error checking of LAPB is unnecessary. By using a simplified framing with no error correction mechanisms, Frame Relay can send Layer 2 information very rapidly, compared to these other WAN protocols.

Point-to-Point Protocol (PPP)

PPP, described by RFC 1661, was developed by the IETF. PPP contains a protocol field to identify the network-layer protocol.

Integrated Services Digital Network (ISDN)

ISDN is a set of digital services that transmits voice and data over existing phone lines.

SUMMARY

This chapter concludes an overview of all the layers of the OSI reference model except one: the network layer. The next chapter focuses on the network layer, where routing functionality and capabilities are defined.

Chapter Three Test
Physical and Data Link Layers

Estimated Time: 15 minutes

Complete all the exercises to test your knowledge of the materials contained in this chapter. Answers are listed in Appendix A, "Chapter Test Answer Key."

Question 3.1

What are the Cisco router interface abbreviations for Ethernet, Token Ring, and FDDI?

_____ Ethernet

_____ Token Ring

_____ FDDI

Question 3.2

Which frame types use an Ethernet type field to define the protocol in use?

_____

_____

Question 3.3

What are the two sublayers defined within the data link layer?

_____

_____

Question 3.4

Write the letter identifying the correct statement in column 2 that describes the given protocol or standard in column 1.

Write Letters	Column 1 Protocol or Standard	Column 2 Topology, Function, or Characteristic
_____	SDLC	A) Equivalent of IEEE 802.5
_____	EIA/TIA-232	B) Voice-grade access, formerly a recommended standard
_____	802.3	C) Uses primary and secondary roles for IBM data links
_____	Frame Relay	D) From original Xerox work; uses field for protocol type
_____	Ethernet II	E) From IEEE efforts; uses field for length rather than type
_____	FDDI	F) Uses simplified HDLC for higher-speed communication
_____	Token Ring	G) Uses five-bit symbols rather than octets in its framing

Network Layer and Path Determination

This chapter discusses the network layer of the OSI reference model. It covers basic information such as how network-layer addressing works with different protocols. It explains the difference between routing and routed protocols and contrasts static and dynamic routes. It explains how routers track the distance between locations.

The chapter then covers distance vector, link-state, and hybrid routing approaches. It explains the strengths of each approach and describes how each resolves common routing problems.

NETWORK LAYER BASICS

The network layer interfaces to networks and provides best effort end-to-end packet delivery services to its user, the transport layer. The network layer sends packets from the source network to the destination network.

First, this chapter examines general performance of the network layer, including how it determines and communicates the chosen path to a destination, how protocol addressing schemes work and vary, and how routing protocols work.

Path Determination

Which path should traffic take through a cloud of networks? Path determination occurs at Layer 3, the network layer. The path determination function enables a router to evaluate the available paths to a destination and to establish the preferred handling of a packet.

Routing protocols use network topology information when evaluating network paths. This information can be configured by the network administrator or collected through dynamic processes running in the network.

After the router determines which path to use, it can proceed with switching the packet: taking the packet it accepted on one interface and forwarding it to another interface or port that reflects the best path to the packet's destination.

For example, Figure 4–1 depicts a mesh network. There are several possible paths between host A and host C. The path determination process is used to find the best path possible.

Figure 4–1
Host A and host C are connected through multiple paths.

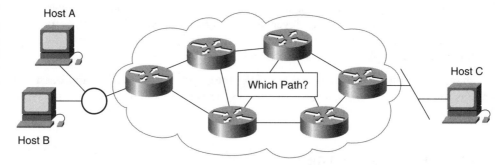

To enable path determination, the routing services provide:

- Routing table initialization and maintenance
- Routing update processes and protocols
- Routing domains and address specifications
- Route metric assignment and control

Communicating Path Information

Routers exchange information about available paths through an internetwork. To identify a path, a name must be assigned to each of the networks along a path. Network addresses are used to identify each network link. Path information contains the names of all the networks that must be crossed along the path.

In Figure 4–2, each line between the routers has a number that the routers use as a network address. These addresses convey information about the path of media connections used by the routing process to pass packets from a source toward a destination.

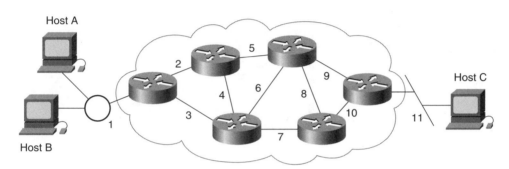

Figure 4–2
Addresses represent the path of media connections.

Routers use path information along with path determination mechanisms, path switching mechanisms, and route processing functions to determine the best path along an internetwork.

The consistency of Layer 3 addresses across the entire internetwork also improves the use of bandwidth by preventing unnecessary broadcasts. By using consistent end-to-end addressing to represent the path of media connections, the network layer can find a path to the destination without unnecessarily burdening the devices or links on the internetwork with broadcasts.

Routers contain broadcasts by allowing the network layer to find the destination.

Key Concept

Addressing: Network and Host

Network addresses consist of a network portion and a host portion. Both are needed to deliver packets from source to destination. In Figure 4–3, the two-part numbers—1.2, 1.3, 2.1, and so on—represent sets of network and host addresses. The first digit is the network portion of the number, and the second digit is the host portion.

The router uses the network portion of the address to identify the source or destination network of a packet within an internetwork. Figure 4–3 shows three network numbers—1, 2, and 3—known by the router.

Figure 4–3
An internet-work address consists of a network por-tion and a host portion.

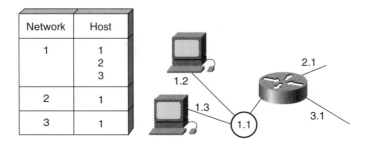

The host portion of the address refers to a specific port or device on the network. For instance, in Figure 4–3 three hosts—1, 2, and 3—are shown sharing the network number 1. The host or node address in a packet identifies that the packet is on its source or destination port or device on the network. For LANs, this port or device address can reflect the real Media Access Control (MAC) address of the device.

However, unlike a MAC address that has a preestablished and usually fixed relationship to a device, a network address has a logical relationship with a specific device. Some network addressing systems, such as IP addressing systems, require you to manually assign a unique host address to each device. Other network addressing systems, such as Novell's IPX addressing system, borrow the MAC address to use as a unique host address. This latter technique is a feature that is being incorporated into IPv6 addressing.

Key Concept The network and host portions of a network address convey different levels of location specificity. A network address is similar to a street address, such as Main Street. Host addresses are similar to the building number, such as 123 Main Street.

Routing processes on internetworks typically are concerned only with the network portion of an address; that is, the information required to deliver data to the appropriate network. After the destination network has been reached, however, the final router in the path must use the host address portion to send the packet to the appropriate device's hardware address on the final network.

Internetworking devices can have more than one network address. Different addresses must be assigned for each network-layer protocol supported by a particular device. For example, a device connected to both an AppleTalk and a DECnet internetwork must be assigned two network addresses.

For some network-layer protocols, a network administrator assigns network addresses according to a preconceived internetwork addressing plan. For other network-layer protocols, assigning addresses is partially or completely dynamic; that is, the protocol automates the process.

Not all network protocols use the host address in the manner shown in Figure 4–3. For example:

- Novell IPX uses MAC addresses as host addresses; for interfaces that do not have a MAC address, Novell IPX may apply a duplicate MAC address for the node address.

- DECnet modifies the MAC address to contain a computed node address.

Protocol Addressing Variations

The two-part network addressing scheme extends across all the protocols covered in this book. How do you interpret the meaning of the address parts? What authority allocates the addresses? These answers vary from protocol to protocol.

Figure 4–4 shows three sample addressing schemes.

In the TCP/IP sample IP address, dotted decimal numbers show a network part and a host part. The network 10 uses the first of the four numbers as the network part and the last three sets of numbers—8.2.48—as a host address.

The Novell IPX example uses a variation of the two-part address. The network address 01ac.eb0b is a hexadecimal (base 16) number that cannot exceed a fixed maximum number of digits. The host address 0000.0c00.6e25 (also a hexadecimal number) is a fixed 48 bits long. This host address derives automatically from information in the hardware of the specific LAN device.

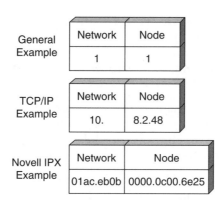

IP and IPX are the two most common Layer 3 address types. You will learn more about these and other protocol addressing rules in the next few pages and in subsequent chapters.

TCP/IP Network Addressing

TCP/IP networks represent addresses as 32-bit entities, divided into a network portion and a host portion, as shown in Figure 4–5.

The Internet Request For Comments (RFC) 1117 divides the network portion into classes. All classes of specific Internet-legal network addresses come from a central authority: the InterNIC (Internetwork Information Center). The most common of these classes follow:

- Class A—Using 8 bits for the network, with the remaining 24 bits for host addressing

- Class B—Using 16 bits for the network, with the remaining 16 bits for host addressing

- Class C—Using 24 bits for the network, with the remaining 8 bits for host addressing

- Class D—Used for IP multicast addresses

IP networks typically are subdivided into subnetworks. When an IP address has been subnetted, the network part of the address is described by two elements: the network number, still assigned by the NIC, and the subnetwork number, assigned by the local network administrator. IP addressing is covered in more detail in Chapter 9, "IP Addressing."

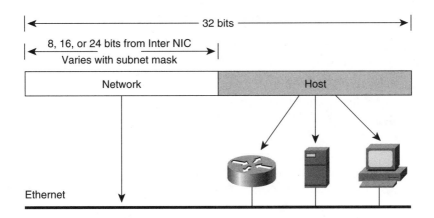

Figure 4–5
IP addresses are 32 bits (4 bytes) long.

Other Protocol Addressing

A router can handle many other protocol addressing schemes besides IP addressing. Table 4–1 summarizes the main details about three of the most common of these schemes.

Protocol	Network Address	Node Address
Novell IPX	Up to 32 bits (hex); refers to the media (for example, Ethernet)	48 bits (hex); usually the MAC address of a LAN interface
AppleTalk	Up to 16 bits (dec); refers to one or one of many nets in cable range on media	Up to 8 bits added to network number; usually dynamically assigned on LAN
X.25 (X.121)	4 (dec) digits of DNIC with 2- or 3-digit Data Country Code and 1 network digit	Up to 10 or 11 digits of Network terminal Number; usually assigned by WAN service provider

Table 4–1
Addressing details for IPX, AppleTalk, and X.25.

The addressing schemes listed in Table 4–1 are covered in greater detail in Chapter 11, "Configuring Novell IPX," Chapter 12, "Configuring AppleTalk," and Chapter 15, "Configuring X.25."

Cisco routers can handle these and many other protocol-specific Layer 3 addressing schemes. Two other protocols, DECnet and Banyan VINES, are covered in Appendixes B and C.

Routing Uses Network Addresses

Routers relay a packet from one data link to another. To relay a packet, a router uses two basic functions: a path determination function and a switching function.

Figure 4–6 illustrates how routers use the addressing for routing and switching functions. When a packet destined for network 10.1.0.0 arrives at Router 1, the router knows that the packet should be sent out port E0.

Figure 4–6
The network portion of the address is used to make path selections.

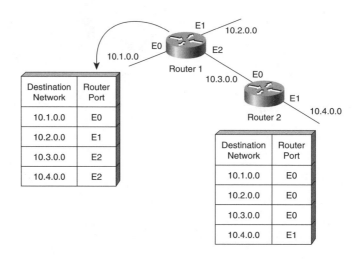

Although the path determination function sometimes is capable of calculating the complete path from the router to the destination, a router is responsible only for passing the packet to the best network along the path. This best path is represented as a direction to a destination network. For example, in Figure 4–6, if a packet that is destined for network 10.4.0.0 arrives at Router 1, the router knows that the best direction to send the packet is out interface E2. Router 2 is the next hop, or router, along the path. The router uses the network portion of the address to make these path selections.

The switching function allows a router to accept a packet on one interface and forward it on a second interface. The path determination function enables the router to select the most appropriate interface for forwarding a packet.

Key Concept With respect to Layer 3, the term *switching* is used to describe moving packets in from one port and out another port. This is different from Layer 2 switching functionality, which refers to forwarding a packet from one port to another port based on the MAC address only.

Rout*ed* Versus Rout*ing* Protocol

Confusion often exists between the similar terms *routing protocol* and *routed protocol*.

Routed Protocol

A routed protocol is a protocol that contains enough network-layer addressing information for user traffic to be directed from one network to another network. Routed protocols define the format and use of the fields within a packet. Packets that use a routed protocol are conveyed from end system to end system through an internetwork.

The Internet protocol IP and Novell's IPX are examples of routed protocols.

Routing Protocol

A routing protocol supports a routed protocol by providing mechanisms for sharing routing information. Routing protocol messages move between the routers. A routing protocol allows the routers to communicate with other routers to update and maintain routing tables. Routing protocol messages do not carry end-user traffic from network to network. A routing protocol uses the routed protocol to pass information between routers. TCP/IP examples of routing protocols are Routing Information Protocol (RIP), Interior Gateway Routing Protocol (IGRP), and Open Shortest Path First (OSPF).

Usually routing protocols function only between routers, but because some routing protocols are unaware of other routers, they rely on data-link broadcast messages to provide information to other routers.

At times, these broadcast messages are used by end systems for their own purposes. For example, an end system receiving a router's update broadcast can record the existence of the router and use the router at a later time if it needs to acquire information about the topology of the internetwork.

For example, AppleTalk's address discovery mechanism—AARP—relies on end systems learning about neighboring routers.

Communications that use routed protocols, such as IP, can be forwarded from one network to another network. Routing protocols, such as IP RIP, are used to make decisions on the best path that those packets should travel. **Key Concept**

Network-Layer Protocol Operations

When a host application sends a packet to a destination on a different network, a data-link frame is received on one of a router's interfaces. The router strips off the MAC header and examines the frame's network-layer header, such as an IP or IPX header, to make a forwarding decision, as shown in Figure 4–7.

Figure 4–7
Each router provides its services to support upper-layer functions.

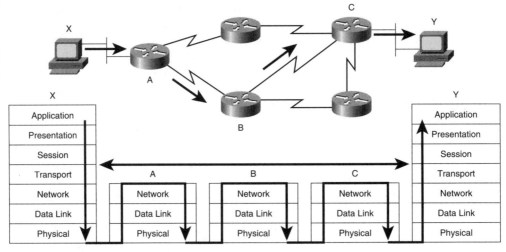

The network-layer data is sent to the appropriate network-layer process, and the data-link layer frame itself is discarded.

The network-layer process examines the header to determine the destination network and then references the routing table that associates networks to outgoing interfaces.

The packet is again encapsulated in the data-link frame for the selected interface and queued for delivery to the next hop in the path.

This process occurs each time the packet switches through another router. At the router connected to the network containing the destination host, the packet is again encapsulated in the destination LAN's data-link frame type for delivery to the protocol stack on the destination host.

Multiprotocol Routing

Routers are capable of supporting multiprotocol routing. That is, routers can support multiple independent routing algorithms and maintain associated routing tables for several routed protocols concurrently. This capability allows a router to interleave packets from several routed protocols over the same data links.

Each routed and routing protocol has no knowledge of other protocols. This concept is called *ships-in-the-night routing.*

For example, in Figure 4–8, router 1 and router 2 handle IP, IPX, AppleTalk, and DECnet traffic. The routing information for each environment is not absorbed by and does not affect the other protocols.

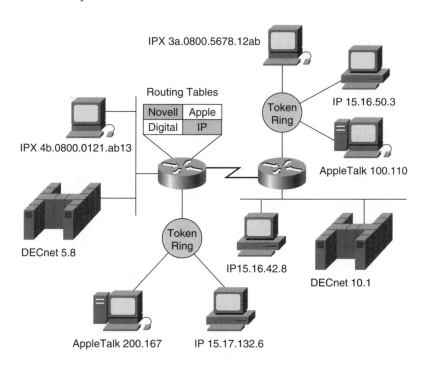

Figure 4–8
Routers pass traffic from all routed protocols over the internetwork.

Later, this chapter discusses an alternative to multiprotocol routing: integrated routing in balanced protocols such as Enhanced IGRP.

Static Versus Dynamic Routes

Static knowledge is administered manually: A network administrator enters it into the router's configuration. The administrator must manually update this static route entry whenever an internetwork topology change requires an update. Static knowledge can be private; by default, it is not conveyed to other routers as part of an update process. You can, however, configure the router to share this knowledge.

Dynamic knowledge works differently. After the network administrator enters configuration commands to start dynamic routing, route knowledge is updated automatically by a routing process whenever new topology information is received from the internetwork. Changes in dynamic knowledge are exchanged between routers as part of the update process.

Static Route Example

Static routing has several useful applications when it reflects a network administrator's special knowledge about network topology. One such application is security. Dynamic routing tends to reveal everything known about an internetwork to sources outside it. For security reasons, it might be appropriate to conceal parts of an internetwork. Static routing allows an internetwork administrator to specify what is advertised about restricted partitions.

Another application is when an internetwork partition is accessible by only one path. In such a case, a static route to the partition can be sufficient. This type of partition is called a *stub network*. Configuring static routing to a stub network avoids the overhead of dynamic routing.

For example, in Figure 4–9, router A is configured with a static route to the remote stub network; there is no reason to allow periodic routing updates across the WAN link between router A and router B, as would occur with dynamic routing.

Figure 4–9
Static routing entries can eliminate the need to allow route updates across the WAN link.

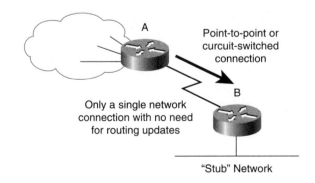

For example, in a network that supports dial-on-demand routing (DDR), the link is not up all the time. Therefore, routing updates should not occur due to cost constraints of the WAN link. Static routing entries can be used for this type of WAN link.

Static routing generally is not sufficient for large or complex networks because of the time required to define and maintain static route table entries. Dynamic routing is used to enable routers to build their routing tables automatically and make the appropriate forwarding decisions.

Default Route

A default route is a path on which a router should forward a packet if it does not have specific knowledge about the packet's destination.

Figure 4–10 shows a use for a default route—a routing table entry that is used to direct frames for which the next hop is not explicitly listed in the routing table. Default routes can be set manually by the administrator (static configuration), or they can be set by dynamic routing with knowledge of many protocols. Figure 4–10 shows the use of a default route.

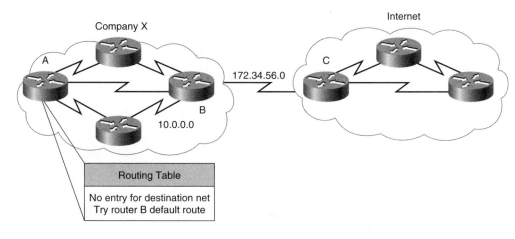

Figure 4–10
A default route is used if the next hop is not explicitly listed in the routing table.

In Figure 4–10, company X routers possess specific knowledge of the topology of the company X internetwork, but not of other networks. Maintaining knowledge of every other internetwork accessible by way of the Internet cloud is unnecessary and unreasonable, if not impossible.

Instead of maintaining specific internetwork knowledge, each router in company X is informed by the default route that it can reach any unknown destination by directing the packet to the Internet.

Adapting to Topology Change

The internetwork shown in Figure 4–11 adapts differently to topology changes depending on whether it uses statically or dynamically configured knowledge.

Static knowledge allows the routers to properly route a packet from network to network. In Figure 4–11, router A refers to its routing table and follows the static knowledge there to relay the packet to router D. Router D does the same and relays the packet to router C. Router C delivers the packet to the destination host.

Figure 4–11
*Dynamic rout-
ing enables
routers to
automatically
use backup
routes when-
ever neces-
sary.*

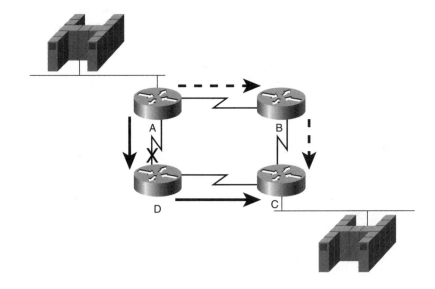

But what happens if the path between router A and router D fails? Obviously, router A will not be able to relay the packet to router D with a static route. Until router A is manually reconfigured to relay packets by way of router B, communication with the destination network is impossible.

Dynamic knowledge offers more automatic flexibility. According to the routing table generated by router A, a packet can reach its destination over the preferred route through router D. However, a second path to the destination is available by way of router B. When router A recognizes the link to router D is down, it adjusts its routing table, making the path through router B the preferred path to the destination. The routers continue sending packets over this link.

When the path between routers A and D is restored to service, router A can once again change its routing table to indicate a preference for the counterclockwise path through routers D and C to the destination network.

Dynamic Routing Operations

The success of dynamic routing depends on two basic router functions:

- Maintenance of a routing table
- Timely distribution of knowledge—in the form of routing updates—to other routers

Dynamic routing relies on a routing protocol to disseminate knowledge. A routing protocol defines the set of rules used by a router when it communicates with neighboring routers. For example, a routing protocol describes:

- How updates are conveyed
- What knowledge is conveyed
- When to convey knowledge
- How to locate recipients of the updates

In Figure 4–12, router 1 uses IP's RIP protocol to pass routing information from its routing table to router 2.

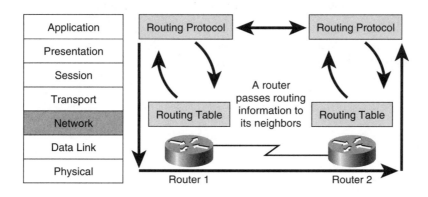

Figure 4–12
Routing protocols maintain and distribute routing information.

Representing Distance with Metrics

When a routing algorithm updates the routing table, its primary objective is to determine the best information to include in the table. Each routing algorithm interprets "best" in its own way. The algorithm generates a number—called the *metric*—for each path through the network. Typically, the smaller the metric, the better the path.

Metrics can be calculated based on a single characteristic of a path or by combining several characteristics. The metrics most commonly used by routing protocols follow (see also Figure 4–13):

- *Hop count*—Refers to the number of routers a packet must go through to reach a destination. The lower the hop count, the better the path. Path length is used to indicate the sum of the hops to a destination.

- *Ticks*—Used with Novell IPX RIP to reflect delay. Each tick is 1/18th of a second.

- *Cost*—Path cost is the sum of the costs associated with each link to a destination. Costs are assigned (automatically or manually) to the process of crossing a network. Slower networks typically have a higher cost than faster networks. The lowest "cost" route is the one believed to be the fastest route available.

- *Bandwidth*—The rating of a link's maximum throughput. Routing through links with greater bandwidth does not always provide the best routes. For example, if a high-speed link is busy, sending a packet through a slower link might be faster.

- *Delay*—Depends on many factors, including the bandwidth of network links, the length of queues at each router in the path, network congestion on links, and the physical distance to be traveled. A conglomeration of variables that change with internetwork conditions, delay is a common and useful metric.

- *Load*—Dynamic factor that can be based on a variety of measures, including CPU use and packets processed per second. Monitoring these parameters on a continual basis can be resource intensive.

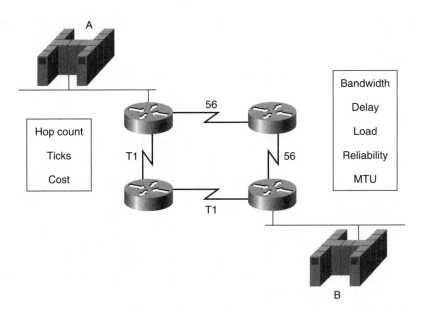

Figure 4–13
A variety of metrics can be used to define the best path.

- *Reliability*—Reflects the propensity of network links to fail and the speed with which they are repaired. You can take multiple reliability factors into account when assigning reliability ratings. Reliability ratings are usually assigned by the network administrator but can be calculated dynamically by the protocol.

- *MTU* (maximum transmission unit)—The maximum message length in octets that is acceptable to all links on the path. It would be considered the fastest path to travel along a route that supports larger MTUs and allows maximum packet sizes to be used end-to-end.

Although not used directly by the router, expense is another important metric influence. Some organizations might not care about performance as much as operating expenses. For instance, even though the bandwidth is less and the delay is longer, sending packets over leased lines rather than through more expensive public lines may be preferable to some enterprises.

ROUTING PROTOCOLS

Most routing algorithms can be classified as conforming to one of two basic algorithms: *distance vector* or *link state*.

The *distance vector* routing approach determines the direction (vector) and distance to any link in the internetwork.

The *link-state* (also called shortest path first) approach learns the exact topology of the entire internetwork (or at least the partition in which the router is situated).

A third approach, the balanced hybrid approach, combines aspects of the link-state and distance vector algorithms.

The rest of this chapter covers procedures and problems for each of these routing algorithms and presents techniques for minimizing the problems.

Key Concept **There is no single best routing algorithm for all internetworks. Network administrators must weigh technical and nontechnical aspects of their network to determine the best algorithm. Cisco IOS software can configure whatever routing choices best fit the administrator's internetwork. Distance vector protocols are generally less computationally intensive than link-state methods, but typically do not scale well to large networks due to several factors including hop count limitations, time to converge, etc.**

The routing algorithm is fundamental to dynamic routing. Whenever the topology of the internetwork changes because of growth, reconfiguration, or component failure, the router's knowledge base of the network must also change.

Knowledge of the network topology needs to be accurate and consistent from router to router. This accurate, consistent view is called *convergence*. When all routers in an internetwork are operating with the same knowledge, the internetwork is said to have converged.

Fast convergence is a desirable internetwork feature because it reduces the period of time that routers have outdated knowledge for making routing decisions that could be incorrect, wasteful, or both.

Distance Vector Routing

Distance vector-based routing algorithms (also known as Bellman-Ford algorithms) periodically pass copies of a routing table from router to router. Updates between routers also communicate topology changes immediately when they occur.

Each router receives a routing table from other routers connected to the same network (its direct neighbors), as shown in Figure 4–14. For example, in the figure, router B receives information from router A, its router neighbor across the WAN link. Router B adds a distance vector number (such as a number of hops) increasing the distance vector, and then passes the routing table to its other neighbor, router C. This step-by-step process occurs in all directions between direct-neighbor routers.

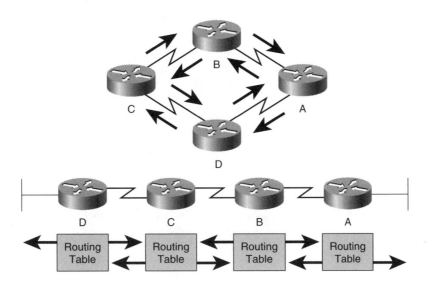

Figure 4–14
Distance vector routers periodically pass copies of their routing table to neighbor routers and accumulate distance vectors.

In this way, the algorithm accumulates network distances so it can maintain a database of internetwork topology information. Distance vector algorithms do not allow a router to know the exact topology of an internetwork.

Distance vector information is somewhat analogous to the information found on signs at a highway intersection. A sign points toward a road leading away from the intersection and indicates the distance to the destination. Further down the highway, another sign also points toward the destination, but now the distance to the destination is shorter. So long as each successive point on the path shows that the distance to the destination is successively shorter, the traffic is following the best path.

Examples of distance vector routing protocols are IPX RIP and IP RIP.

Distance Vector Network Discovery

Each router using distance vector routing begins by identifying its own directly connected networks. In Figure 4–15, the interface to each directly connected network is shown in the routing tables as having a distance of 0.

As the distance vector network discovery process proceeds, routers discover the best path to destination networks based on accumulated metrics from each neighbor.

Figure 4–15
*Distance vec-
tor routers
discover the
best path to
destinations
from each
neighbor.*

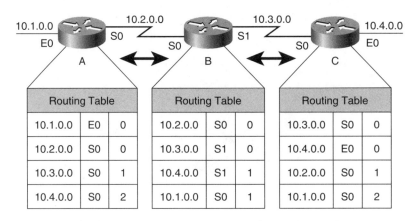

For example, router A learns about other networks based on information it receives from router B. Each of the other network entries learned from router B is placed in router A's routing table and has an accumulated distance vector to show how far away that learned network is in the given direction.

Distance Vector Topology Changes

As mentioned earlier, routing table updates communicate topology changes. As with the network discovery process, topology change updates proceed step by step from router to router, as shown in Figure 4–16.

Figure 4–16
*Updates pro-
ceed step by
step from
router to
router.*

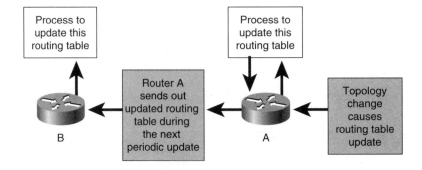

Distance vector algorithms call for each router to send its entire routing table to each of its adjacent neighbors. Distance vector routing tables include information about the total path cost (defined by its metric) and the logical address of the first router on the path to each network it knows about. In Figure 4–15, the metric of each path is shown in the third column of the routing tables.

When a router receives an update from a neighboring router, it compares the update to its own routing table. If it learns about a better route (smaller metric) to a network from its neighbor, the router updates its own routing table. To calculate the new metric, the router adds the cost of reaching the neighbor router to the path cost reported by the neighbor. The new metric is entered into the router's routing table.

For example, if router B in Figure 4–16 is one unit of cost from router A, router B would add 1 to all costs reported by router A when router B runs the distance vector processes to update its routing table.

Typically, a router sends updates by multicasting or broadcasting its table on each configured port; but other methods, such as sending the table only to preconfigured neighbors, are employed by some routing algorithms.

Multicast is used by the RIP2, OSPF, and EIGRP routing protocols. RIP and IGRP use broadcast.

Problem: Routing Loops

Routing loops can occur if the internetwork's slow convergence on a new configuration causes inconsistent routing entries. Figure 4–17 uses a simplistic network design to show how a routing loop can develop. Later, this chapter looks at how routing loops occur and are corrected on more complex network designs.

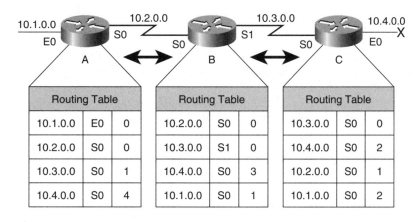

Figure 4–17
Router A updates its table to reflect the new but erroneous hop count.

Routing Table		
10.1.0.0	E0	0
10.2.0.0	S0	0
10.3.0.0	S0	1
10.4.0.0	S0	4

Routing Table		
10.2.0.0	S0	0
10.3.0.0	S1	0
10.4.0.0	S0	3
10.1.0.0	S0	1

Routing Table		
10.3.0.0	S0	0
10.4.0.0	S0	2
10.2.0.0	S0	1
10.1.0.0	S0	2

In Figure 4–17, network 10.4.0.0 has failed, initiating a routing loop between routers A, B, and C. The following steps describe the process of the loop:

- Just before the failure of network 10.4.0.0, all routers have consistent knowledge and correct routing tables. The network is said to have converged. For this example, the cost function is hop count so the cost of each link is 1. Router C is directly connected to network 10.4.0.0 with a distance of 0. Router A's path to network 10.4.0.0 is through router B, with a hop count of 2.

- When network 10.4.0.0 fails, router C detects the failure and stops routing packets out its E0 interface. However, router A has not yet received notification of the failure and still believes it can access 10.4.0.0 through router B. Router A's routing table reflects a path to network 10.4.0.0 with a distance of 2.

- Because router B's routing table indicates a path to network 10.4.0.0, router C believes it now has a viable path to network 10.4.0.0 through router B. Router C updates its routing table to reflect a path to network 10.4.0.0 with a hop count of 2.

- Router A receives the new routing table from router B, detects the modified distance vector to network 10.4.0.0, and recalculates its own distance vector to 10.4.0.0 as 3.

Because routers A, B, and C conclude that the best path to network 10.4.0.0 is through each other, packets destined to network 10.4.0.0 continue to bounce between the three routers.

Symptom: Counting to Infinity

The invalid updates about network 10.4.0.0 continue to loop, and the hop count increments each time the update packet passes through another router. This process of continually incrementing the hop count is called *counting to infinity*. Without countermeasures to stop the process, the loop and the process of counting to infinity will continue indefinitely.

Solution: Defining a Maximum

The countermeasure to counting to infinity is that distance vector protocols define infinity as some maximum number. Such a maximum can be defined for any routing metric, including hop count. For example, RIP has a maximum hop count of 16.

With this approach, the routing protocol permits the routing loop until the metric exceeds its maximum allowed value. Once the metric value exceeds the maximum, network 10.4.0.0 in Figure 4–17 is considered unreachable. The routers will designate it as unreachable in their routing tables and stop circulating update information indicating the network is reachable.

By defining a maximum, distance vector routing algorithms are self-correcting in response to incorrect routing information, although not immediately so. A loop may occur for some finite period of time, until the maximum metric value is exceeded.

A related concept is the Time To Live (TTL) parameter. The TTL is a packet parameter that decreases each time a router processes the packet. When the TTL reaches zero, a router discards or drops the packet without forwarding it. A packet caught in a routing loop is removed from the internetwork when its TTL expires. IP uses a *Time To Live* counter to stop "count to infinity" problems. When the TTL reaches 0, a router discards the packets, thereby preventing the packets from looping forever.

Solution: Split Horizon

One way to eliminate routing loops and speed up convergence is through the technique called *split horizon*. The logic behind split horizon is that it is never useful to send information about a route back in the direction from which the information originally came. In Figure 4–18, for example, router B learns that network 10.4.0.0 is down through the following steps:

- Router B has access to network 10.4.0.0 through router C. It makes no sense for router B to announce to router C that router B has access to network 10.4.0.0 through router C, as router C will always have the best information about 10.4.0.0.

- Given that router B passed the announcement of its route to network 10.4.0.0 to router A, it makes no sense for router A to announce its distance from network 10.4.0.0 to router B.

- Having no alternative path to network 10.4.0.0, router B concludes that network 10.4.0.0 is inaccessible.

In its basic form, the split horizon technique simply does not allow update information to flow out the same interface it arrived on.

Another form of split horizon, called poison reverse, is discussed next.

Solution: Poison Reverse

Poison reverse is a variation of split horizon. Poison reverse attempts to eliminate routing loops caused by inconsistent updates. With this technique, a router that discovers

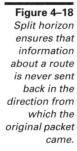
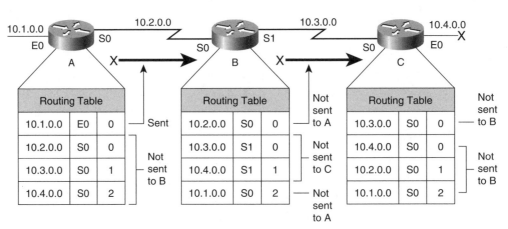

Figure 4–18
*Split horizon
ensures that
information
about a route
is never sent
back in the
direction from
which the
original packet
came.*

an inaccessible route sets a table entry that keeps the network state consistent while other routers gradually converge correctly on the topology change. Used with hold-down timers, which are described in the next section, route "poisoning" is a solution to long loops.

For example, when network 10.4.0.0 goes down, as shown in Figure 4–19, router C can poison its link to network 10.4.0.0 by recording a table entry for that link as having infinite cost (that is, being unreachable). By poisoning its route to network 10.4.0.0, router C is not susceptible to other incorrect updates about network 10.4.0.0 coming from neighboring routers that might claim to have a valid alternate path.

When an update shows the metric for an existing route to have increased sufficiently, there is a loop. The route should be removed (poisoned) and put into holddown. Currently, the rule is that a route is removed if the composite metric increases more than a factor of 1.1. It is not safe for just any increase in the composite metric to trigger removal of the route, because small metric changes can occur due to changes in channel occupancy or reliability. This rule is needed only to break very large loops, because small ones will be prevented by split horizon, triggered updates, and holddowns.

Solution: Hold-Down Timers

Hold-down timers are used to prevent regular update messages from inappropriately reinstating a route that may have gone bad. Holddowns tell routers to hold any changes that might affect routes for some period of time. The hold-down period is usually calculated to be just greater than the period of time necessary to update the entire network with a routing change.

Hold-down timers work as follows:

- When a router receives an update from a neighbor indicating that a previously accessible network is now inaccessible, the router marks the route as inaccessible and starts a hold-down timer, as shown in Figure 4–19. If at any time before the hold-down timer expires an update is received from the same neighbor indicating that the network is again accessible, the router marks the network as accessible and removes the hold-down timer.

- If an update arrives from a different neighboring router with a better metric than originally recorded for the network, the router marks the network as accessible and removes the hold-down timer.

- If at any time before the hold-down timer expires an update is received from a different neighboring router with a poorer metric, the update is ignored. Ignoring an update with a poorer metric when a holddown is in effect allows more time for the knowledge of the change to propagate through the entire network.

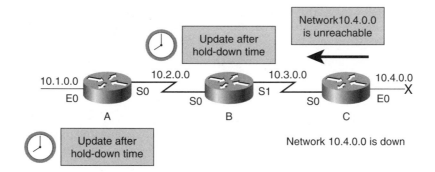

Figure 4–19
A router keeps an entry for the "network down" state, allowing time for other routers to recompute for this topology change.

Solution: Triggered Updates

In the previous examples of routing loops, the loops were caused by erroneous information calculated as a result of inconsistent updates, slow convergence, and timing of updates. If routers wait for their regularly scheduled updates before notifying neighboring routers of network catastrophes, serious problems can occur.

Normally, new routing tables are sent to neighboring routers on a regular basis. For example, IP RIP updates occur every 30 seconds. IPX RIP updates occur every 60 seconds. A *triggered update* is an update that is sent immediately in response to some change in the routing table. The router that detects a topology change immediately sends an update message to adjacent routers that, in turn, generate triggered updates notifying their adjacent neighbors of the change. This wave of updates will propagate throughout that portion of the network where routes connect to the faulty link.

In Figure 4–20, for example, router C immediately announces that network 10.4.0.0 is unreachable. Upon receipt of this information, router B announces through interface S0 that network 10.4.0.0 is down. In turn, router A sends an update out interface E0.

Figure 4–20
With the triggered update approach, nodes send messages as soon as they notice a change in their routing table.

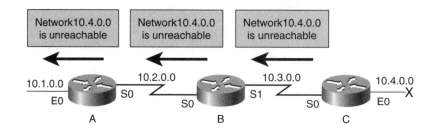

Triggered updates would be sufficient if you could guarantee that the wave of updates reached every appropriate router immediately. However, there are two problems:

- Packets containing the update message can be dropped or corrupted by some link in the network.

- The triggered updates do not happen instantaneously. It is possible that a router that has not yet received the triggered update will issue a regular update at just the wrong time, causing the bad route to be reinserted in a neighbor that had already received the triggered update.

Coupling triggered updates with holddowns is designed to get around these problems. The hold-down rule says that when a route is removed, no new route to the same destination will be accepted for a certain period of time. Thus, the triggered update has time to propagate throughout the network.

Implementing Solutions in Multiple Routes

The individual solutions discussed so far work together to prevent routing loops in a more complex network design. In this scenario, the routers have multiple routes to each other. Consider the design shown in Figure 4–21. Routers A, D, and E each have two routes to network 10.4.0.0.

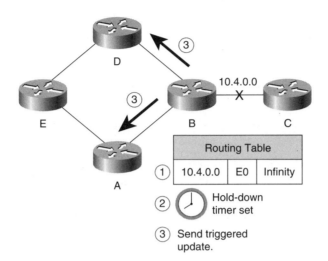

Figure 4–21
As soon as router B detects that network 10.4.0.0 is down, it poisons its route entry in its routing table.

When network 10.4.0.0 fails, the following steps must occur:

1. Poison route—As soon as router B detects the failure of network 10.4.0.0, router B poisons its route to that network by indicating an infinite hop count to that network.
2. Set hold-down timer—Once router B poisons its route to network 10.4.0.0, router B then sets its hold-down timer.
3. Send triggered update—Router B also sends a triggered update to routers D and A, indicating that network 10.4.0.0 is "possibly down." New route information propagates through the rest of this network as the series of connected routers set hold-down timers and trigger updates (steps 2 and 3). Routers D and A receive the triggered update and set their own hold-down timers to suppress any route changes for a specific period of time. Routers D and A, in turn, send a triggered update to router E indicating the possible inaccessibility of network 10.4.0.0.

Finally, router E receives the triggered update about the status of network 10.4.0.0 from routers D and A. Router E then sets its hold-down timer and waits until one of the following events occurs:

- The hold-down timer expires. In this case, router E knows that network 10.4.0.0 is definitely unavailable.

- Another update is received indicating the network status has changed. In this case, Router E updates its tables with the new information.

- Another update is received indicating a new route with a better metric. In this case, Router E updates its tables with the new route information.

During the hold-down period, Router E assumes the network status is unchanged from its original state and will attempt to route packets to network 10.4.0.0.

Link-State Routing

The second basic algorithm used for routing is the *link-state algorithm.*

Link-state routing algorithms—also known as *shortest path first (SPF) algorithms*—maintain a complex database of topology information. Whereas the distance vector algorithm has entries for distant networks and a metric value to reach those networks but no knowledge of distant routers, a link-state routing algorithm maintains full knowledge of distant routers and how they interconnect. Examples of link-state routing protocols are NLSP, OSPF, and IS-IS.

Link-state routing uses link-state packets (LSPs), a topological database, the SPF algorithm, the resulting SPF tree, and, finally, a routing table of paths and ports to each network. The following pages cover these processes and databases in more detail.

Link-State Network Discovery

Link-state network discovery mechanisms are used to create a common picture of the entire internetwork. All link-state routers share this common view of the internetwork. This is similar to having several identical maps of a town. In Figure 4–22, four networks (W, X, Y, and Z) are connected by three link-state routers (A, B, and C).

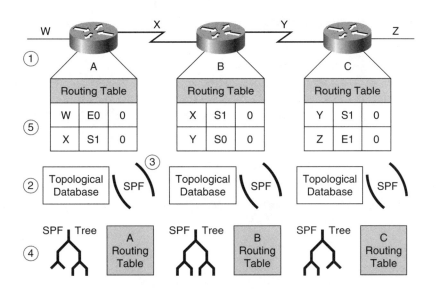

Figure 4–22
In link-state routing, routers calculate the shortest path to destinations in parallel.

Network discovery for link-state routing uses the following processes:

- Routers learn of their neighbors; that is, other routers that are on directly connected networks with them. This process is often referred to as neighbor notification. In link-state routing, each router connected to a network keeps track of its neighbors.

- Routers transmit LSPs onto the network. The LSPs contain information about which networks the routers are connected to.

- Next, routers construct their topological databases consisting of all the LSPs from the internetwork.

- The SPF algorithm computes network reachability, determining the shortest path from a router to each other network in the link-state protocol internetwork. The router uses the Dijkstra algorithm to construct this logical topology of shortest paths as an SPF tree with itself as root. The SPF tree expresses paths from the router to all destinations.

- The router lists its best paths and the ports to these destination networks in the routing table.

After the routers dynamically discover the details of their internetwork, they can use the routing table for switching packet traffic.

Link-State Topology Changes

Link-state algorithms rely on routers having a common view of the network. Whenever a link-state topology changes, the routers that first become aware of the change send information to other routers or to a designated router that all other routers can use for updates. This action entails the propagation of common routing information to all routers in the internetwork. To achieve convergence, each router does the following:

- Keeps track of its neighbors: the neighbor's name, whether the neighbor is up or down, and the cost of the link to the neighbor.

- Constructs an LSP that lists the names and link costs of its neighbor routers. This information includes new neighbors, changes in link costs, and links to neighbors that have gone down.

- Sends out this LSP so that all other routers receive it.

- When it receives an LSP, records the LSP in its database so that it can store the most recently generated LSP from each other router.

- Using accumulated LSP data to construct a complete map of the internetwork topology, it proceeds from this common starting point to rerun the SPF algorithm and compute routes to every network destination.

Each time an LSP causes a change to the link-state database, the link-state algorithm recalculates the best paths and updates the routing table. Then every router takes the topology change into account as it determines the shortest paths to use for packet switching.

Unlike distance vector algorithms, link-state routing algorithms are immediately self-correcting. A loop is terminated by a link-state routing algorithm as soon as the link-state database and routing table are updated.

Link-State Concerns

No routing protocol is perfect, of course. Network administrators need to keep in mind two primary concerns about link-state routing:

- Processing and memory requirements

- Bandwidth requirements

Running link-state routing protocols in most situations requires that routers use more memory and perform more processing. Network administrators must ensure that the routers they select are capable of providing these resources for routing.

Routers keep track of their neighbors and the networks they reach through other routing nodes. For link-state routing, memory must hold information from various link-state advertisements, the topology tree, and the routing table.

The processing complexity of computing the shortest path first is proportional to the number of links in the internetwork times the number of routers in the network.

Another cause for consideration is the bandwidth consumed for initial link-state packet flooding. During the initial discovery process, all routers using link-state routing protocols send LSPs to all other routers. This action floods the internetwork as routers make their peak demand for bandwidth and temporarily reduces the bandwidth available for routed traffic that carries user data.

After this initial flooding, link-state routing protocols generally require internetwork bandwidth only to send infrequent or event-triggered LSPs that reflect topology changes.

Problem: Link-State Updates

The most complex and critical aspect of link-state routing is making sure that all routers get all the LSPs necessary. Routers with different sets of LSPs will calculate routes based on different topological data. Then routes become unreachable as a result of the disagreement among routers about a link. Figure 4–23 provides an example of inconsistent path information.

Consider the following sequence of events in Figure 4–23:

- Suppose that network 1 between routers C and D goes down. As discussed earlier, both routers construct an LSP to reflect this unreachable status.

- Soon afterward, network 1 comes back up; another LSP reflecting this next topology change is needed.

- If the original "Network 1, Unreachable" update message from router C uses a slow path, it may arrive at router A after router D's "Network 1, Back Up Now" LSP.

- With unsynchronized LSPs, router A faces a dilemma about which SPF tree to construct: Does it use paths with or without network 1, which was most recently reported as unreachable?

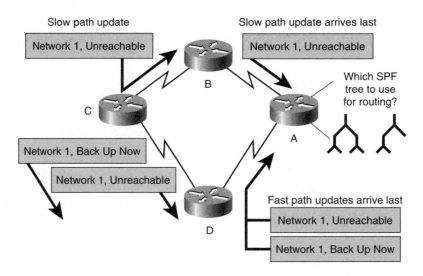

Figure 4–23
*Unsynchro-
nized updates
create incon-
sistent path
decisions.*

If LSP distribution to all routers is not synchronized correctly, link-state routing can result in invalid routes.

Utilizing link-state protocols on very large internetworks can intensify the problem of faulty LSPs being distributed.

For example, if one part of the internetwork comes up first with other parts coming up later, as often occurs when a network is in the process of growing, the order for sending and receiving LSPs will vary. This variation can alter and impair convergence. Routers might learn about different versions of the topology before they construct their SPF trees and routing tables.

Also, on a large internetwork, there is more likely to be variation in transmission speed in different parts of the network. Parts that update more quickly can cause problems for parts that update more slowly. Eventually a partition can split the internetwork into a fast updating part and a slow updating part. Then network administrators must troubleshoot the link-state complexities to restore acceptable connectivity.

A kind of chicken-and-the-egg problem exists for link-state routing that is exacerbated on large internetworks. Specifically, correct delivery of LSPs depends on correct routing table entries, but correct routing table entries depend on accurate LSPs. Routers sending out LSPs cannot assume they will be correctly transported because existing routing table entries might not reflect the current topology.

With faulty updates, LSPs can multiply as they propagate through the internetwork, unproductively consuming more and more bandwidth.

Solution: Link-State Mechanisms

Link-state routing has several techniques for preventing or correcting potential problems arising from resource requirements and LSP distribution:

- A network administrator can reduce the periodic distribution of LSPs so that updates occur only after some long, configurable duration. Reducing the rate of periodic updates does not interfere with LSP updates triggered by topology changes.

- LSP updates can go to a multicast group rather than in a flood to all routers. On interconnected LANs, you can use one or more designated routers as the target depository for LSP transmissions. Other routers can use these designated routers as a specialized source of consistent topology data.

- In large networks, you can set up a hierarchy made up of different areas. A router in one area of the hierarchical domain does not need to store and process LSPs from other routers not located in its area.

- For problems of LSP coordination, link-state implementations can allow for LSP time stamps, sequence numbers, aging schemes, and other related mechanisms to help avoid inaccurate LSP distribution or uncoordinated updates.

Comparing Distance Vector Routing to Link-State Routing

You can compare distance-vector routing to link-state routing in several key areas, as listed in Table 4–2.

Distance Vector	Link-State
Views net topology from neighbor's perspective	Gets common view of entire network topology
Increments metrics as an update passes from router to router	Calculates the shortest path to other routers
Frequent, periodic updates: slow convergence	Event-triggered updates: faster convergence
Passes copies of routing table to neighbor routers	Passes link-state routing updates to other routers

Table 4–2
Comparing distance vector and link-state operational qualities.

The key differences can be summarized as follows:

- Distance vector routing gets all topological data from the perspective it receives from processing the routing table information of its neighbors. Link-state routing obtains a wide view of the entire internetwork topology by accumulating all necessary LSPs.

- Distance vector routing determines the best path by adding to the metric value of each route for each router that must be crossed to get to a network as tables are exchanged from router to router. The larger the metric, the farther away a network is, the less suitable the path is. For link-state routing, each router works simultaneously to calculate its own shortest path to destinations.

- With most distance vector routing protocols, updates for topology changes come in periodic table updates. The entire tables pass incrementally from router to router, usually resulting in slower convergence than in link-state routing. With link-state routing protocols, updates are usually triggered by topology changes. Relatively small LSPs passed to all other routers, or a multicast group of routers, usually result in faster convergence.

Designing your network's routing characteristics to meet technical goals—that is, to use the quickest, shortest, cheapest, or most reliable path—is not always your only goal as a network administrator. Business concerns can also influence routing policy. Conformance with the policies, priorities, and partnerships of an organization impacts routing choices. For example, one routing selection might be considered more desirable because it uses the facilities of a partner or avoids the facilities of a competitor. Multivendor support or standards conformity might outweigh technical superiority.

Operational issues such as the concern for network simplicity are also important. For the chosen routing protocol to properly fit some organizations, it must be easy to set up and manage. It must handle several routed protocols without requiring multiple inconsistent and complex configuration templates.

Finally, avoiding the risk of unproven technologies can also be a factor in designing routing policies and in sustaining a network administrator's career.

Hybrid Routing

This chapter so far has presented the two major types of routing protocols: distance vector and link state.

An emerging third type of routing protocol combines aspects of both. This third type is called *balanced hybrid*.

The balanced hybrid routing protocol uses distance vectors with more accurate metrics to determine the best paths to destination networks. However, it differs from most distance vector protocols by using topology changes to trigger routing database updates.

The balanced hybrid routing type converges relatively quickly, like the link-state protocols. However, it differs from these protocols by emphasizing economy in the use of required resources such as bandwidth, memory, and processor overhead.

Examples of balanced hybrid protocols are OSI's Intermediate System-to-Intermediate System (IS-IS) routing and Cisco's Enhanced Interior Gateway Routing Protocol (Enhanced IGRP).

BASIC ROUTING PROCESSES

Regardless of whether a network uses distance vector or link-state routing mechanisms, its routers must perform the same basic routing functions. The network layer must relate to and interface with various lower layers. Routers must be capable of seamlessly handling packets encapsulated into different lower-level frames without changing the packets' Layer 3 addressing.

LAN-to-LAN Routing

Figure 4–24 shows an example of network layer interfacing in LAN-to-LAN routing. In this example, packet traffic from source host 4 on Ethernet network 1.0 needs a path to destination host 5 on network 2.0. The LAN hosts depend on the router and its consistent network addressing to find the best path.

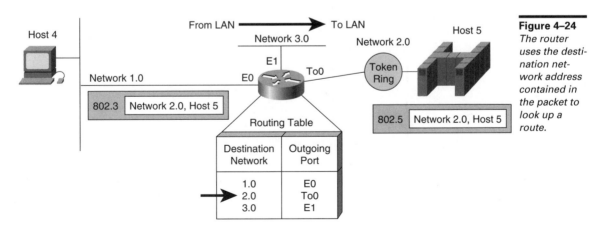

Figure 4–24
The router uses the destination network address contained in the packet to look up a route.

When the router checks its router table entries, it discovers that the best path to destination network 2.0 uses outgoing port To0, the interface to a Token Ring LAN.

Although the lower-layer framing must change as the router switches packet traffic from the Ethernet on network 1.0 to the Token Ring on network 2.0, the Layer 3 addressing for source and destination remains the same. In Figure 4–24, the destination address remains network 2.0, host 5 despite the different lower-layer encapsulations.

LAN-to-WAN Routing

As an internetwork grows, the path taken by a packet might encounter several relay points and a variety of data-link types beyond the LANs. For example, in Figure 4–25, a packet from the top workstation at address 1.3 must traverse three data links to reach the file server at address 2.4 shown on the bottom.

Figure 4–25
Routers maintain the end-to-end address information as they forward the packet.

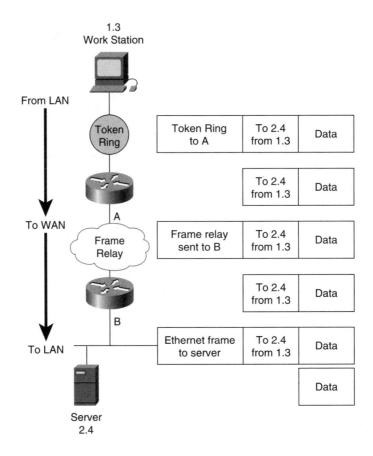

The routed communications follows these basic steps.

1. The workstation sends a packet to the file server by encapsulating the packet in a Token Ring frame addressed to router A at the data link layer and the file server at the network layer.
2. When router A receives the frame, it removes the packet from the Token Ring frame, encapsulates it in a Frame Relay frame, and forwards the frame to router B.
3. Router B removes the packet from the Frame Relay frame and forwards the packet to the file server in a newly created Ethernet frame.
4. When the file server at 2.4 receives the Ethernet frame, it extracts and passes the packet to the appropriate upper-layer process.

The routers enable LAN-to-WAN packet flow by keeping the network layer source and destination addresses constant while encapsulating the packet at the interface to a data link that is appropriate for the next hop along the path.

SUMMARY

Routers are devices that implement network layer services, including path determination and switching. Typically, routers are required to support multiple protocol stacks, each with their own routing protocols, and to allow these different environments to operate in parallel.

This chapter concludes the overview of background concepts, especially the OSI reference model, that you need to understand before configuring routers. The rest of this book focuses on operations and techniques for configuring Cisco routers to operate a variety of protocols and media types. In particular, the next chapter covers the steps required to boot up a Cisco router, and to enter and use the various operating modes.

Chapter Four Test
Network Layer and Path Determination

Estimated Time: 15 minutes

Complete all the exercises to test your knowledge of the materials contained in this chapter. Answers are listed in Appendix A, "Chapter Test Answer Key."

Question 4.1

T   F   Link-state routers build a common view of internetwork topology.

Question 4.2

T   F   Link-state routing uses periodic updates, resulting in relatively slow convergence.

Question 4.3

T   F   Distance vector routing develops a view from neighbor routers' perspectives.

Question 4.4

T   F   Distance vector routing passes an updated routing table from neighbor to neighbor.

Question 4.5

T   F   In link-state routing, events trigger updates for relatively fast convergence.

Question 4.6

T   F   Distance vector routers process updates in parallel with other routers.

Question 4.7

TCP/IP routers forward packets based on the contents of the packet's _____ header.

Question 4.8

On a 10BaseT network, the _____ frame is placed directly in front of the network layer header.

Questions 4.9–4.11

Write the letter of the header that could contain the following data:

A. Data-link header
B. Network-layer header
C. Transport-layer header

Question 4.9 _____ Destination 10.4.0.6
Question 4.10 _____ Source 00.00.OB.A4.26.39
Question 4.11 _____ Destination port 20

Basic Router Operations

This chapter discusses basic startup procedures, the various command modes, and status commands of a router. You will develop a model of the router based on components that are configurable, and then build on this model to understand how configuration commands work.

The screens in this section reflect Cisco IOS Release 11.2(6). If you are running a different version, your screens may vary from those shown.

STARTING UP A ROUTER

This section covers the router components that play a key role in the configuration process. It also examines the router startup routines. Knowing which components are involved in the configuration process gives you a better understanding of how the router stores and uses your configuration commands. Being aware of the steps that take place during router initialization will help you determine what and where problems may occur when you start up your router.

External Configuration Sources

The router can be configured from many locations (see Figure 5–1), including

- *Console Port*—Upon initial installation, you configure the router from a console terminal, which is connected to the router via the console port.

- *Auxiliary Port*—You can also configure a router using the auxiliary port.

- *Virtual Terminals*—You can configure a router from virtual terminals 0 through 4 after the router is installed on the network. Note that you can access a VTY typically via Telnet.

- *TFTP Server*—You can also download configuration information from a TFTP server on the network. The TFTP server can be a UNIX or PC workstation that acts as a central depository for files. You can keep configuration files on the TFTP server and then download them to the router.

- *Network Management Station*—You can manage router configuration from a remote system running network management software such as CiscoWorks or HP OpenView.

Figure 5–1
Router config-uration infor-mation can come from many sources.

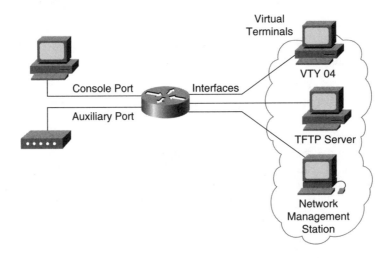

To send and receive configuration information to and from a virtual terminal, TFTP server, or network management station, the router must be configured to support IP traffic.

Cisco routers can be accessed for configuration purposes from the console port, the auxiliary port, and five VTY lines at the same time; up to seven people can configure the router at once.

Because of this, security of the router should be strictly observed through password protection to avoid unauthorized access of the router configuration files.

Internal Configuration Components

The internal architecture of the Cisco router supports components that play an important role in the startup process, as shown in Figure 5–2. These components are RAM/DRAM, NVRAM, Flash memory, ROM, interfaces, and ports.

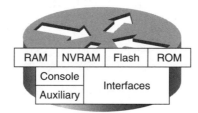

Figure 5–2
The internal configuration components consist of several elements.

RAM/DRAM

RAM/DRAM (Random Access Memory/Dynamic Random Access Memory) is the main storage component for the router. RAM is also called working storage and contains the dynamic configuration information.

NVRAM

NVRAM (nonvolatile RAM) contains a backup copy of your configuration. If the power is lost or the router is turned off for a period of time, the backup copy of the configuration enables the router to return to operation without needing to be reconfigured.

Flash Memory

Flash memory is a special kind of erasable, programmable read-only memory. This memory contains a copy of Cisco Internetwork Operating System (Cisco IOS) software. Flash memory has a structure that enables it to store multiple copies of the Cisco IOS software, allowing you to load a new level of the operating system in every router in your network and then, at some convenient time, to upgrade the whole network to that new level. Flash memory content is retained when you power down or restart.

ROM

ROM contains an initializing bootstrap program and a small monitoring system that can be used for recovery from a catastrophe. The Cisco 2500, 4000, and 4500 router series have a subset of the Cisco IOS software in ROM. The Cisco 7000 and 7500 router series have full Cisco IOS software in ROM. You can upgrade the ROM software by replacing pluggable chips on the CPU.

Interfaces

Interfaces are the network connections through which packets enter and exit the router.

Interface support is available for Token Ring, Ethernet, Fast Ethernet, and GigaBit Ethernet. Additionally there is support for ISDN, BRI, ATM, DDDI, and other physical connections.

Auxiliary Ports

Cisco IOS software also allows the auxiliary port to be used for asynchronous routing as a network interface.

System Startup Overview

The startup routines for Cisco IOS software have the goal of starting router operations. Cisco routers are designed to deliver reliable service for the connecting user networks. To succeed, the startup routines must perform three tasks:

1. Check hardware and conduct a power-on self-test (POST).
2. Find and load the Cisco IOS software image. The image is the data the router uses for its operating system.
3. Find and apply the router configuration information. This information includes statements about router-specific attributes, protocol functions, and interface addresses.

First the router makes sure that it comes up with tested hardware. During the POST process, the router executes diagnostics from ROM on all modules. These diagnostics verify the basic operation of the CPU, memory, and interface circuitry.

After verifying the hardware functions, the router proceeds with software initialization. Some startup routines act as fallback operations that are able to perform the router startup should other routines be unable to do so. This flexibility allows Cisco IOS software to start up in a variety of initial situations.

Next in the startup sequence, the router searches its configuration register to determine where to find the Cisco IOS software. If your router does not find a valid system image, or if you interrupt the boot sequence, the system enters ROM monitor mode. From ROM monitor mode, you can also boot the device or perform diagnostic tests. You can configure the router to automatically initiate ROM monitor mode every time the router starts up.

You can also configure the router to boot the Cisco IOS image file from ROM or to look in NVRAM for user-defined instructions on where to locate the image file.

After the Cisco IOS software has been loaded, the router will attempt to load the configuration file if one exists. This file contains all the configuration information you specified for this particular router. The configuration file is stored in NVRAM; however, you can configure the router to load the configuration file from a TFTP server.

If no configuration file exists, the router will revert to setup mode. Setup mode is an interactive dialog that allows you to create a basic configuration for the router. If the router is configured to load the software from a TFTP server and the server cannot be found, the router uses the configuration file existing in NVRAM. If the TFTP server is available, the router loads the alternate configuration file stored on the TFTP server.

Figure 5–3 shows the entire startup sequence of a router.

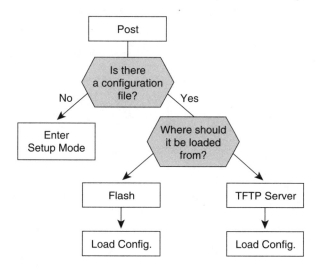

Figure 5–3
The entire startup sequence of a router.

The router can start up from ROM, Flash, or TFTP server. After the router has completed the initialization process, it begins operations. At this point, you can build new configuration parameters or alter an existing one. In either case, you access the router through the user interface commands.

ACCESSING THE USER INTERFACE

This section covers how to manipulate the Cisco IOS software from a router console. The router console can be a workstation running a terminal emulation package, such as Hyperterm, or a remote device running Telnet. Both methods give you access to the Cisco IOS software user interface to log in, log out, and enter commands for the router.

This section also covers accessing context-sensitive help, using editing commands, and reviewing command history.

Router Modes

The Cisco IOS user interface provides access to command modes, as shown in Figure 5–4.

Figure 5–4
*Cisco IOS sup-
ports seven
command
modes.*

User EXEC Mode Limited examination of router. Remote access. **Router >**	**Setup Mode** Prompted dialog used to establish an initial configuration.
Privileged EXEC Mode Detailed examination of router. Debugging and testing. File manipulation. Remote access. **Router#**	**RXBoot Mode** Boot helper software which helps the router boot when it cannot find a valid Cisco IOS image in Flash memory. **Router<boot>**
	Global Configuration Mode Commands that affect the system as a whole. **Router (config)#**
ROM Monitor Mode Used if the operating system does not exist in Flash or the boot sequence was interrupted during startup. **> or rommon >**	**Other Configuration Modes** Complex and multiline configurations. **Router (config-mode)#**

TIPS

Enter a question mark (?) at the system prompt to obtain a list of commands available for each command mode.

User EXEC Mode

Cisco IOS software provides a command interpreter called EXEC, which interprets the commands you type and carries out the corresponding operations.

EXEC has two levels of access to commands: user mode and privileged mode.

After you log in to the router, you are automatically in user EXEC command mode. In general, the user EXEC mode contains nondestructive commands that allow you to connect to remote devices, change terminal settings on a temporary basis, perform basic tests, and list system information. User EXEC mode is indicated by the device host name followed by the angle bracket (>).

Privileged EXEC Mode

The privileged EXEC mode commands set operating parameters. The privileged commands include those commands contained in user EXEC mode, as well as the **configure** command through which you can access the remaining command modes. Privileged EXEC mode also includes high-level testing commands, such as **debug**. To enter privileged EXEC mode, enter **enable** at the user EXEC prompt. The privileged EXEC mode prompt consists of the device host name followed by the pound sign (#).

Privileged EXEC mode contains potentially destructive commands and should be password protected.

From the privileged level, you can access a number of specific configuration modes.

ROM Monitor Mode

ROM monitor mode is a command-line interface (CLI) that allows you to configure your router. ROM monitor mode occurs if your router does not find a valid system image or if you interrupt the boot sequence during startup. The ROM Monitor prompt is the angle bracket (>). On the Cisco 1003, 1600, 2600, 3600, 4500, 7200, and 7500 series, rommon> is the default ROM monitor prompt. The **continue** command takes you from ROM monitor to user EXEC mode.

TIPS

From the Cisco 2000, 2500, 3000, and 4000 series routers, you can also enter ROM monitor mode by entering the reload command and then pressing the **Break** key during the first 60 seconds of startup.

Setup Mode

This mode is an interactive prompted dialog at the console that helps the new user create a first-time basic configuration. You can also enter setup mode by entering **setup** at the privileged EXEC prompt. Setup mode consists of a series of questions and does not exhibit a defining prompt of its own.

RXBoot Mode

This mode is a special mode you enter by altering the settings of the configuration register and rebooting the router. RXBoot mode provides the router with a subset of the Cisco IOS software and helps the router boot when it cannot find a valid Cisco IOS image in Flash memory. The RXBoot mode prompt is the host name followed by <boot>.

Global Configuration Mode

Global configuration commands apply to features that affect the system as a whole. You initiate global configuration mode by entering the **configure** command at the privileged EXEC mode prompt. Global configuration mode is indicated by the device host name (config) followed by the pound sign (#). To exit to privileged EXEC mode, enter **exit,** **end,** or press **Ctrl-Z** at the prompt.

From global configuration mode, you can access a number of other command modes.

Other Configuration Modes

These modes provide more specific multiple-line configurations that target individual interfaces or functionality, such as modifying the operation of an interface, configuring multiple virtual interfaces (called subinterfaces) on a single physical interface, or setting an IP routing protocol.

There are more than 17 different specific configuration modes. To learn more about them, refer to www.cisco.com.

Logging In to the Router

When you first log in to the router, you will automatically be in user EXEC mode. To exit user EXEC mode, type **logout** at the prompt.

The EXEC command interpreter waits for a specified interval of time for you to start input. If no input is detected, the EXEC resumes the current connection, and you log in to the router again. The default interval the router waits for input is 10 minutes; an interval of 0 (zero) specifies the router will not time out. The **no exec-timeout** command removes the timeout definition and is the same as entering the **exec-timeout** 0 command. This command is entered in the line configuration mode, which is discussed later in this book.

Enter privileged EXEC mode by entering **enable** at the user EXEC mode prompt, as shown in Figure 5–5. If privileged EXEC mode has been password protected, you will be prompted to enter the password. Exit privileged EXEC mode by entering the **disable** or **exit** commands.

To log out of the router, type **exit** or **logout**.

Using Passwords

Cisco IOS software supports a variety of security features for controlling access to your routers. The most basic form of security is to control who can log in to your router. This access can be controlled by one or more of the following:

- A line access password
- A privileged EXEC mode password
- Encrypted passwords

Figure 5–5
You must type
enable *to*
enter privi-
leged EXEC
mode after
you log in to
the router.

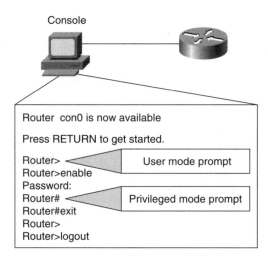

Console

Router con0 is now available

Press RETURN to get started.

Router>   ◁ User mode prompt
Router>enable
Password:
Router#   ◁ Privileged mode prompt
Router#exit
Router>
Router>logout

Passwords are set during initial configuration in setup mode or by issuing the **enable password** command. If an enable password has not been set, enable mode can be accessed only from the router console.

Individual Line Passwords

You can secure access to the router by password protecting individual lines. This level of security requires the users to verify authorization before they can access any line, including the line console.

Privileged EXEC Mode Password

You can also control access to privileged EXEC mode by assigning a password to this mode during the initial setup of the router.

Encrypted Passwords

Cisco provides a feature that allows you to encrypt passwords. If you use this feature, passwords are stored in the router in an encoded form and are masked when you display the router configuration parameters. This encryption is enabled using the **service password-encryption** command.

Using a Password

You can configure your router to have a user password check, as shown in Figure 5–6. The password you enter does not appear onscreen. If you do not enter anything, the login process will time out after a while.

You get three tries to enter the correct password. The router will let you know if the password you entered is incorrect. Press **Return** to acknowledge the message and start over from the idle console.

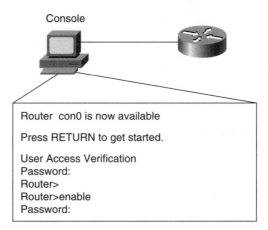

Console

```
Router con0 is now available

Press RETURN to get started.

User Access Verification
Password:
Router>
Router>enable
Password:
```

Figure 5–6
*Password adding is set during initial configuration or by using the **enable password** command.*

User Mode Command List

Once you are in user EXEC mode, you can display all the available commands by typing a question mark (**?**) at the user EXEC mode prompt, as shown in Figure 5–7.

The screen displays 22 lines at one time. The "-- More --" prompt at the bottom of the display indicates that multiple screens are available as output. You can resume output of the next available screen by pressing the spacebar. To display the next line, press the **Return** key (or, on some keyboards, the **Enter** key). Press any other key to return to the user EXEC prompt.

Figure 5–7
Type a question mark (?) to display available commands.

```
Router>?
Exec commands:
    access-enable    Create a temporary Access-List entry
    atmsig           Execute Atm Signalling Commands
    cd               Change current device
    clear            Reset functions
    connect          Open a terminal connection
    dir              List files on given device
    disable          Turn off priviledged commands
    disconnect       Disconnect an existing network connection
    enable           Turn on privileged commands
    exit             Exit from the EXEC
    help             Description of the interactive help system
    lat              Open a lat connection
    lock             Lock the terminal
    login            Log in as a particlar user
    logout           Exit from the EXEC
    mrinfo           Request neighbor and version information
                     from a multicast router
- - more - -
```

Privileged Mode Command List

As in the user EXEC mode, you display the available commands for the privileged EXEC mode by typing a question mark (?) at the privileged prompt, as shown in Figure 5–8. Notice that doing so displays a much larger list of EXEC commands.

Figure 5–8
Typing a question mark lists available commands in privileged EXEC mode.

```
Router#?
Exec commands:
    access-enable    Create a temporary Access-List entry
    access-template  Create a temporary Access-List entry
    bfe              For manual emergency modes setting
    clear            Reset functions
    clock            Manage the system clock
    configure        Enter configuration mode
    connect          Open a terminal connection
    copy             Copy configuration or image data
    debug            Debugging functions (see also 'undebug')
    disable          Turn off privileged commands
    disconnect       Disconnect an existing network connection
    enable           Turn on privileged commands
    erase            Erase flash or configuration memory
    exit             Exit from the EXEC
    help             Description of the interactive help system
    lat              Open a lat connection
    lock             Lock the terminal
    login            Log in as a particular user
    logout           Exit from the EXEC
    mbranch          Trace multicast route down tree branch

- - more - -
```

Context-Sensitive Help

If you know a command but are not sure of the complete command syntax, Cisco IOS software supports context-sensitive help. This feature allows you to get a list of any keywords and arguments associated with a specific command. Both the user and privileged EXEC modes support context-sensitive help.

You can abbreviate commands and keywords to the number of characters that allow a unique abbreviation. For example, you can abbreviate the **clock** command to **clo**.

When using context-sensitive help, the space (or lack of a space) before the question mark (?) is significant. To obtain a list of commands that begin with a particular character sequence, type those characters followed immediately by the question mark (?). Do not include a space. This form of help is called word help, because it completes a word for you.

To list keywords or arguments, enter a question mark (?) in place of a keyword or argument. Include a space before the question mark. This form of help is called command syntax help, because it reminds you which keywords or arguments are applicable based on the command, keywords, and arguments you already have entered.

Using Enhanced Editing Commands

The user interface includes an enhanced editing mode that provides a set of editing key functions. This feature allows you to alter or correct long or complex commands without having to retype them.

Although enhanced editing mode is automatically enabled with the current software release, you can disable it and revert to the editing mode of previous software releases. To disable editing mode, enter the **terminal no editing** command at the user EXEC mode prompt.

You might also want to disable enhanced editing if you have written scripts that do not interact well when enhanced editing is enabled.

The editing command set provides a horizontal scrolling feature for commands that extend beyond a single line on the screen. When the cursor reaches the right margin, the command line shifts 10 spaces to the left. You cannot see the first 10 characters of the line, but you can scroll back and check the syntax at the beginning of the command. The following key combinations help automate scrolling of long lines as described:

- <Ctrl><A>   Move to the beginning of the command line
- <Ctrl><E>   Move to the end of the command line

- <Esc><B>   Move to the beginning of the previous word
- <Ctrl><F>   Move forward one character
- <Ctrl><B>   Move back one character
- <Esc><F>   Move forward one word

To scroll back, press **Ctrl-B** or the **Left** arrow key repeatedly until you are at the beginning of the command entry, or press **Ctrl-A** to return directly to the beginning of the line.

In Figure 5–9, the command entry extends beyond one line. When the cursor first reaches the end of the line, the line is shifted 10 spaces to the left and redisplayed. The dollar sign ($) indicates that the line has been scrolled to the left. Each time the cursor reaches the end of the line, the line is again shifted 10 spaces to the left.

Figure 5–9
The editing command scrolls the line horizontally if the line is too long to display.

> Router> $ value for customers, employees, and partners.

Reviewing Command History

The user interface provides a history or record of commands you have entered. This feature is particularly useful for recalling long or complex commands or entries. With the command history feature, you can complete the following tasks:

- Set the command history buffer size
- Recall commands
- Disable the command history feature

To view the current history settings, type **show history** at the privileged EXEC prompt, as shown in Figure 5–10.

By default, command history is enabled, and the system records 10 command lines in its history buffer. To change the number of command lines the system will record during the current terminal session, use the **terminal history size** command.

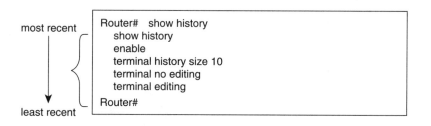

Figure 5–10
Show history
displays a
record of
recent com-
mands.

The following key combinations automate moving through the command history as described:

- <Ctrl><P> or **Up** arrow   Last (previous) command recall
- <Ctrl><N> or **Down** arrow   More recent command recall
- <Tab>   Entry completion

To recall commands in the history buffer beginning with the most recent command, press **Ctrl-P** or the **Up** arrow key. Repeat the key sequence to recall successively older commands.

To return to more recent commands in the history buffer after recalling commands, press **Ctrl-N** or the **Down** arrow key. Repeat the key sequence to recall successively more recent commands.

After you enter the unique characters for a command, press the **Tab** key, and the interface will finish the entry for you.

On most laptop computers, you may also have additional select and copy facilities available. Copy a previous command string, and then paste or insert it as your current command entry and press **Return**.

EXAMINING ROUTER STATUS

This section covers basic commands that you can issue to determine the current status of a router. These commands will help you obtain vital information you need when monitoring and troubleshooting router operations.

It is important to be able to monitor the health and state of your router at any given time. Cisco routers have a series of commands that allow you to determine if the router is functionally correctly or where problems have occurred, as shown in Figure 5–11.

Figure 5–11
*Many com-
mands are
available to
monitor router
configuration.*

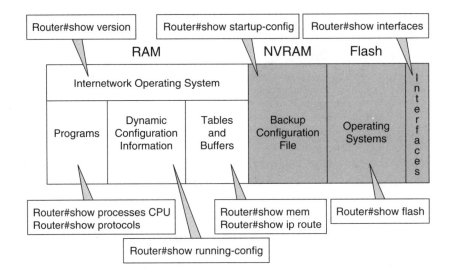

Router status commands are as follows:

- **show version**—Displays the configuration of the system hardware, the software version, the names and sources of configuration files, and the boot images.

- **show processes**—Displays information about the active processes.

- **show protocols**—Displays the configured protocols. This command shows the status of any configured Layer 3 (network) protocol.

- **show memory**—Shows statistics about the router's memory, including memory free pool statistics.

- **show ip route**—Displays the entries in the routing table.

- **show flash**—Shows information about the Flash memory device.

- **show running-config** (**write term** on Cisco IOS Release 10.2 or earlier)— Displays the active configuration parameters.

- **show startup-config** (**show config** on Cisco IOS Release 10.2 or earlier)— Displays the backup configuration file.

- **show interfaces**—Displays statistics for all interfaces configured on the router.

The following sections examine several of these commands in more detail.

show version Command

The **show version** command displays information about the Cisco IOS software version running on the router, as shown in Figure 5–12.

```
Router#show version

IOS (tm) 2500 Software (C2500-JS-L), Version 11.2 (6), RELEASE SOFTWARE (fcl)
Copyright (C) 1986-1997 by cisco Systems, Inc.
Compiled Tue 06-MAY-97 16:17 by Kuong
Image text-base: 0x0303ED8C, data-base: 0x00001000

ROM: System Bootstrap, Version 5.2 (8a), RELEASE SOFTWARE
ROM: 2500-XBOOT Bootstrap Software, Version 10.1(1), RELEASE SOFTWARE (fcl)

router uptime is 1 week, 3 days, 32 minutes
System restarted by reload
System image file is "c2500-js-1", booted via tftp from 171.69.1.129

- - more - -
```

Figure 5–12
Results of the
show version
command.

This information is important to know when you are upgrading the software on your routers or when you are troubleshooting a problem with Cisco support personnel.

Notice that this command not only shows you the version of the router software, but also gives you statistics on how long the system has been up, the name of the system image file, and where the system image file originated. Each time the version of the Cisco IOS software is revised or updated, a revision number is applied to the version. This revision number appears in parentheses directly following the version number. In this example, the version of the Cisco IOS software is 11.2, and the revision number is (6).

TIPS

If several routers are exhibiting the same behavioral problems, use the **show version** command. Perhaps all problem routers obtained the same image file from the same TFTP server, which could indicate that the image file on the TFTP server is corrupted.

show startup-config Command
and show running-config Command

The **show startup-config** and **show running-config** commands (shown in Figure 5–13) are among the most used Cisco IOS software EXEC commands.

Figure 5–13
*The show
startup-
config* and
*show running-
config* are
some of the
most useful
EXEC
commands.

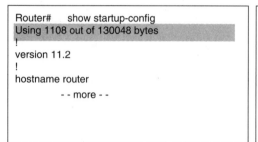

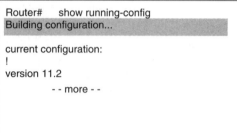

The **show startup-config** command allows an administrator to see the image size and startup configuration commands the router will use on the next restart. This backup file is loaded into memory when the router is initialized and contains all the information you specified about the router interfaces. You will know that you are looking at the startup configuration file when you see a message indicating how much nonvolatile memory has been used.

The **show running-config** command displays the configuration information running in terminal memory. You will know that you are looking at the running configuration when you see the words "Current Configuration" on the screen. You can make changes to the running configuration information; however, those changes will be lost when the router powers down. To record any configuration changes you make to the running configuration, you must copy those changes to the startup configuration file stored in NVRAM. To copy those changes to the startup configuration file, use **copy running-config startup-config** command.

In Cisco IOS Release 10.2 and earlier, the command **write terminal** shows the running configuration, and the command **show config** shows the startup configuration.

show interfaces Command

The **show interfaces** command displays configurable parameters and real-time statistics related to the interfaces on the router. This command is very useful in determining the activity and behavior of a specific interface or in verifying any changes you made to the router interfaces.

A few of the statistics you can obtain from the **show interfaces** command are:

- State of the interface
- MTU size, maximum tranmission unit size.
- Internet address for the interface
- MAC address for the LAN card (for example, Ethernet, Token Ring, or FDDI interface)
- Encapsulation type
- Number of packets received
- Number of input and output packet errors
- Number of collisions detected (on an Ethernet interface)

This command is extremely useful in helping you determine the health and operational history of a router.

Examine Figure 5–14. The top line of the output tells you that the line is up. A few lines down, the output provides the Internet address and below that some link metrics: BW is bandwidth; DLY is delay. On the next line down is HDLC, the default encapsulation protocol for serial lines on Cisco routers.

```
Router#show interfaces
Serial0 is up, line protocol is up
    Hardware is MK5025
    Internet address is 183.8.64.129, subnet mask is 255.255.255.128
    MTU 1500 bytes, BW 56 Kbit, DLY 20000 usec, rely 255/255, load 9/255
    Encapsulation HDLC, loopback not set, keepalive set (10 sec)
    Last input 0:00:00, output 0:00:01, output hang never
    Last clearing of "show interface" counters never
    Output queue 0/40, 0 drops: input queue 0/75, 0 drops
    Five minute input rate 1000 bits/sec, 0 packets/sec
    Five minute output rate 2000 bits/sec, 0 no buffer
        331885 packets input, 62400237 bytes, 0 no buffer
        Received 230457 broadcasts, 0 runts, 0 giants
        3 input errors, 3 CRC, 0 frame, 0 overrun, 0 ignored, o abort
        403591 packets output, 66717279 bytes, 0 underruns
        0 output errors, 0 collisions, 8 interface resets, 0 restarts
        45 carrier transitions
```

Figure 5–14
*The **show
interfaces**
command.*

A runt is a packet that is too small to be legal, and a giant is a packet that is too big.

show protocols Command

Use the **show protocols** EXEC command to display the protocols configured on the router, as shown in Figure 5–15.

This command shows the global and interface-specific status of any configured Layer 3 protocols (for example, IP, DECnet, IPX, and AppleTalk).

Figure 5–15
The show protocols command displays the status of Layer 3 protocols.

```
Router#show protocols
Global values:
    Internet Protocol routing is enabled
DECENT routing is enabled
XNS routing is enabled
Vines routing is enabled
Appletalk routing is enabled
Novell routing is enabled
- - more - -
Ethernet0 is up, line protocol is up
Internet address is 183.8.128.2, subnet mask is 255.255.255.128
Decnet cost is 5
XNS address is 3010.aa00.0400.0284
CLNS enabled
Vines metric is 32
AppleTalk address is 3012.93, zone 1d-e0
Novell address is 3010.aa00.0400.0284
- - more - -
```

The Cisco router often divides configuration file information into global configuration and interface configuration. The global section tells you what traffic this router is capable of passing. The interface section provides more detailed configuration information, such as protocol addresses.

SUMMARY

In this chapter, you've learned in the abstract how a Cisco router is initiated with startup routines. The next chapter expands on this topic from a hands-on viewpoint, including how to load configuration files, name a router, set passwords, and configure an interface. The commands introduced in this chapter to examine the router configuration will be applied in the next chapter to the task of verifying implementations.

Chapter Five Test
Basic Router Operations

Estimated Time: 15 minutes

Complete all the exercises to test your knowledge of the materials contained in this chapter. Answers are listed in Appendix A, "Chapter Test Answer Key."

Question 5.1

In the blank boxes, write the correct command to access each router element in Figure 5–16.

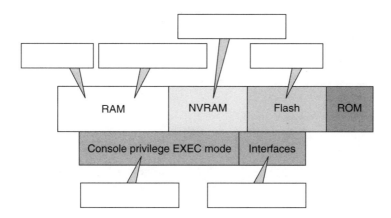

Figure 5–16

Question 5.2

How can you use the context-sensitive help to list privileged EXEC commands?

Question 5.3

What command is used to enter privileged EXEC mode?

Question 5.4

What happens when you enter the **show history** command?

Question 5.5

What happens when you type **exit** at the privileged EXEC mode prompt?

CHAPTER 6

Configuring a Router

This chapter covers loading a configuration file, configuring a serial interface, and defining your configuration environment. The chapter presents the configuration modes available to a privileged user, as well as commands to enter and review changes. Throughout most of this chapter, it is assumed that a configuration file already exists; the focus is on loading and managing the file, not on creating it. However, the last sections of this chapter overview how to use Setup mode to create or change a simple configuration file.

When you configure a router, there are several possible steps you can take to make and save changes. Depending on the particular configuration task, you may not use all the steps all the time. In this chapter, you will

- Load an existing configuration file
- Change the router identification
- Assign a password to both the user and privileged EXEC modes
- Configure a serial interface
- Save the changes to NVRAM

The router uses information from the configuration file when it starts up. The configuration file contains commands to customize router operation. If no configuration file is available, the system configuration dialog setup guides you through creating one. For details on the router startup sequence, refer to Chapter 5, "Basic Router Operations."

LOADING CONFIGURATION FILES

Because the configuration variables affect the router as a whole, you must be in global configuration mode before you can create, load, or alter any existing configuration information. To enter global configuration mode, type the **configure** command at the privileged EXEC mode.

Router configuration information can be generated by several means, as shown in Figure 6–1. Configuration commands can come from a terminal, nonvolatile memory (NVRAM), or a file stored on a network server. The default is to enter commands from the terminal console.

Figure 6–1
There are four ways to load router configuration files.

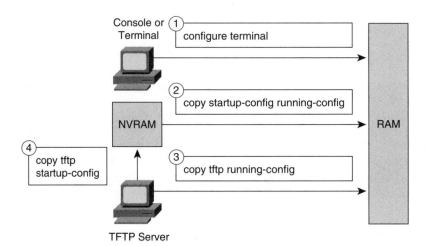

For IOS release 10.3 and later, you can specify the source of the configuration commands directly when you enter global configuration mode by typing one of the following commands:

- **configure terminal**—Executes configuration commands from the terminal
- **copy tftp running-config**—Copies a file from a TFTP server to RAM

- **copy tftp startup-config**—Loads a configuration file from a TFTP server directly into NVRAM

- **copy startup-config running-config**—Copies the configuration information in NVRAM to RAM. The router acts as a command-line compiler and reads the NVRAM configuration file line by line, overwriting only lines that already exist in RAM (a process called *gentle overlay*). If there is a conflict between the two sets of parameters, the router will not turn off processes.

Loading Configuration Files—Releases Before 10.3

The following commands are used with Cisco IOS Release 10.0 and earlier:

- **configure terminal**—Executes configuration commands from the terminal

- **configure memory**—Executes the commands stored in NVRAM

- **configure network**—Copies the configuration file from a network server to RAM

- **configure overwrite**—Loads a configuration file directly into NVRAM without affecting the running configuration. Be careful not to load a file that is larger than NVRAM.

The **configure network** command supports only TFTP servers.

Loading Configuration Files from a TFTP Server

If you have a network consisting of many routers, you can maintain the consistency of your configuration files and reduce your workload by repeatedly using one or two backup configurations. Using a network server to store backups of your configuration files can save you time and keystrokes. The file is then downloaded when you need it. The following example copies a configuration file from a TFTP server to the router, as shown in Figure 6–2.

To retrieve the configuration file stored on your TFTP network server, follow these steps:

1. Enter configuration mode by entering the **copy tftp running-config** command (or **configure network** if you are using Release 10.0 or earlier).

Figure 6–2
*A TFTP server
can store
backup config-
uration files to
be down-
loaded when
needed.*

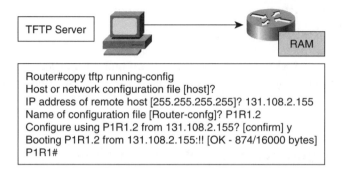

```
Router#copy tftp running-config
Host or network configuration file [host]?
IP address of remote host [255.255.255.255]? 131.108.2.155
Name of configuration file [Router-confg]? P1R1.2
Configure using P1R1.2 from 131.108.2.155? [confirm] y
Booting P1R1.2 from 131.108.2.155:!! [OK - 874/16000 bytes]
P1R1#
```

2. Enter the type of configuration file. The prompt gives you the option to load one of two configuration file types from the server. These two files are:

   • **Host configuration file**—This configuration file contains commands that apply to a router in particular. This file type is the default.

   • **Network configuration file**—This configuration file contains commands that apply to all routers and terminal servers on the network.

3. Enter the network address of the remote host from which you are retrieving the configuration file.
4. Enter the name of the configuration file or accept the default name.

FUNDAMENTAL CONFIGURATION TASKS

This section looks at configuration commands that individualize and secure a router. It also discusses commands that alter router interfaces, focusing on the serial interface. Finally, it covers the steps required to save the altered configuration file. It first looks at fundamental configuration tasks by reviewing the router modes that will be utilized.

Overview of Router Modes

Chapter 5 introduced the different router modes that affect configuration. Now it's time to take a look at those modes in greater detail.

Type **configure terminal** to enter global configuration mode. Global configuration mode recognizes commands that affect the whole router. For example, because the router has one enable password and one host name, these commands are accepted in global configuration mode. Also, in global configuration mode the router recognizes one-line commands. Some of these commands cause the router to enter other configuration modes, where it recognizes more complicated and detailed commands.

An example of another configuration mode is the interface configuration mode. You enter this mode by specifying a particular interface. Some commands, such as **ring-speed**, are appropriate to Token Ring but would not make any sense on a serial interface. Therefore, specifying the interface tells the command executive more about the nature of the configuration commands you are about to enter.

The prompt always identifies the current active mode, including global configuration mode, as shown in Figure 6–3.

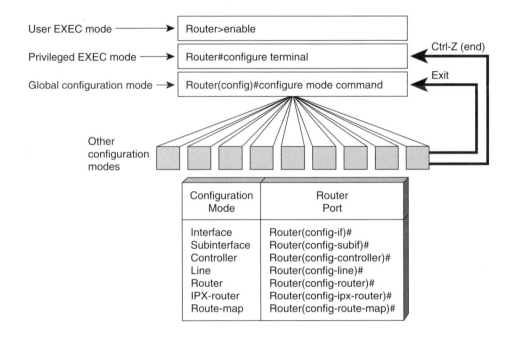

Figure 6–3
Each router mode has a distinct prompt so you always know what the current active mode is.

From the global configuration mode, you have access to the specific configuration modes, including

- *Interface mode*—Supports commands that configure operations on a per-interface basis

- *Subinterface mode*—Supports commands that configure multiple virtual interfaces on a single physical interface

- *Controller mode*—Supports commands that configure channelized T1

- *Line mode*—Supports commands that configure the operation of a terminal line

- *Router mode*—Supports commands that configure an IP routing protocol
- *IPX-Router mode*—Supports commands that configure the Novell network-layer protocol
- *Route-Map mode*—Supports commands that configure routing tables and source and destination information

For a complete list of the router configuration modes, refer to "Understanding the User Interface" at www.cisco.com.

If you type **exit,** the router will back out one level, eventually allowing you to log out. In general, typing **exit** from one of the specific configuration modes will return you to global configuration mode. Pressing **Ctrl-Z** leaves configuration mode completely and returns the router to the privileged EXEC mode.

TIPS

A common mistake when in a nonglobal configuration mode is to forget to exit back to the global configuration mode when done. Checking the prompt verifies that you are in the correct configuration.

Configuring Router Identification

One of the first tasks in configuring your router is to name it. Naming your router helps you to better manage your network by uniquely identifying each router within the network. The name of the router is considered to be the host name and is the name displayed at the system prompt. If no name is configured, the system default router name is Router. You assign the router name in global configuration mode. In Figure 6–4, the router name is set to P1R1.

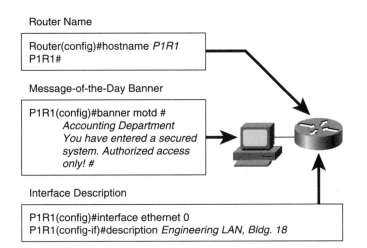

Router Name

```
Router(config)#hostname P1R1
P1R1#
```

Message-of-the-Day Banner

```
P1R1(config)#banner motd #
    Accounting Department
    You have entered a secured
    system. Authorized access
    only! #
```

Interface Description

```
P1R1(config)#interface ethernet 0
P1R1(config-if)#description Engineering LAN, Bldg. 18
```

Figure 6–4
*Router identi-
fication sets
local identity
or message
for the
accessed
router or
interface.*

TIPS

Router identification is very convenient when you are remotely configuring routers, because it is a quick reminder of which router you are accessing. Name is also used by Network Management Services such as Cisco Works.

You can configure a message-of-the-day banner to be displayed on all connected terminals. This banner is displayed at login and is useful for conveying messages. When you enter the **banner motd** command, follow the command with one or more blank spaces and a delimiting character of your choice, for example, the pound sign (#). After you add the banner text, terminate the message with the delimiting character. (The **motd** keyword stands for "message of the day.")

You can have many lines in the banner—a full screen—and if you know how to enter the special commands for VT extended modes, you can get elongated and highlighted characters. If you are in a secure network, a banner is a good place to put network maintenance information such as the description of users who depend on this router and where the router is located.

TIPS

Remember that anyone can see banner information. You should be very careful about the wording of your banner message. Including the word "Welcome" is an explicit invitation to anyone, including hackers, to enter your network.

Some other banners are available, including an idle banner, which is displayed on a terminal or console when it is not in use. Some people use the idle banner to display the corporate logo.

You can add a description to an interface to help you remember specific information about that interface, such as what network that interface services. This description is meant solely as a comment to help identify how the interface is being used. The description will appear in the output when you display the configuration information that exists in router memory.

Description commands are very helpful for identifying links to a WAN.

The description function is easy to implement using the syntax **description** (*string*). Consider the following example:

```
PIR1(config-if)#description Network Lab, Tyler Building
```

Password Configuration

As discussed in Chapter 5, you can secure your router by using passwords to restrict access. Passwords can be established both on individual lines and to the privileged EXEC mode. Passwords are case sensitive.

The console password is set independently from other line passwords, such as virtual terminal passwords. If your console is in your office under lock and key, you may not want a console password; just press **Return,** and you will be in user mode automatically. However, if your console is out where anyone has access to it, you might want to put a password on the console line.

Console Password

The **line console 0** login password xxxx establishes a login password on the console terminal.

The **line vty 0 4** login password xxxx establishes a login password on incoming Telnet sessions.

The **enable password** command restricts access to the privileged EXEC mode.

Virtual Terminal Password

The virtual terminal password must be set for remote configuration support. Telnet requires a password check. The numbers 0 and 4 are a range—that is, VTY lines zero through four, which equals five lines and equates to five incoming Telnet sessions.

The same password can be used for all five lines. However, you might want to set one of the virtual terminal passwords to be unique. This setting is often used in a large network with many network administrators. You set four identical VTY passwords so that everyone can get into the router, and set one VTY password to something else. This way, if a catastrophic problem occurs on the network and all common VTY lines are in use, the one unique line is in reserve for recovery.

Privileged EXEC Mode Password

The router has one **enable** password. Whoever owns this password can do anything with the router, so be careful about communicating this password to others.

To provide an additional layer of security, particularly for passwords that cross the network or are stored on a TFTP server, you can use the **enable secret** command. Both **enable password** and **enable secret** commands allow you to establish an encrypted password that users must enter to access enable mode (the default) or any privilege level you specify, but the **enable secret** command offers an improved encryption algorithm.

TIPS

Cisco recommends that you use the **enable secret** command whenever possible because of its encryption algorithm. Use the **enable password** command only if you boot an older image of the Cisco IOS software or if you boot from older boot ROMs that do not recognize the **enable secret** command.

If you configure the enable secret password, it is used instead of the enable password, not in addition to it.

Password Encryption

Passwords can be further protected from display through the use of the **service password-encryption** command. To set password encryption, enter the following command in the global configuration mode:

```
Router(config)#service password-encryption
```

TIPS

To disable passwords, use the **no** form of the specific password command in global configuration command mode.

If an encrypted password is lost, you must erase the configuration on the router and create a new file from setup mode.

Configuring and Managing an Interface

A router's main function is to relay packets from one data link to another. To do that, the characteristics of the interfaces through which the packets are received and sent must be defined. These characteristics include, but are not limited to, the address of the port, the data encapsulation method, media-type, bandwidth, and direct memory access buffering.

Many features are enabled on a per-interface basis. Interface configuration mode contains commands that modify the operation of an Ethernet, Token Ring, FDDI, or serial port. When you issue the **interface** command, you must define the interface type and number. The number is assigned to each interface at the factory and is used to identify each interface, which is particularly useful when you have multiple interfaces of the same type on a single router.

An example of an interface type and number is:

```
Router(config)#interface serial 0
Router(config)#interface ethernet 1
```

Some series routers can accept multiple interface cards with multiple ports on each card. In this case, the first number is the number of the card, or slot number. The second number is the port on the card. For example, on the second interface card, the first Ethernet interface is specified as Ethernet 2/0.

You define an interface in the Cisco series routers by slot and port number:

```
Router(config)#interface ethernet 1/0
```

You define an interface in the Cisco 7000 and 7500 series routers with VIP cards by slot, port adapters, and port numbers, in that order:

```
Router(config)#interface ethernet 1/0/0
```

To quit the interface configuration mode, type **exit** at the system prompt.

The show interfaces Command

The **show interfaces** command displays all the statistics for all the interfaces on the router, as shown in Figure 6–5.

```
Router#show interface
    Ethernet0 is up, line protocol is up
    Hardware is Lance, address is 0060.4740.c2b6 (bia 0060.4740.c2b6)
    MTU 1500 bytes, BW 10000 Kbit, DLY 1000 usec, rely 255/255, load 1/255
    Encapsulation ARPA, loopback not set, keepalive set (10 sec)
    ARP type: ARPA, ARP Timeout 04:00:00
                          •
                          •
                          •
Serial1 is up, line protocol is down
    Hardware is MK5025
    MTU 1500 bytes, BW 1544 Kbits, DLY 20000 usec, rely 255/255, load 9/255
    Encapsulation HDLC, loopback not set, keepalive set (10 sec)
                          •
                          •
                          •
```

Figure 6–5
The show interfaces command is useful for configuring and monitoring routers.

If you want to view the statistics for a specific interface, enter the **show interfaces** command followed by the specific interface and port number. The following example uses the interface command to display the statistics for the serial interface, port 1:

```
Router#show interfaces serial 1
```

To view the statistics for the Ethernet interface, port 0, enter

```
Router#show interfaces ethernet 0
```

If you use the **show interfaces** command on the Cisco 3640, 7000, and 7200 series routers without the *slot/port* arguments, information for all interface types will be shown. For example, if you type **show interfaces ethernet,** you will receive information for all Ethernet, serial, Token Ring, and FDDI interfaces. The only way to specify a particular interface is by adding the type *slot/port* argument.

In Figure 6–5, the top line of the output tells that the line is up. Below that are some characteristics. For the Ethernet interface, you see the MAC address of the card, the maximum transmission units, and the bandwidth. For the serial interface, the default bandwidth for a serial line is T1. On the next line down, you find HDLC, which is the default encapsulation protocol for serial lines on Cisco routers.

Interpreting Interface Status

One of the most important elements of the **show interface** command output is display of the line and data-link protocol status. Figure 6–6 indicates the key summary line to check and the status meanings.

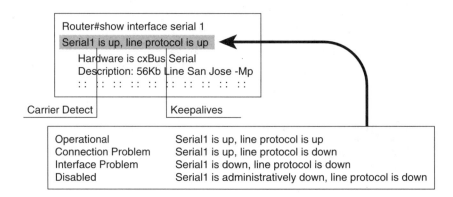

Figure 6–6
*Use the **show interface serial** command to identify line and protocol problems.*

The first parameter ("Serial1 is up," in Figure 6–6) refers to the hardware layer and essentially reflects whether the interface is receiving the Carrier Detect signal from the other end. The second parameter ("line protocol is up" in Figure 6–6) refers to the data link layer. This parameter reflects whether the data link layer protocol keepalives are being received.

The following parameter combinations are possible:

- If both the interface and the line protocol are up, the connection is operational.

- If the hardware is up and the line protocol is down, a Layer 2 problem exists, such as:

  No keepalives

  No clock rate

  Wrong connector

  Encapsulation mismatch

  In a back-to-back connection, the other end of the connection is "administratively down"

- If both the line protocol and the interface are down, a cable might never have been attached when the router was powered up.

- If the information says "administratively down," you have manually disabled this interface. Later in this chapter, you learn how to disable interfaces.

Configuring a Serial Line

One of the more common interface configurations is a serial interface configuration. This chapter will use a serial interface configuration as a sample task. (Later chapters will cover configuration of other kinds of interfaces, such as Ethernet, Token Ring, and subinterfaces.) A serial interface can be configured from the console or through a virtual terminal line. Figure 6–7 shows how to configure a serial line.

The steps of serial line configuration are as follows:

1. Enter global configuration mode. In this example, you are configuring the interface from the console terminal.
2. Once in global configuration mode, you must identify the specific interface against which you will be issuing commands.
3. If you are configuring an interface that will act as a DCE, you must specify a clock rate for it. (See the next section, "Determining DCE/DTE Status," for an explanation of DCEs and DTEs in serial link environments.) Desired clock rate in bits per second are: 1200, 2400, 4800, 9600, 19200, 38400, 56000, 64000, 72000, 125000, 148000, 500000, 800000, 1000000, 1300000, 2000000, or 4000000. The default clock rate for serial lines is T1.

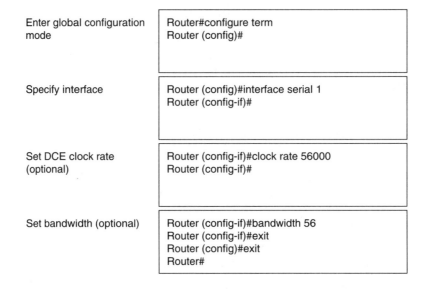

Figure 6–7
You must use global configuration mode to set up a serial line.

Enter global configuration mode

```
Router#configure term
Router (config)#
```

Specify interface

```
Router (config)#interface serial 1
Router (config-if)#
```

Set DCE clock rate (optional)

```
Router (config-if)#clock rate 56000
Router (config-if)#
```

Set bandwidth (optional)

```
Router (config-if)#bandwidth 56
Router (config-if)#exit
Router (config)#exit
Router#
```

Issue the **clock rate** command with the desired speed. Be sure to type the complete clock speed. For example, a clock rate of 56000 cannot be abbreviated to 56.

4. Enter the desired bandwidth for the interface. The **bandwidth** command overrides the default bandwidth that is displayed in the **show interfaces** command and is used by some routing protocols such as IGRP.

If you change the interface from a DCE to a DTE, use the **no clock rate** command to remove the clock rate.

TIPS

When running the EIA/TIA-232 line at high speeds and long distances, data can phase shift with respect to the clock. To prevent this shifting, use the **dce-terminal-timing enable** command.

Determining DCE/DTE Status

On serial links, one side of the link acts as the DCE and the other side of the link acts as the DTE. By default, Cisco routers are DTE devices but can be configured as DCE devices. In a "back-to-back" cable configuration where a modem is not used, the DCE must provide a clocking signal.

Before you begin to configure or alter your serial interface, you need to know if the interface is configured as DTE or DCE. The **show controllers serial** command displays information specific to the interface hardware, as shown in Figure 6–8.

```
Router#show controllers serial 1
HD unit 1, idb = 0xBFD3C, driver structure at 0xC39A0
buffer size 1524 HD unit 1, V.35 DCE cable, clockrate 56000
cpb = 0x83, eda = 0x800, cda = 0x814
RX ring with 16 entries at 0x830800
00 bd_ptr=0x0800   pak=0x0C54F0   ds=0x836938   status=80 pak_sizes=22
01 bd_ptr=0x0814   pak=0x0C5158   ds=0x835BC8   status=80 pak_sizes=22
02 bd_ptr=0x0828   pak=0x0C4F8C   ds=0x835510   status=80 pak_sizes=269
03 bd_ptr=0x083C   pak=0x0C4DC0   ds=0x834E58   status=80 pak_sizes=22
04 bd_ptr=0x0850   pak=0x0C6184   ds=0x839840   status=80 pak_sizes=22
05 bd_ptr=0x0864   pak=0x0C4BF4   ds=0x8347A0   status=80 pak_sizes=22
                             •
                             •
                             •
```

Figure 6–8
*The **show controllers serial** command indicates if the interface is a DTE or DCE.*

Most of the information displayed is proprietary and is used by Cisco technical support personnel for diagnostic purposes. The **show controllers serial** command shows if the interface is cabled as a DCE or DTE. If the interface is configured as a DCE, the **show controllers serial** display also reflects the current clock rate.

You can display information pertinent to all controllers, to a controller type, or to a specific controller. The command

        Router#**show controllers**

displays information about all controllers installed in the router. The command

        Router#**show controllers serial**

displays information about all controllers of a specific type. The command

        Router#**show controllers serial 1**

displays information about a specific controller.

Verifying Your Changes

Taking a moment or two to verify the changes you have made to an interface can ensure no mistakes were made during the process of saving your changes. The **show interfaces** command enables you to view the current interface configuration and status. For example, Figure 6–9 shows the result of issuing the **show interfaces serial 1** command.

Figure 6–9
*Always verify
your changes
using the
show inter-
faces com-
mand.*

```
Router#show interfaces serial 1
Serial1 is up, line protocol is up
Hardware is MK5025
MTU 1500 bytes, BW 56 Kbit, DLY 20000 usec, rely 255/255, load 9/255
Encapsulation HDLC, loopback not set, keepalive set (10 sec)
Last input 0:00:00, output 0:00:01, output hang never
Last clearing of "show interface" counters never
Output queue 0/40, 0 drops; input quere 0/75, 0 drops
Five minute input rate, 1000 bits/sec, 0 packets/sec
Five minute output rate 2000 bits/sec, 0 packets/sec
331885 packets input, 62400237 bytes, 0 no buffer
Received 230457 broadcasts, 0 runts, 0 giants
3 input errors, 3 CRC, 0 frame, 0 overrun, 0 ignored, 0 abort
403591 packets output, 66717279 bytes, 0 underruns
0 output errors, 0 collisions, 8 interface resets, 0 restarts
45 carrier transitions
```

Shutting Down an Interface

At some point, you might want to disable an interface, for example, to perform hardware maintenance on a specific interface or segment of a network. You might also want to disable an interface if a problem exists on a particular segment of the network and you need to isolate that segment from the rest of the network until the problem is detected or repaired.

The **shutdown** command, as shown in Figure 6–10, administratively turns off an interface. To reinstate the interface, use the **no shutdown** command.

```
Router#configure term
Router(config)#interface serial 1
Router(config-if)#shutdown
%LINEPROTO-5-UPDOWN:  Line Protocol on Interface Serial1, changed state to down
%LINK-5-CHANGED:  Interface Serial, changed state to administratively down
```

Figure 6–10
*The com-
mands **shut-
down** and **no
shutdown**
change the
interface state.*

```
Router#configure term
Router(config)#interface serial 1
Router(config-if)#no shutdown
%LINK-3-UPTOWN:  Interface Serial, changed state to up down
%LINEPROTO-5-UPDOWN:  Line Protocol on Interface Serial1, changed state to up
```

Verifying Configuration Changes—Release 10.3 and Later

After you make the changes to the running configuration variables, you should verify
your new environment. Figure 6–11 shows the procedures you can use when working
with Cisco IOS Release 10.3 or later.

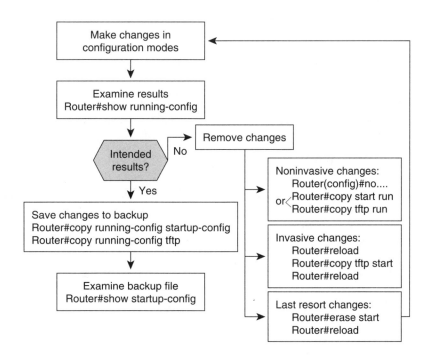

Figure 6–11
*Changing con-
figurations on
IOS 10.3 and
later.*

To verify your changes, use the **show running-config** command. This command displays the current configuration variables in memory.

If the variables displayed are not what you intended, you can correct the environment by:

- Issuing the **no** form of a configuration command or by copying a new set of configuration parameters into RAM

- Restarting the system and automatically loading a new configuration file from NVRAM

- Removing the startup configuration file with the **erase startup-config** command and replacing it with a new configuration file from an alternate source

If you have already copied the erroneous configuration information to the startup configuration file in NVRAM, you can

- Create new configuration variables in the running configuration and copy those new variables to the startup configuration file. To do this, enter

  ```
  Router#copy running-config startup-config
  ```

- Remove the saved configuration with the **erase** command and restart the system. In this case, the router configuration will revert to the factory defaults.

CAUTION
_____

After you enter the **reload** command, the system asks if you want to save the current configuration. If you do not want to save the running configuration, respond **no** at the prompt.

Verifying Configuration Changes—Releases Before 10.3

Figure 6–12 shows the procedures to verify configuration changes when you are working with Cisco IOS releases before 10.3.

To verify your changes, use the **show configuration** command. This command displays the current configuration variables in memory.

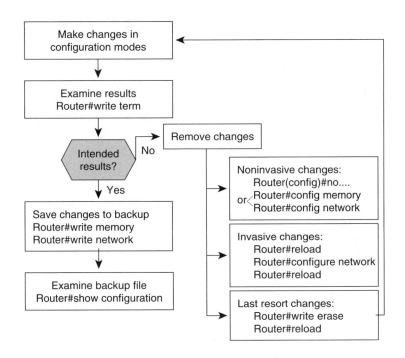

Figure 6–12
*Changing configura-
tions in earlier
versions of
IOS.*

If the variables displayed are not what you intended, you can correct the environment by:

- Issuing the **no** form of a configuration command or by copying a new set of configuration parameters into RAM

- Restarting the system and automatically loading a new configuration file from NVRAM

- Removing the startup configuration file with the **write erase** command and replacing it with a new configuration file from an alternate source

If you have already copied the erroneous configuration information to the startup configuration file in NVRAM, you can:

- Create new configuration variables in the running configuration and copy those new variables to the startup configuration file. To do this, enter

```
Router#write memory
```

Saving Configuration Changes—Release 10.3+

When you have determined that the new variables are correct, you must save your changes to the startup configuration file, as shown in Figure 6–13.

Figure 6–13
You can save configuration changes to NVRAM and (optionally) a remote server.

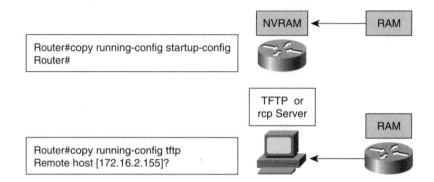

Saving the changes ensures the router uses the new variables when you copy the startup configuration file into memory or perform a reload.

To save the configuration variables to the startup configuration file in NVRAM, enter the following command at the privileged EXEC system prompt:

```
Router#copy running-config startup-config
```

To save the configuration variables to a remote server on the network, enter the following command at the privileged EXEC system prompt:

```
Router#copy running-config tftp
```

where the variable **tftp** represents the target server type.

TIPS

If you make a change on the router and then upload it to the TFTP server, it will overwrite the older file—except that this new copy will not have any comments in it. For this reason, the **copy running-config tftp** command is often used to create an initial file on the server. After the file is created, changes are made first on the TFTP server using your favorite editor and then downloaded into the router. This method preserves the comments in the configuration file.

You can include comments in the configuration file by preceding a line of text with an exclamation point (!). Including comments may help define the purpose of any commands you have placed in the configuration file.

Saving Configuration Changes—Releases Before 10.3

To save the configuration variables to the startup configuration file in NVRAM using pre-Release 10.3 software, enter the following command at the privileged EXEC system prompt:

```
Router#write memory
```

To save the configuration variables to a remote server on the network using pre-Release 10.2 software, enter the following command at the privileged EXEC system prompt:

```
Router#write network
```

The **write network** command supports only TFTP servers.

MANAGING THE CONFIGURATION ENVIRONMENT

As your network grows, there may come a time when you want to store your Cisco IOS software and configuration files on a central server, which would allow you to control the number and revision level of software images and configuration files you must maintain. This section discusses the alternative sources for Cisco IOS software and how to direct the router to locate the software. It also looks at how to modify the existing Configuration Register setting to reflect a new location for the system image.

Locating the Cisco IOS Software

The Configuration Register boot field determines whether the router loads an operating system image, and if so, where it obtains this system image. You can modify the Configuration Register boot field to tell the router how to load a system image on startup. Instead of using the default system image and configuration file to start up, you can specify a particular system image and configuration file located elsewhere on the network.

The default source for Cisco IOS software depends on the hardware platform, but most commonly the router looks to the configuration commands saved in NVRAM, as shown in Figure 6–14. Settings can be placed in the Configuration Register to enable alternatives for where the router will bootstrap Cisco IOS software.

Figure 6–14
If the router cannot locate the IOS software, it enters ROM monitor mode.

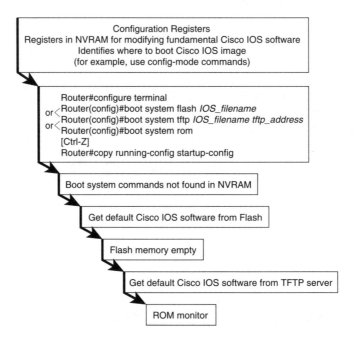

You specify **boot system** commands to define the sequence for fallback sources. You save these statements in NVRAM with the **copy running-config startup-config** command. The router will use the new sequence during the next startup. If the **boot system** command is not used, however, the system has its own fallback alternatives (refer to Figure 6–14). The router defaults to the Cisco IOS software in Flash memory.

If Flash memory is empty, however, the router will try a network alternative. The router uses the Configuration Register value to form a filename from which to boot a default system image stored on a network server.

Finally, if the router has exhausted all alternative paths and cannot locate the Cisco IOS software, the router enters ROM monitor mode.

The order in which the router looks for system bootstrap information depends on the boot field setting in the Configuration Register.

Determining the Current Configuration Register Value

Before you alter the Configuration Register, you should determine how the router is loading the software image. Use the **show version** command to obtain the current Configuration Register value, as shown in Figure 6–15. The last line of the resulting display contains the Configuration Register values.

```
Router#show version

IOS (tm) 2500 Software (C2500-JS-L), Version 11.2(6), RELEASE SOFTWARE(fcl)
Copyright (c) 1986-1997 by Cisco Systems, Inc.
Complied Tue 06-May-97 16:17 by Kuong
Image text-base: 0x0303ED8C, data-base; 0x00001000

ROM: System Bootstrap, Version 5.2(8a), RELEASE SOFTWARE
ROM: 3000 Bootstrap Software (IGS-RXBOOT), version 10,2 (8a),
RELEASE SOFTWARE (fcl)
Router uptime is 21 hours, 13 minutes
System restarted by reload
System image file is "flash:c2500-js-1.112-6.bin", booted via flash
    •
    •
    •
Configuration register is 0x2102
```

Figure 6–15
The Configuration Register indicates that NVRAM is examined for boot system commands.

Note that the **show version** command displays over two screen pages. You need to go to the second screen page to display the current Configuration Register values.

TIPS

If you are in ROM monitor mode, use the **o** command to list the Configuration Register settings.

You can change the default Configuration Register setting with the enabled config-mode **config-register** command. Type **configure terminal** in privileged EXEC mode to enter configuration mode, as shown in Figure 6–16.

Figure 6–16
Enter configuration mode to change the default Configuration Register.

```
Router#configure terminal
Router(config)#config-register 0x2102
[Ctrl-Z]
Router#reload
```

The Configuration Register is a 16-bit register. The lowest four bits of the Configuration Register (bits 3, 2, 1, and 0) form the boot field. A hexadecimal number is used as the argument to set the value of Configuration Register. Table 6–1 shows the hexadecimal options and their meanings.

Table 6–1
Boot field options in the Configuration Register.

Configuration-Register Value	Meaning
0x0	Use ROM monitor mode (Manually boot using the *b* command)
0x1	Automatically boot from ROM (Provides IOS subset)
0x2 to 0xF	Examine NVRAM for *boot system* commands (0x2 default if router has Flash)

To change the boot field and leave all other bits set to their default values, follow these guidelines:

- Set the boot field to 0 to automatically enter ROM monitor mode. (This value sets the boot field bits to 0-0-0-0.) The router displays the **>** or the **rommon>** prompt in this mode.

- Set the boot field to 1 to configure the system to boot automatically from ROM. (This value sets the boot field bits to 0-0-0-1.) The router displays the **router(boot)>** prompt in this mode.

- Set the boot field to any values, 2 to F, to configure the system to use the boot system commands in NVRAM. This is the default. (These values set the boot field bits to 0-0-1-0 through 1-1-1-1.)

Use the **show version** command to verify the changes in the boot field setting.

TIPS

Password recovery is one of the possible reasons to change the Configuration Register. For more information, refer to Appendix F, "Password Recovery."

Boot System Options in Software

As mentioned earlier, you can define a fallback sequence for the router to use during startup. Figure 6–17 shows the **boot system** commands used to specify one fallback sequence for booting Cisco IOS software.

The boot system commands in Figure 6–17 specify that a Cisco IOS image will load first from Flash memory, next from a network server, and finally from ROM.

Loading from Flash memory allows you to copy a system image without changing electrically erasable programmable read-only memory (EEPROM). Information stored in Flash memory is not vulnerable to network failures that can occur when loading system images from network servers.

Flash

```
Router#configure terminal
Router(config)#boot system flash c2500-js-1
[Ctrl-Z]
Router#copy running-config startup-config
```

Network

```
Router#configure terminal
Router(config)#boot system tftp test.exe 172.16.13.111
[Ctrl-Z]
Router#copy running-config startup-config
```

ROM

```
Router#configure terminal
Router(config)#boot system rom
[Ctrl-Z]
Router#copy running-config startup-config
```

In case Flash memory becomes corrupted, the boot system entries in Figure 6–17 next specify that a system image should be loaded from a network server.

Finally, if Flash memory is corrupted and the network server fails to load the image, booting from ROM is the final bootstrap option in software in Figure 6–17. Note that the system image in ROM likely will be a subset of Cisco IOS software, lacking the protocols, features, and configurations of full Cisco IOS software. It may also be an older version of Cisco IOS software if you have updated software since you purchased the router.

The command **copy running-config startup-config** saves your desired command sequence in NVRAM. The router will execute the **boot system** commands in the order in which they were entered into configuration mode.

Preparing for a Network Backup Image

Retaining a backup copy of your software image ensures that you always have a copy of the Cisco IOS software in case the system image in your router becomes corrupted.

Geographically distributed routers need a source or backup location for software images. Using a network server allows image and configuration uploads and downloads over the network. The network server can be another router, a workstation, or a host system, as shown in Figure 6–18.

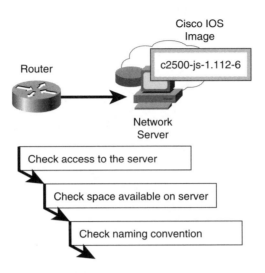

Figure 6–18
The network server can be another router, a workstation, or a host system.

Before you copy software between the network server and Flash memory in the router, you should check for preliminary conditions:

- Make sure that you have access to the network server.

- Verify that the server has sufficient room to accommodate the Cisco IOS software image.

- Check the filename requirements and file space of the network server.

Viewing Available Memory and the Image Filename

The **show flash** command is an important tool for gathering information about your router memory and image file, as shown in Figure 6–19.

```
Router#show flash

System flash directory:
Files   Length        Name/status
  1     7871172     c2500-js-1.112-6.bin

[7871236 bytes used, 517372 available, 8388608 total]
8192K bytes of processor board System flash (Read ONLY)
```

Figure 6–19
*Use the **show flash** command to verify that Flash memory has room for the Cisco IOS image.*

With the **show flash** command, you can determine the following:

- Total amount of memory on your router

- Amount of memory available

- Name of the system image file (such as c2500-js-1.112-6.bin) used by the router

The name for the Cisco IOS image file contains multiple parts, each with a specific meaning:

- The first part of the image name contains the platform on which the image runs. In the example in Figure 6–19, the platform is C2500.

- The second part of the name identifies the special capabilities of the image file. A letter or series of letters identifies the feature sets supported in that image. In this example, the *j* indicates this is an enterprise image, and the *s* indicates it contains extended capabilities.

- The third part of the name specifies where the image runs and if the file is compressed. In this example, l indicates the file is relocatable and not compressed.

- The fourth part of the name indicates the version number. In this example, the version number is 11.2 (6).

- The final part of the name is the file extension. The .bin extension indicates this file is a binary executable file.

The Cisco IOS software naming conventions, name part field meaning, image content, and other details are subject to change. Refer to your sales representative, distribution channel, or Cisco Connection Online (CCO) for updated details.

CAUTION

You should be careful in reading the Cisco IOS image filename. Some fonts display the lowercase letter *l* and the number 1 as the same character. How you type the characters will impact the capability of the router to load the files correctly.

Creating Software Image Backup

You create a software backup image file by copying the image file from a router to a network server, as shown in Figure 6–20.

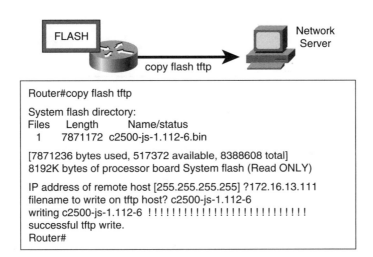

Figure 6–20
Back up current files prior to updating Flash.

```
Router#copy flash tftp

System flash directory:
Files    Length        Name/status
  1      7871172   c2500-js-1.112-6.bin

[7871236 bytes used, 517372 available, 8388608 total]
8192K bytes of processor board System flash (Read ONLY)

IP address of remote host [255.255.255.255] ?172.16.13.111
filename to write on tftp host? c2500-js-1.112-6
writing c2500-js-1.112-6 !!!!!!!!!!!!!!!!!!!!!!!!!!!!!!
successful tftp write.
Router#
```

To copy the current system image file from the router to the network server, use the **copy flash** command in the privileged EXEC mode.

When using a TFTP server, enter

```
Router#copy flash tftp
```

The **copy flash** command also requires you to enter the IP address of the remote host and the name of the source and destination system image file. The **copy flash** command automatically displays the contents of Flash, including the image filename. Methods to obtain IP addresses are discussed in Part 2 of this book, "Networking Protocol Suites."

To stop the copy process, press the **Control** and **Shift** and **6** keys (**Ctrl-^**).

TIPS

The routers that come by default with Flash memory have a preloaded copy of the Cisco IOS software. Although Flash is extremely reliable—good for 65 years and 100,000 rewrites—it is a good idea to make a backup copy of the Cisco IOS software if you have a network server available. If you had to replace Flash memory for some reason, you would have a backup copy at the revision level currently running on your network.

Upgrading the Image from the Network

You must load a new system image file on your router if the existing image file has become damaged or if you are upgrading your system to a newer software version. You can download the new image from the network server using the following commands.

When using a TFTP server, enter

```
Router#copy tftp flash
```

The command will prompt you for the IP address of the remote host and the name of the source and destination system image file. Enter the correct and appropriate filename of the update image as it appears on the server, as shown in Figure 6–21.

Figure 6–21
You must assign a source location and filename for the upgraded image file.

```
Router#copy tftp flash
IP address of remote host [255.255.255.255] ?172.16.13.111
Name of tftp filename to copy into flash [ ]?c4500-js-l
copy c4500-js-1.112-6 from 172.16.13.111 into flash memory? [confirm] <Return>
xxxxxxxx bytes available for writing without erasure.
erase flash before writing? [confirm]<Return>
Clearing and intializing flash memory (please wait) ####...##
Loading from 172.16.13.111: !!!!!!!!!!!!!!!!!!!!!!!!!!!!!!
!!!!!!!!!!!!!! (text omitted) [OK - 324572/524212 bytes]
Verifying checksum...
v v v v v v v v v v v v v v v v v v v v v v v v v v v v v v v v v v v v v v v v v v v
v v v v v v (text omitted)
Flash verification successful. Length = 1804637, checksum = 0xA5D3
```

Before performing this procedure, make sure you are able to reach your TFTP server and that the proper Cisco IOS software image is available. Be sure to use the **show flash** command to view the file and to compare its size with that of the original on the server before changing the **boot system** commands and rebooting the machine on the new level.

Adequate preparation and providing a backup before update operations offers the most secure administration of Cisco IOS image loading.   **Key Concept**

After you confirm your entries, the procedure asks if you want to erase Flash. Erasing Flash makes room for the new image. You perform this task if there is not sufficient Flash memory for more than a single Cisco IOS image.

If no free Flash memory space is available, or if the Flash memory has never been written to, the erase routine is required before new files can be copied. The system informs you of these conditions and prompts you for a response.

As shown in Figure 6–21, each exclamation point (!) means that one User Datagram Protocol (UDP) segment has successfully transferred. The series of Vs indicates successful checksum verification of a segment.

TIPS

You can put comments into a configuration file on a TFTP server by starting the comment line with an exclamation point (!). When a file is downloaded into the router, the comments are stripped off.

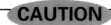

For a period of time after Flash erasure, you will not have a copy of the operating system on the router. This situation is risky. Some routers run the Cisco IOS software from Flash. If you erase the Flash memory on these types of routers, such as the 2500, the router will cease to function.

Overwriting an Existing Image

If you attempt to copy a filename that already exists in Flash memory, the system notifies you, as shown in Figure 6–22.

Figure 6–22
*The response
indicates that
an image
already exists.*

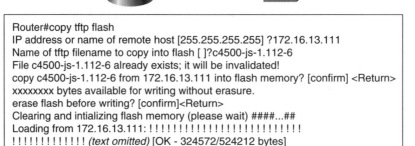

```
Router#copy tftp flash
IP address or name of remote host [255.255.255.255] ?172.16.13.111
Name of tftp filename to copy into flash [ ]?c4500-js-1.112-6
File c4500-js-1.112-6 already exists; it will be invalidated!
copy c4500-js-1.112-6 from 172.16.13.111 into flash memory? [confirm] <Return>
xxxxxxxx bytes available for writing without erasure.
erase flash before writing? [confirm]<Return>
Clearing and intializing flash memory (please wait) ####...##
Loading from 172.16.13.111: !!!!!!!!!!!!!!!!!!!!!!!!!!!!!!
!!!!!!!!!!!!!! (text omitted) [OK - 324572/524212 bytes]
Verifying checksum...
v v v v v v v v v v v v v v v v v v v v v v v v v v v v v v v v v v v v v v v v v v v v
v v v v v v (text omitted)
Flash verification successful. Length = 1204637, checksum = 0x95D9
```

System image filenames are not case sensitive; both upper- and lowercase versions of the filename are viewed as the same filename.

The existing file is "deleted" when you copy the new file into Flash. The first copy of the file still resides within Flash memory but is rendered unusable in favor of the newest version, and will be listed with the [deleted] tag when you use the **show flash** command.

If more than one file of the same name is copied to Flash, regardless of case, the last file copied becomes the valid file.

If you abort the copy process, the newer file is marked [deleted] because the entire file was not copied and is not valid. In this case, the original file still resides within Flash memory and is the image file used.

Use the **show flash** command to view the file and to compare its size with that of the original on the server before changing the **boot system** commands to use the new image.

Up to now, you've seen how to locate and store the Cisco IOS software image. Next you learn how to set up the configuration in NVRAM.

Creating or Changing a Configuration Using Setup Mode

If your device is a new router or the contents of NVRAM are corrupted, the router cannot find the critical configuration information needed to start up. In either of these cases, the router software enters setup mode and prompts you for configuration information. You can also force setup mode by entering the following command in privileged EXEC mode:

```
Router#setup
```

The primary purpose of the setup mode is to rapidly bring up a minimal-feature configuration, which is accomplished through the System Configuration Dialog program.

The first thing System Configuration Dialog asks is if you want to continue with the setup program, as shown in Figure 6–23.

```
Router#setup

    - - - System Configuration Dialog - - -

At any point you may enter a question mark '?' for help.
Use ctrl-c to abort configuration dialog at any prompt.
Default settings are in square brackets '[ ]'.

Continue with configuration dialog? [yes/no]:     yes

First, would you like to see the current interface summary? [yes]:     no
```

Figure 6–23
The System Configuration Dialog prompts you through the setup.

You can exit by entering **No** at the prompt. To begin the initial configuration process, enter **Yes**. You can press **Ctrl-C** to terminate the process and start over at any time. When you are using the command form of **setup** (Router#setup), **Ctrl-C** returns you to the privileged EXEC prompt (Router#).

TIPS

The System Configuration Dialog facility provides help text for each prompt. To access help, press the question mark (**?**) key at the target prompt.

For many of the prompts in the System Configuration Dialog of the **setup** command facility, default answers appear in square brackets ([ ]) following the question. Pressing the **Return** key allows you to use the defaults. If you are configuring the system for the first time, the factory defaults are provided. If there is no factory default, as in the case of passwords, no value is displayed in the brackets.

Refer to the Cisco *Using ClickStart, AutoInstall, and Setup* documentation at www.cisco.com for more information on setup mode.

Setup Mode Global Parameters

If you choose to continue with the System Configuration Dialog, you are first prompted to set the global parameters for the router.

The first global parameter allows you to define the router host name. This host name will precede the Cisco IOS prompts for all configuration modes. At initial configuration, the router name default is shown between the square brackets as [Router], as shown in Figure 6–24.

The next global parameters show how to set the various passwords used on the router. These parameters are where you define the enable secret password discussed in the "Privileged EXEC Mode Password" section. When you enter a string of password characters for this prompt, the characters are processed by Cisco-proprietary encryption. This process enhances the security of the password string. Whenever anyone lists the contents of the router configuration file, this enable password appears as a meaningless string of characters. Cisco recommends you use different passwords for the enable and secret password to maintain an enhanced level of security.

When you answer "yes" to a prompt, additional subordinate questions may appear about that protocol.

```
Configuring global parameters:

  Enter host name [Router]:   P1R1

The enable secret is a one-way cryptographic secret used
instead of the enable password when it exists.

  Enter enable secret [<Use current secret>]:

  Enter enable password [sanfran]:
% Please choose a password that is different from the enable secret
  Enter enable password [sanfran]:    cisco
  Enter virtual terminal password [sanjose]:
  Configure SNMP Network Management? [no]:
  Configure IP? [yes]:
    Configure IGRP routing? [yes]:
      Your IGRP autonomous system number [1]:
                               •
                               •
                               •
```

Figure 6–24
The default router name at initial configuration is shown in brackets.

Setup Mode Interface Parameters

After you have configured the global parameters, you are prompted for parameters for each installed interface, as shown in Figure 6–25.

```
Configuring interface parameters:

Configuring interface TokenRing0
  Is this interface in use? [no]:     <Return>

Configuring interface Serial0
  Is this interface in use? [yes]:
  Configure IP on this interface? [yes]:
  Configure IP unnumbered on this interface? [no]
    IP address for this interface:   172.16.97.67
    Number of bits in subnet field [0]:
    Class B network is 172.16.0.0,0 subnet bits; mask is 255.255.0.0
  Configure Novell on this interface? [yes]:    no

Configuring interface Serial1:
  Is this interface in use? [yes]:    no
```

Figure 6–25
Setup prompts you for parameters for each interface.

If you choose not to configure an interface, the System Configuration Dialog will bypass any subsequent prompts relating to that interface. If you choose to configure an interface, enter your configuration values at each prompt.

You can configure every protocol that you enabled in the global section for each interface. So you can see that some interfaces can be enabled for multiple protocols, while others may be running only one protocol.

For the serial line Serial0, the system prompts you for an IP address on this port and whether to configure IP unnumbered with the default setting. "Yes" indicates the common practice of enabling IP processing on that interface without assigning an explicit IP address to the interface. Usually the reason for using IP unnumbered is to conserve IP addresses. It can be unnecessary to use an explicit IP address for a serial link that connects point-to-point with another serial interface.

Setup Mode Script Review

When you complete your changes, the setup command facility shows you the configuration command script that was created during the setup session, as shown in Figure 6–26.

Figure 6–26
Setup prompts you to save the configuration.

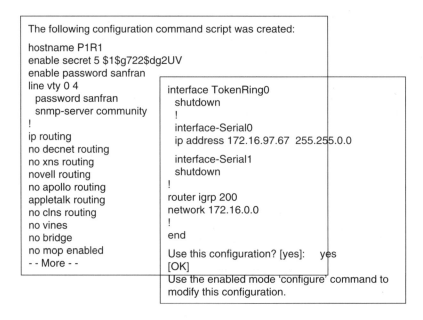

```
The following configuration command script was created:

hostname P1R1
enable secret 5 $1$g722$dg2UV
enable password sanfran
line vty 0 4
   password sanfran
   snmp-server community
!
ip routing
no decnet routing
no xns routing
novell routing
no apollo routing
appletalk routing
no clns routing
no vines
no bridge
no mop enabled
- - More - -

interface TokenRing0
   shutdown
!
interface-Serial0
   ip address 172.16.97.67  255.255.0.0

interface-Serial1
   shutdown
!
router igrp 200
network 172.16.0.0
!
end

Use this configuration? [yes]:     yes
[OK]
Use the enabled mode 'configure' command to
modify this configuration.
```

This script allows you to view your changes before they are saved to NVRAM.

The commands are divided into global and interface sections. Notice in Figure 6–26 that both interface TokenRing0 and Serial1 are shut down—that is, administratively disabled. To use such interfaces, you must enter interface configuration mode and type

no shutdown to turn them back on. If these interfaces have never been configured, you must use the **configure** command to enter the appropriate values to make the interfaces operational.

At the end, setup will ask you if you want to use this configuration. If you enter "yes," the configuration file is loaded into working storage, and a copy is stored in backup memory in NVRAM. This is the only time the router will automatically store a configuration file in NVRAM; after this initial configuration, you must explicitly tell the router to make a backup copy.

The script tells you to use configuration mode to modify the command after setup has been used. The script file generated by setup is additive; you can turn on features with setup, but you cannot turn them off. Also, setup does not support many of the advanced features of the router or those features that require a more complex configuration such as access lists. These complex configurations are covered further in Chapter 13, "Basic Traffic Management with Access Lists."

SUMMARY

This chapter has defined how to create a running and startup configuration, and how to make global configuration changes. You've also configured a serial interface and determined the load location of the Cisco IOS image. The next chapter focuses on using the Cisco Discovery Protocol to locate other routers.

Chapter Six Test
Configuring a Router

Estimated Time: 15 minutes

Complete all the exercises to test your knowledge of the materials contained in this chapter. Answers are listed in Appendix A, "Chapter Test Answer Key."

Question 6.1

T   F   Configuration files can come from the console, NVRAM, or a network server.

Question 6.2

T   F   The router loads the image file based on the values set in the Configuration Register.

Question 6.3

T   F   Multiple source options provide flexibility and fallback alternatives.

Question 6.4

T   F   If the router cannot find a configuration file, the router enters shutdown mode.

Question 6.5

T   F   The router has a single configuration mode to handle all configuration areas.

Question 6.6

T   F   Cisco routers provide multiple levels of password protection.

Question 6.7

T   F   Cisco routers support configuration parameters that aid in router identification.

Question 6.8

T   F   Cisco IOS Release 10.3 or later uses the **copy configure** command to save configuration files.

Question 6.9

What command is used to enter global configuration mode and specify that configuration commands will originate from the terminal?

_____

Question 6.10

What command defines a login banner?

_____

Question 6.11

What command is used to create a message-of-the-day banner?

_____

Question 6.12

What must you enter after the message-of-the-day banner command to indicate the end of the banner message?

_____

Question 6.13

What command string puts you in interface configuration mode for interface Serial1?

_____

Question 6.14

If you set both the enable and secret passwords, which one is used to enter privileged EXEC mode?

_____

Question 6.15

What command do you use to verify that interface Serial1 is cabled as a DCE interface?

_____

Question 6.16

What command do you use to enable an interface?

_____

Question 6.17

In what mode must the router be before you can issue the **no shutdown** command?

_____

Question 6.18

What command do you issue to obtain the current Configuration Register setting?

_____

Question 6.19

What commands boot the system file from Flash? ROM? TFTP server?

Flash: _____

ROM: _____

TFTP server: _____

Question 6.20

What mode must the router be in before you can issue a **boot system** command?

_____

Discovering and Accessing Other Cisco Routers

This chapter discusses the Cisco Discovery Protocol (CDP) and how you can use it to view interface and CDP configuration parameters on your local router. However, CDP is most commonly used to obtain protocol addresses and platform information about neighboring routers.

The screens in this section reflect Cisco IOS Release 11.2(6). If you are running a different version, your screens may vary from those shown.

CISCO DISCOVERY PROTOCOL OVERVIEW

CDP provides a single proprietary command that enables you to access a summary of the multiple protocols and addresses configured on other directly connected routers.

CDP runs over a data link layer connecting lower physical media and uppernetwork-layer protocols, as shown in Figure 7–1. Because CDP operates at this level, two or more CDP devices that support different network-layer protocols can learn about each other.

Physical media that support the Subnetwork Access Protocol (SNAP) connect CDP devices. These physical media can include all LANs, Frame Relay and SMDS WANs, and ATM networks.

When a Cisco device running Cisco IOS Release 10.3 or later boots up, CDP starts up by default and automatically discovers neighboring Cisco devices running CDP. In the past, the ability to obtain information about remote devices required tools provided by TCP/IP. With CDP, discovery of devices now extends beyond those devices running

Upper-Layer Entry Addresses	TCP/IP	Novell IPX	AppleTalk	Others
Cisco Proprietary Data-Link Protocol	CDP discovers and shows information about directly connected Cisco devices			
Media Supporting SNAP	LAN's	Frame Relay	ATM	Others

TCP/IP. Because CDP is protocol independent, it will discover directly connected Cisco devices regardless of which protocol suite they are running.

CDP runs on all Cisco manufactured equipment, including:

- Routers
- Access servers
- Workgroup switches

Once CDP discovers a device, it can display any of the various upper-layer protocol address entries used on the discovered device's port: IPX, AppleTalk Datagram Delivery Protocol (DDP), DECnet CLNS, and others.

CDP uses an assigned HDLC protocol type value. To use this proprietary type value, CDP must run on media that support Layer 2 encapsulation that uses SNAP. If you run an analyzer on your network, the analyzer may inform you that an "unknown" protocol is on your network. You should be able to recognize the format for CDP exchanges so you don't spend time trying to eliminate this traffic. The SNAP format for CDP exchanges is hexadecimal aaaa03.00000c.2000, where *aaaa03* is the LLC, *00000c* is the Cisco organization ID, and *2000* is the HDLC protocol type for CDP.

The CDP process sends and receives advertisements about neighboring CDP devices on a specific multicast address (0100.0ccc.cccc). By periodically sending and receiving hello-type updates, each CDP device learns about other CDP devices and can determine if any medium in the data link has gone down or come up.

Cisco IOS Releases 10.3 and later include CDP Management Information Base (MIB) objects within the Cisco proprietary SNMP MIB extension. This capability gives variables in CDP frames the potential to extend into SNMP network management applications.

USING CDP ON A LOCAL ROUTER

CDP is enabled by default at the global level and on each interface in order to send or receive CDP information. Advertisement and discovery using CDP involves data-link frame exchanges, and only directly connected neighbors exchange CDP frames.

To disable CDP on a router, enter the following command in global configuration mode:

```
Router(config)#no cdp run
```

To disable CDP on a specific interface, enter the following command in interface configuration mode:

```
Router(config-if)#no cdp enable
```

The **show cdp interface** command displays information about your own router interfaces, as shown in Figure 7–2. The interface values include the CDP timers, the interface status, and the encapsulation used by CDP for advertisement and discovery frame transmission.

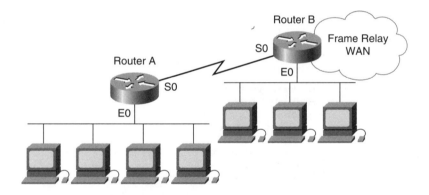

Figure 7–2
Using the **show cdp interface** *command on router A reveals information about the router's own interfaces.*

The sending time indicates the interval at which CDP frames are sent. The holdtime indicates a Time To Live (TTL) for what CDP sends. Neighbors that receive a holdtime value must discard the CDP information about the device if the time specified elapses before the neighbor receives another transmitted TTL value. Also, to prevent obsolete information, prior to losing power a router transmits a TTL frame with a value of zero. CDP devices receiving this frame discard information about the disabled device.

ALTERING CDP PARAMETERS

Routers use CDP to constantly update neighboring devices about the state of the router and its interfaces. The CDP timer regulates how often updates are transmitted. The default value for the CDP timer is 60 seconds. You can alter the CDP timer to send updates using a shorter or longer time interval, as shown in Figure 7–3. Decreasing the timer interval provides faster updates to neighboring routers but increases your bandwidth usage.

Figure 7–3
You can alter CDP's timer and holdtime parameters.

```
Router A# (config)# cdp timer 30
Router A# (config)# exit
Router A# show cdp interface
Serial0 is up, line protocol is up
    Encapsulation is HDLC
    Sending CDP packets every 30 seconds
    Holdtime is 180 seconds
```

```
Router A# (config)# cdp holdtime 90
Router A# (config)# exit
Router A# show cdp interface
Serial0 is up, line protocol is up
    Encapsulation is HDLC
    Sending CDP packets every 30 seconds
    Holdtime is 90 seconds
```

To alter the CDP timer, enter the following command in global configuration mode:

```
Router(config)#cdp timer [seconds]
```

where [seconds] represents the interval between updates.

The holdtime indicates how long the CDP packets being sent from your router should be held by the receiving device before being discarded. The default value is 180 seconds. If the device receives a more recent update or if this holdtime value expires, the device must discard the CDP entry. You might want to set the holdtime lower than the default setting if information about your router changes often and you want the receiving devices to purge this information more quickly.

To alter the CDP holdtime, enter the following command in global configuration mode:

```
Router(config)#cdp holdtime [seconds]
```

where [seconds] represents the amount of time a receiving device should hold the information sent by your device before discarding it.

SHOWING CDP NEIGHBORS

The primary use of CDP is to discover platform and protocol information on your neighboring devices. Use the **show cdp neighbors** command to display the CDP updates received on the local router, as shown in Figure 7–4.

```
Router A# show cdp neighbors
Capability Codes:  R - Router, T - Trans Bridge,
                   B - Source Route Bridge,
                   S - Switch, H - Host, I - IGMP

Device ID          Local Intrfce    Holdtime    Capability    Platform    Port ID
RouterB.cisco.com      Eth 0          151          R T           AGS        Eth 0
RouterB.cisco.com      Ser 0          165          R T           AGS        Ser 3

Router A# show cdp neighbors detail
─────────────────────────────────────
Device ID: routerB.cisco.com
Entry address(es):
    IP address: 198.92.68.18
    CLNS address: 490001.1111.1111.1111.00
    Appletalk address: 10.1
Platform: AGS, Capabilities: Router Trans-Bridge
Interface: Ethernet0, Port ID (outgoing port): Ethernet0
Holdtime: 143 sec
```

Figure 7–4
The command **show cdp neighbors** *displays; the results of the CDP discovery process.*

Notice that for each local port, the display shows the following:

- *Neighbor device ID*—For example, the router's configured host name and domain name, if any

- *Local port type and number*—Interface that is connected to the neighbor

- *Decremental holdtime value in seconds*

- *Neighbor's device capability code*—For example, if the device acts as a source-route bridge as well as a router

- *Hardware platform of the neighbor*
- *Neighbor's remote port type and number*

Key Concept The show cdp neighbors detail command displays a complete profile of neighbor routers and is the most widely used of the show cdp commands.

In the example in Figure 7–4, a neighbor's device name contains a domain name; therefore, the Device ID column for router B displays a domain-name entry in the form *company.com*. To check the device as a single target, include the domain by entering the command variation **show cdp entry routerB.cisco.com**.

Showing CDP Entries for a Neighboring Device

The **show cdp entry** *device-name* command displays a CDP entry for a specific neighboring router. Use the asterisk (*) in place of the device name as a wildcard to display information for all directly attached devices.

As shown in Figure 7–5, the output from this command includes all the Layer 3 addresses present in the neighbor router B. You can see the IP, CLNS, and DECnet network addresses of the targeted CDP neighbor with the single command entry on router A.

Figure 7–5
Enter the specific Cisco router device name to view its CDP information.

```
Router A# show cdp entry routerB

Device ID: routerB
Entry address(es):
   IP address: 198.92.68.18
   CLNS address: 490001.1111.1111.1111.00
   Appletalk address: 10.1
Platform: AGS, Capabilities: Router Trans-Bridge
Interface: Ethernet0, Port ID (outgoing port): Ethernet0
Holdtime: 155 sec

Version:
IOS (tm) GS Software (GS3),  11.2(13337)  [asastry 161]
Copyright (c) 1986-1996 by Cisco Systems, Inc.
Compiled Tue 14-May-96 1:04
```

The holdtime value for a neighboring router indicates how long ago the CDP frame arrived with this information. The display also includes abbreviated version information about router B.

CDP was designed and implemented as a very simple, low-overhead protocol. A CDP frame can be as small as 80 octets, mostly made up of the ASCII strings that represent information like that shown in Figure 7–5.

If the device name has been configured with a domain name, you must enter it as part of the target device name. For example, if router B were running with a configuration containing a domain-name entry in the form company.com, you would include the domain by entering the CDP command variation **show cdp entry routerB.cisco.com**.

Note that the CDP entry device name is case sensitive. In the example in Figure 7–5, the **show cdp entry** command will display only CDP information for device name **routerB**. The command will not display CDP information for any other derivation of the device name, such as **RouterB**, **routerb**, or **ROUTERB**.

SUMMARY

CDP is one way to learn about other routers on the network. In Part 2, "Networking Protocol Suites," you learn other methods, such as ping and Telnet, to discover remote router information. Ping and Telnet require that you know the address of the router you are trying to contact. Use CDP to determine that address.

Chapter Seven Test
Discovering and Accessing Other Cisco Routers

Estimated Time: 15 minutes

Complete all the exercises to test your knowledge of the materials contained in this chapter. Answers are listed in Appendix A, "Chapter Test Answer Key."

Question 7.1

T   F   CDP is a protocol and media independent tool.

Question 7.2

T   F   CDP displays information about directly connected routers, hubs, and management consoles.

Question 7.3

T   F   CDP sends and receives updates at regular intervals.

Question 7.4

T   F   CDP can be enabled by router or by interface.

Question 7.5

T   F   CDP is off by default.

Question 7.6

T   F   CDP uses broadcasts for discovery.

Question 7.7

T   F   CDP is used only for discovery on LANs.

Question 7.8

What is the purpose of the CDP timer setting?

Question 7.9

Why would you consider disabling CDP?

Question 7.10

What is the purpose of CDP holdtime parameter?

Question 7.11

How can you identify CDP packets on your network?

PART 2

Networking Protocol Suites

TCP/IP Overview

The *Transmission Control Protocol/Internet Protocol* (TCP/IP) suite of protocols was developed as part of the research done by the Defense Advanced Research Projects Agency (DARPA). It was originally developed to provide communication between devices connected through DARPA. Now TCP/IP is the de facto standard for internetwork communications and serves as the transport protocol for the Internet, enabling millions of computers to communicate globally. Originally, TCP/IP was included with the Berkeley Software Distribution of UNIX to connect UNIX hosts that were remote to each other, as shown in Figure 8–1.

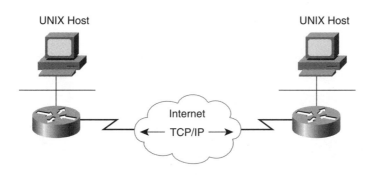

Figure 8–1
TCP/IP is the transport used on the Internet.

In the mid-1970s, DARPA established a *packet-switched* network to provide electronic communication between research institutions in the United States. DARPA and other government organizations understood the potential of packet-switched technology and were just beginning to face the problem that virtually all companies with networks now have: how to establish communication between dissimilar computer systems. Packet-switched networks enable companies to connect to an internetwork cloud that handles getting information from one location to another without a direct connection or link between two end devices.

For more information about the history of TCP/IP, refer to the following World Wide Web address:

```
http://www.ietf.cnrl.reston.va.us/home.html
```

This book focuses on TCP/IP for several reasons:

- TCP/IP is a universally available protocol that you will most likely use at work.

- TCP/IP is a useful reference for understanding other protocols because it includes elements that are representative of other protocols.

- TCP/IP is important because the router uses it as a configuration tool. The router uses Telnet for remote configuration, TFTP to transfer configuration files and operating system images, and SNMP for network management.

TCP/IP PROTOCOL STACK

The TCP/IP Internet protocols can be used to communicate across any set of interconnected networks. They are equally well suited for both LAN and WAN communication. The TCP/IP Internet protocol suite includes not only Layer 3 and 4 specifications (such as IP and TCP), but also specifications for such common applications as e-mail, remote login, terminal emulation, and file transfer.

The TCP/IP protocol stack maps closely to the OSI reference model in the lower layers. All standard physical and data-link protocols are supported. The TCP/IP stack's model, the DARPA model, consists of only four layers that are closely associated to the OSI model:

- *Application layer*—Defines the upper-layer functionality included in the application, presentation, and session layers of the OSI model. This includes support for the communicating component of an application, code formatting and conversion, and session establishment and maintenance between applications.

- *Transport layer*—Defines connectionless and connection-oriented transport functionality

- *Internet layer*—Defines internetworking functionality for routed protocols

- *Network interface layer*—Defines the data-link properties and media access methods

The four-layer TCP/IP protocol stack is shown in Figure 8–2.

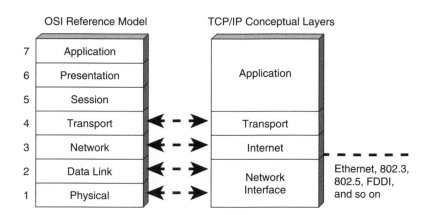

Figure 8–2
The four-layer model that TCP/IP is mapped to is similar to the OSI model in defined functionality.

TCP/IP information is transferred in a sequence of datagrams. One message may be transmitted as a series of datagrams, which are then reassembled into the message at the receiving location. The terms *packet* and *datagram* are nearly interchangeable. However, a datagram is a unit of data, whereas a packet is a physical entity that appears on a network. In most cases, a packet contains a datagram. In some protocols, though, a datagram is divided into a number of packets to accommodate a requirement for smaller transmittable pieces.

Creation and documentation of the Internet protocols closely resembles an academic research project. The protocols are specified in documents called *RFCs (Request for Comments)*. RFCs are published, reviewed, and analyzed by the Internet community. One of the most interesting RFCs is RFC 1700, which lists the assigned numbers for the Internet community. Another important RFC is RFC 791, which covers IP functionality.

APPLICATION LAYER OVERVIEW

Application protocols exist for file transfer, e-mail, and remote login and map to the functionality defined in the application layer of the TCP/IP model. Network management is also supported at the application layer, as shown in Figure 8–3.

Figure 8–3
*Some applica-
tions can
reside on a
router.*

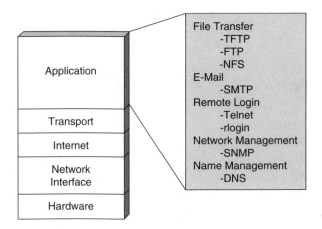

Many of these applications can reside on a router or host device, as well. For example, TFTP (Trivial File Transfer Protocol) and Telnet are applications that may be used to access and configure a router. If a router supports SNMP (Simple Network Management Protocol), the router can be managed by another SNMP device with sufficient permissions. Finally, DNS (Domain Name System) enables a router to reply to DNS host name queries.

TRANSPORT LAYER OVERVIEW

The transport layer performs several functions, including:

- Flow control provided by sliding windows
- Reliability provided by sequence numbers and acknowledgments

Two protocols are provided at the transport layer: TCP and UDP (see Figure 8–4).

- TCP is a connection-oriented, reliable protocol. It is responsible for breaking messages into segments, reassembling them at the destination station, resending anything that is not received, and reassembling messages from the segments. TCP supplies a virtual circuit between end-user applications.

- UDP is connectionless and unacknowledged. Because it has eliminated all of the acknowledgment mechanisms, UDP is fast and efficient. UDP does not divide application data into pieces. Reliability is assumed to be handled by the upper-layer protocols, by a reliable lower-layer protocol, or by an error-tolerant application.

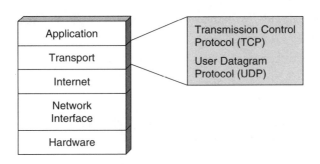

Figure 8–4
Application developers can select a connection-oriented (TCP) or connection-less (UDP) transport.

Typically, TCP connections are slower to establish, but they provide data streaming and very fast, guaranteed delivery of information. UDP, on the other hand, does not use a setup routine, but it also does not provide guaranteed delivery. If packets are lost in a UDP communication, the upper layer must have timers and timeout mechanisms in place to retry the transmission.

TCP Segment Format

Figure 8–5 shows the TCP segment format.

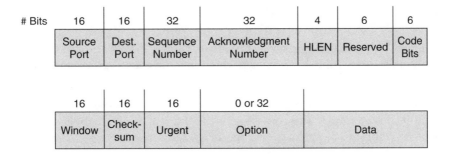

Figure 8–5
TCP segment format.

Field definitions in the TCP segment include the following:

- *Source Port*—Number of the sending port; identifies upper-layer protocol in sending host

- *Destination Port*—Number of the destination port; identifies upper-layer protocol in destination host

- *Sequence Number*—Position in the senders' byte stream of the data in the segment; used to establish reliability

- *Acknowledgment Number*—Next expected TCP octet; used to establish reliability

- *HLEN*—Number of 32-bit words in the header; indicates where the data begins

- *Reserved*—Set to zero

- *Code Bits*—Control functions (such as SYN bits for setup and FIN bits for termination of a session)

- *Window*—Number of octets that the sender is willing to accept; size of receive buffers

- *Checksum*—Calculated checksum of the header and data fields; verifies datagram arrives in tact

- *Urgent Pointer*—Indicates the end of the urgent data; used to signify out-of-band data

- *Option*—One currently defined: maximum TCP segment size; used by vendors to enhance their protocol offering

- *Data*—Upper-layer protocol data

Port Numbers

Both TCP and UDP use *port* (or socket) *numbers* to pass information to the upper layers. Port numbers are used to keep track of different conversations crossing the network at the same time.

Application software developers agree to use well-known port numbers that are defined in RFC1700. For example, any conversation bound for the FTP application uses the standard port number 21, as shown in Figure 8–6.

Figure 8–6
Port numbers indicate the upper-layer protocol that is using the transport.

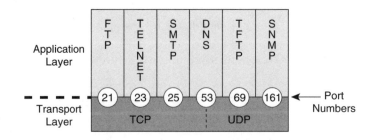

Conversations that do not involve an application with a well-known port number are assigned port numbers randomly chosen from within a specific range instead. These port numbers are used as source and destination addresses in the TCP segment.

It is possible to filter on TCP port numbers. In fact, this is one type of security offered by many firewalls.

Some ports are reserved in both TCP and UDP, but applications might not be written to support them. Port numbers have the following assigned ranges:

- Numbers below 256 are for public applications.

- Numbers from 256 to 1023 are assigned to companies for saleable applications.

- Numbers above 1023 are dynamically assigned by the host application.

The TCP port number, combined with other information, is what UNIX C language developers call a *socket*. However, work sockets have different meanings in XNS and Novell, where they are service access point abstractions or programming interfaces rather than service access point identifiers.

TCP Port Numbers

End systems use port numbers to select the proper application. Originating source port numbers are dynamically assigned by the source host, usually some number greater than 1023. See Figure 8–7.

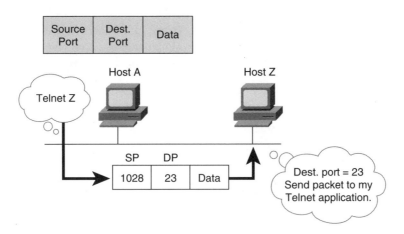

Figure 8–7
The source port and destination ports do not need to be the same.

In most cases, the TCP port number on one side of a conversation is the same on the other side. For example, when a file transfer takes place, the software on one host is communicating with a peer application on another host.

In the example in Figure 8–7, you see a Telnet (TCP port 23) session. It is possible to have multiple Telnet sessions running simultaneously on a host or router. Telnet selects an unused port number above 1023 to represent the source port for each independent session. Notice that the destination port is still 23.

Port numbering is important to understand in order to configure IP extended access lists. Extended access lists can block or forward data based on these numbers. Extended access lists are covered in greater detail in Chapter 13, "Basic Traffic Management with Access Lists."

TCP Handshake/Open Connection

For a connection to be established or initialized, the two TCPs use processes or end stations instead of "TCPs" and must synchronize on each other's *initial sequence numbers* (ISN). Sequence numbers are used to track the order of communications and to ensure that there are no missing pieces of data in a communication that requires multiple packets. The initial sequence number is the starting number used when the TCP connection is established.

Synchronization is accomplished by exchanging segments carrying the ISNs and a control bit called *SYN* (for synchronize). (As a shorthand, segments carrying the SYN bit are also called *SYNs*.) Successful connection requires a suitable mechanism for picking an initial sequence number and a slightly involved handshake to exchange the ISNs.

Synchronization requires that each side send its own ISN and receive a confirmation and ISN from the other side of the connection. Each side must receive the other side's ISN and send a confirming acknowledgment in a specific order, outlined in the following steps:

1. A [→] B SYN my sequence number is X
2. A [←] B ACK your sequence number is X
3. A [←] B SYN my sequence number is Y
4. A [→] B ACK your sequence number is Y

Because the second and third steps can be combined in a single message, this exchange is called the *three-way handshake*, as shown in Figure 8–8.

This sequence is like two people talking. The first person wants to talk to the second, so she says, "I would like to talk with you." (SYN.) The second person responds,

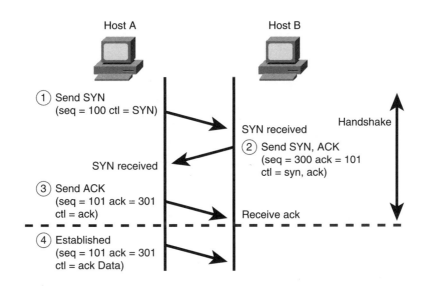

Figure 8–8
*Data cannot
be exchanged
until the
three-way
handshake
has been suc-
cessfully com-
pleted.*

"Good. I want to talk with you." (SYN, ACK.) The first person then says, "Fine; let's talk." (ACK.)

A three-way handshake is necessary because sequence numbers are not tied to a global clock in the network, and TCPs may have different mechanisms for picking the ISN. The receiver of the first SYN has no way of knowing if the segment was an old delayed one, unless it remembers the last sequence number used on the connection (which is not always possible), and so it must ask the sender to verify this SYN.

At this point, either side can begin communicating, and either side can break the connection. TCP is a peer-to-peer (balanced) communication method.

TCP Simple Acknowledgment and Windowing

The window size determines how much data the receiving station can accept at one time. With a window size of 1, each set of bytes must be acknowledged before another set of bytes is transmitted, which results in inefficient use of bandwidth by the hosts.

The purpose of windowing is to improve flow control and reliability. Unfortunately, with a window size of 1, you see a very inefficient use of bandwidth, as shown in Figure 8–9.

To govern the flow of data between devices, TCP uses a flow control mechanism. The receiving TCP reports a window to the sending TCP. This window specifies the number

Figure 8–9
With a window of 1, the sender must wait for an acknowledgment before sending more data.

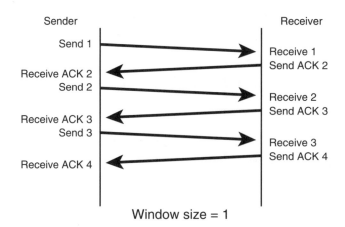

Figure 8–10
A larger window increases flow efficiency.

of bytes, starting with the acknowledgment number, that the receiving TCP is currently prepared to receive, as shown in Figure 8–10.

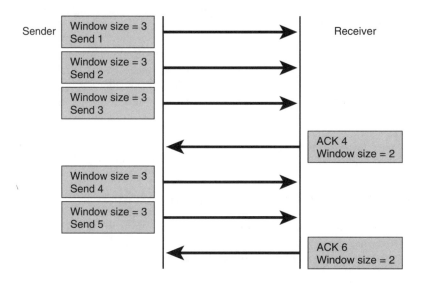

TCP window sizes are variable during the lifetime of a connection. Each acknowledgment contains a window advertisement that indicates how many bytes the receiver can accept. TCP also maintains a congestion control window, which is normally the same size as the receiver's window but is cut in half when a segment is lost (such as when there is congestion). This approach permits the window to be expanded or contracted as necessary to manage buffer space and processing. A larger window size allows more data to be processed.

You can find more information on windowing in RFCs 793 and 813. Also, you can find an explanation of basic TCP windowing in *TCP/IP Illustrated, Vol. 1*, by Richard W. Stevens (pp. 282–284. New York, New York: Addison-Wesley, 1994).

TCP Sequence and Acknowledgment Numbers

TCP provides sequencing of segments by numbering each datagram before transmission. At the receiving station, TCP reassembles the segments into a complete message. If a sequence number is missing in the series, that segment is retransmitted. Segments that are not acknowledged within a given time period result in retransmission.

The sequence and acknowledgment numbers are directional, which means that the communication occurs in both directions. Figure 8–11 highlights the communication going in one direction. The sequence and acknowledgments take place with the sender on the right. TCP provides full-duplex communication. Acknowledgments provide reliability. Another example of a sequenced, connection-oriented protocol is SPX (Sequenced Packet Exchange) for NetWare.

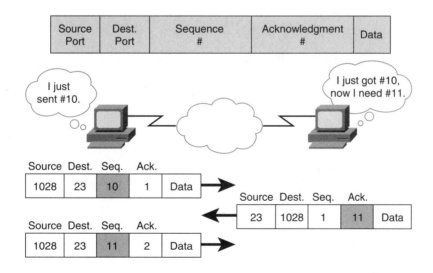

Figure 8–11
The receiver asks for the next datagram in the sequence.

UDP Segment Format

UDP uses no windowing or acknowledgments. Application-layer protocols can provide reliability. UDP is designed for applications that do not need to put sequences of segments together.

Protocols that use UDP include TFTP, SNMP, Network File System (NFS), and Domain Name System (DNS). As you can see in Figure 8–12, a UDP header is relatively small (only eight bytes). However, DNS can use TCP, as well.

Figure 8–12
UDP has no sequence or acknowledgment fields.

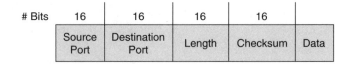

An example that shows how reliability is handled when using UDP is TFTP (Trivial File Transfer Protocol). TFTP uses a checksum. At the end of the transfer, if the checksum does not match, the file did not make it. The user is notified and must type the command again. As a result, the user has become the reliability mechanism.

INTERNET LAYER OVERVIEW

The Internet layer of the TCP/IP stack corresponds to the network layer of the OSI model. Each layer is responsible for getting packets through an internetwork using software addressing.

As shown in Figure 8–13, two protocols operate at the TCP/IP Internet layer, which corresponds to the OSI network layer. These include:

- IP, which provides connectionless, best-effort delivery routing of datagrams. It is not concerned with the content of the datagrams. Instead, it looks for a way to move the datagrams to their destinations.

- Internet Control Message Protocol (ICMP), which provides control and messaging capabilities.

Note that routing protocols are usually considered layer-management protocols that support the network layer. OSPF is totally contained within the network layer.

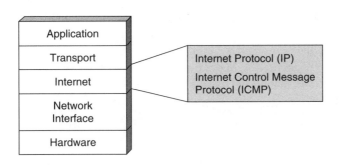

Figure 8–13
The OSI network layer corresponds to the TCP/IP Internet layer.

IP Datagram

Figure 8–14 shows the format of an IP datagram. IP datagrams contain an IP header and data and are surrounded by the MAC layer header and MAC layer trailer.

# Bits	4	4	8	16	16	3	13	8
	VERS	HLEN	Type of Service	Total Length	Identi-fication	Flags	Frag Offset	TTL

# Bits	8	16	32	32	Variable	Variable	Variable
	Protocol	Header Checksum	Source IP Address	Destination IP Address	IP Options (if any)	Padding	Data

Figure 8–14
The IP header is variable in length because of the IP options field.

Field definitions within this IP datagram are as follows:

- *VERS*—Version number; the current popular version is IP version 4; the next generation of IP (called IPng) is IP version 6. IPng is covered in RFC 1752.

- *HLEN*—Header length in 32-bit words; indicates where the transport header starts.

- *Type of Service*—How the datagram should be handled; specifies reliability, precedence, delay, and throughput parameters.

- *Total Length*—Total length (header and data); this includes all upper-layer headers.

- *Identification, Flags, Frag Offset*—Provides fragmentation and reassembly of datagrams to allow differing MTUs (Maximum Transmission Units) or frame sizes in the internetwork.

- *TTL*—Time To Live; Time To Live is a countdown field. Every station must decrement this number by one or by the number of seconds it holds onto the packet. When the counter reaches zero, the Time To Live expires, and the packet is dropped. TTL keeps packets from endlessly wandering the Internet in search of nonexistent destinations.

- *Protocol*—Identifies the upper-layer (Layer 4) protocol that should receive the datagram. Although most IP traffic uses TCP, there are other protocols that can use IP. Transport-layer protocols are numbered similarly to port numbers.

  Note that protocol numbers connect, or *multiplex*, IP to the transport layer. These numbers are standardized in RFC 1700. Cisco uses these numbers in filtering with extended access lists.

- *Header Checksum*—Integrity check on the header.

- *Source and Destination IP Addresses*—32-bit IP addresses that identify the end devices involved in the communication.

- *IP Options*—Network testing, debugging, security, and others.

Internet Control Message Protocol (ICMP)

ICMP resides at the Internet layer, as shown in Figure 8–15.

Figure 8–15
ICMP provides error and control mechanisms.

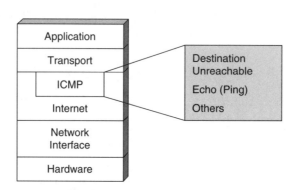

The ICMP is implemented by all TCP/IP hosts. ICMP messages are carried in IP datagrams and use myriad types of defined messages. The following list shows some of the most common and useful ICMP message types:

- *Destination Unreachable*—Report unreachable destination
- *Time Exceeded*—Detect circular pockets
- *Parameter Problem*—Faulty IP packet structure
- *Source Quench*—Flow control
- *Redirect*—Route change
- *Echo*—Test reachability
- *Echo Reply*—Test reachability
- *Timestamp*—Clock sync—Transit time estimation
- *Timestamp Reply*—Clock sync—Transit time estimation
- *Information Request*—Obtain a network address
- *Information Reply*—Obtain a network address
- *Address Mask Request*—Obtain a subnet mask

Refer to RFC 1700 for a more complete list of ICMP messages.

ICMP Testing

If a router receives a packet that it is unable to deliver to its ultimate destination, the router sends an ICMP host unreachable message to the source, as shown in Figure 8–16.

The message might be undeliverable because there is no known route to the destination.

ICMP is even simpler than UDP. ICMP does not use port numbers in its header because all ICMP messages are interpreted by the network software itself; therefore, no port numbers are needed to determine the destination of the message. ICMP does, however, include a type field, and it identifies the ICMP message type echo request.

Figure 8–16
*ICMP indi-
cates that the
desired desti-
nation is
unreachable.*

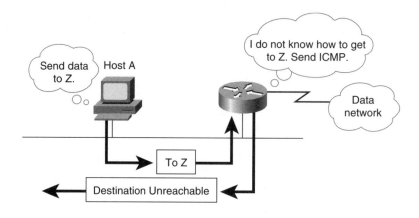

TIPS

ICMP is a tremendous network design and troubleshooting tool.

An echo reply is a successful reply to a **ping** command; however, results could include other ICMP messages, such as unreachables and timeouts, that indicate the **ping** request couldn't reach the destination. Figure 8–17 shows host A sending an echo request to host B. Upon receipt of the echo request, host B sends an echo reply back to host A.

Figure 8–17
*An ICMP echo
request is
generated by
the **ping**
command.*

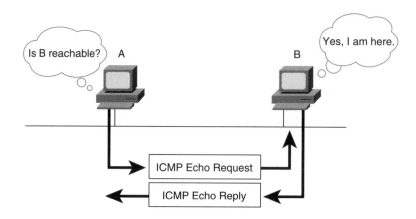

Another tool that uses ICMP is traceroute. Traceroute provides a list of the routers along the path between two devices.

Address Resolution Protocol

Address Resolution Protocol (ARP) is used to resolve or map a known IP address to a MAC sublayer address to allow communication on a multiaccess medium such as Ethernet. To determine a destination address for a datagram, the ARP cache table is checked. If the address is not in the table, ARP sends a broadcast looking for the destination station, as shown in Figure 8–18.

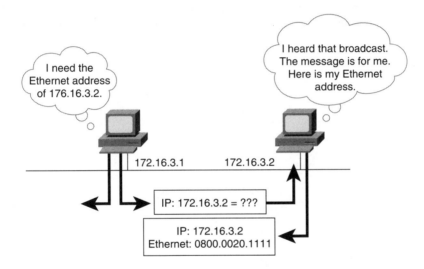

Figure 8–18
ARP is used to get the MAC address.

Every station on the network receives the broadcast.

The term *local ARP* refers to resolving an address when both the requesting host and the destination host share the same media or wire. In the case of a packet destined to a device on another network, the local host will ARP only for the local router's MAC address.

Reverse ARP

Reverse ARP (RARP) is used to obtain an IP address using an RARP broadcast. RARP relies on the presence of an RARP server with a table that contains information used to respond to these requests.

On the local segment, RARP can be used to initiate a remote operating system load sequence. For example, RARP can be used to boot diskless workstations over a network. ARP and RARP are implemented directly on top of the data link layer.

SUMMARY

In this chapter, you read a brief overview of the TCP/IP protocol suite. You learned all about the TCP/IP protocol stack, as well as the four TCP/IP conceptual layers. You also reviewed the protocols maintained at the network layer, including IP (and the IP header structure) and ICMP. Finally, this chapter covered MAC to IP address resolution with ARP and RARP. The following two chapters, Chapter 9, "IP Addressing," and Chapter 10, "IP Routing Configuration," delve further into configuring an IP internetwork.

Chapter Eight Test
TCP/IP Overview

Estimated time: 15 minutes

Complete all the exercises to test your knowledge of the materials contained in this chapter. Answers are listed in Appendix A, "Chapter Test Answer Key." Use the information contained in this chapter to answer the following questions.

Question 8.1

What are the four TCP/IP conceptual layers?

_____

_____

_____

_____

Question 8.2

Which OSI layer does IP map to?

_____

Question 8.3

Which layer does UDP map to?

_____

Question 8.4

For each statement, write the name of the protocol being described.

Protocol Name	Statement
A. _____	Maps a known IP address to a MAC sublayer address
B. _____	Includes Layer 4 protocol ID in header
C. _____	Used to send Destination Unreachable messages
D. _____	Breaks messages into datagrams
E. _____	Uses sequence numbers
F. _____	Relies on application-layer reliability
G. _____	Provides best-effort delivery
H. _____	Reassembles datagrams into messages
I. _____	Handshakes with receiving device
J. _____	Used to send error and control messages
K. _____	Provides connectionless transmission
L. _____	Sends acknowledgments
M. _____	Uses no windowing

IP Addressing

This chapter presents the details of IP address classes, network and node addresses, and subnet masking. The test at the end of the chapter lets you evaluate your understanding of IP address configuration.

TCP/IP ADDRESS OVERVIEW

In a TCP/IP environment, end stations communicate seamlessly with servers or other end stations. This communication occurs because each node using the TCP/IP protocol suite has a unique 32-bit logical address.

Often traffic is forwarded through the internetwork based on the name of an organization rather than an individual person or host. If names are used instead of addresses, the names must be translated to the numeric address before the traffic can be delivered. Location of the organization will dictate the path the data follows through the internetwork.

Each company listed on the internetwork is treated as a single network that must be reached before an individual host within that company can be contacted. Each company network has an address; the hosts that populate that network share those same bits, but each host is identified by the uniqueness of the remaining bits.

IP Addressing Overview

An IP address is 32 bits in length and has two parts:

- Network number
- Host number

The 32 bits are divided into four octets (an octet is eight bits; that is, one byte). Although computers have no difficulty dealing with a 32-bit number, humans do; therefore, you must translate the binary value of each octet into a decimal equivalent to create an address format known as *dotted-decimal notation*, as shown in Figure 9–1. A sample dotted-decimal address is 172.16.122.204.

Figure 9–1
IP addresses are four octets (or four bytes) long and contain a network and host portion.

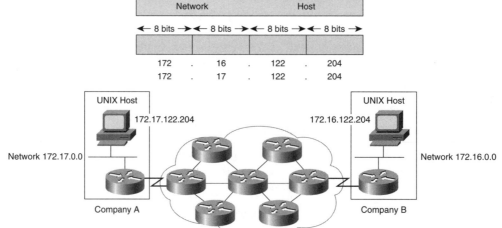

The minimum value for an octet is 0; it contains all zeros. The maximum value for an octet is 255; it contains all ones. Later in this chapter, you'll learn how the binary format is used in IP addressing.

The allocation of addresses is managed by a central authority, the Internet Assigned Numbers Authority (IANA).

This most common form of addressing reflects the widely used IP version 4. Faced with the problem of depleting available addresses, Internet Engineering Task Force (IETF) work is under way for a backward-compatible next generation of IP (IPng, now referred to as IPv6).

IPv6 will offer expanded routing and addressing capabilities with 128-bit addresses rather than the 32-bit addresses shown in Figure 9–1. Addresses from both IP versions will coexist. Initial occurrences of the IPv6 addresses will probably be at locations with address translator software and firewalls.

In some sections of this book, you will work with the addresses on the bit level, so you will convert these addresses into binary, make changes to them, and convert them back to decimal form. Refer to Appendix E, "Decimal to Hexadecimal and Binary Conversion Table," for a binary conversion table.

IP Address Classes

When IP was first developed, there were no classes of addresses. Now, for ease of administration, the IP addresses are broken into classes, as shown in Figure 9–2.

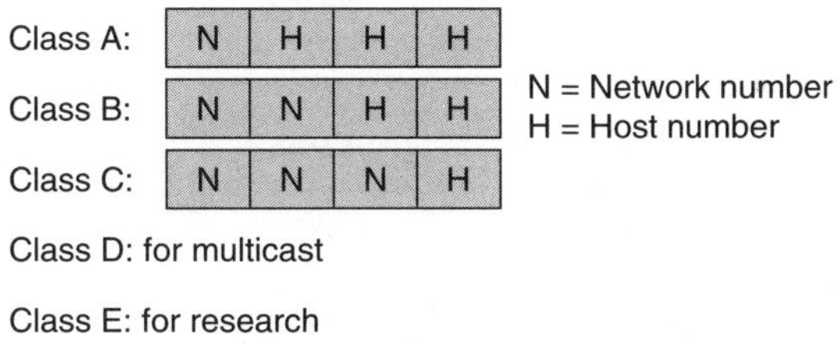

Class A: | N | H | H | H |

Class B: | N | N | H | H |

Class C: | N | N | N | H |

N = Network number
H = Host number

Class D: for multicast

Class E: for research

Figure 9–2
Class A, B, and C addresses are used for IP networks.

There are only 126 Class A addresses, but each address can support up to 16 million hosts. There are 64,000 Class B addresses, which can support up to 64,000 hosts. There are more than 16 million Class C addresses possible, each supporting up to 254 hosts. (Class D and E addresses are also defined. Class D addresses start at 224.0.0.0 and are used for multicast purposes. Class E addresses start at 240.0.0.0 and are used for experimental purposes by Internet designers and engineers.)

To define the number of host addresses possible, perform the following equation, where *n* is the number of bits in the host portion:

$$(2^n - 2) = \text{available host addresses}$$

For example, consider the network address 200.99.44.0 (a Class C address). There are eight bits available in the host portion. The formula is $(2^8 - 2) = 254$.

The reason you must subtract two from the number is to reserve two special host ID numbers: all zeros and all ones. All zeros is used to indicate the network (for example, 200.99.44.0), and all ones in the host ID portion is reserved for a network broadcast (for example, 200.99.44.255).

This addressing scheme allows the addresses to be assigned based on the size of the network. This address design was based on the assumption that there would be many more small networks than large networks in the world.

As the number of networks grows, classes may eventually be replaced by another addressing mechanism, such as classless interdomain routing (CIDR). RFC 1467, "Status of CIDR Deployment in the Internet," presents information about CIDR. RFC 1817, "CIDR and Classful Routing," also presents CIDR information.

IP Address Bit Patterns

Figure 9–3 shows the structure of Class A, B, and C addresses.

Figure 9–3
The first byte's value can be used to indicate the class of the address.

High-Order Bits	Octet in Decimal	High-Order Bits
0	1-126	A
10	128-191	B
110	192-223	C

	#Bits	1	7	24
Class A:		0	network#	host#

	#Bits	1	1	14	16
Class B:		1	0	network#	host#

	#Bits	1	1	1	21	8
Class C:		1	1	0	network#	host#

The value of the first byte of an address determines the class of the address, as well as how many bits make up the network portion of the address. The scope of each class is as follows:

- Class A

  Range of network numbers: 1.0.0.0 to 126.0.0.0

  Number of host addresses: 16,777,214 (16,777,216 − 2)

- Class B

  Range of network numbers: 128.0.0.0 to 191.255.0.0

  Number of host addresses: 65,534 (65,536 − 2)

- Class C

  Range of network numbers: 192.0.0.0 to 223.255.255.0

  Number of host addresses: 254 (256 − 2)

- Class D addresses include

  Range of network numbers: 224.0.0.0 to 239.255.255.0

The First Octet Rule

The *first octet rule* states that the class of `n address can be determined by the numerical value of the first octet.

Once the first octet rule is applied, the router identifies how many bits it must match to interpret the network portion of the address (based on the standard address class), as shown in Figure 9–3. If there is no further identification of additional bits to use as part of the network address, the router can make a routing decision using this address.

The range of class addresses is as follows, with the first octet represented in decimal:

- 1 to 126—Class A address
- 128 to 191—Class B address
- 192 to 223—Class C address
- 224 to 239—Class D address
- 240 to 255—Class E address

Note that the number 127.0.0.0 is reserved for the *loopback address*. The loopback address is used by a device to address itself internally. This technique is used to test the local device's TCP/IP stack and identify possible stack corruption.

CONCEPTS OF IP ADDRESS CONFIGURATION

This section focuses on basic concepts you need to understand before configuring an IP address. By examining various network requirements, you can select the correct class of address and define how to establish IP subnets.

Host Addresses

Each device or interface must have a non-zero host number. A host address of all ones is reserved for an IP broadcast into that network, as shown in Figure 9–4.

Figure 9–4
All hosts must have non-zero IP addresses.

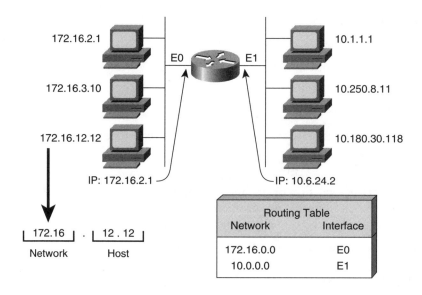

A value of zero refers to "this network" or "the wire itself" (for example, 172.16.0.0). (It was also used for IP broadcasts in some early TCP/IP implementations, although it is rarely found now.)

The use of all zeros is called *subnet zero*. By convention, subnet zero is now reserved and cannot be used to represent interfaces, but refers to the wire itself. The routing table contains entries for network or wire addresses; it usually does not contain any information about hosts.

An IP address and subnet address on an interface achieve three purposes:

- Enable the system to process the receipt and transmission of packets
- Specify the device's local address
- Specify a range of addresses that share the cable with the device

The router is able to distinguish the network portion from the host portion of the address by using a mask that you configure on the interface of the router. Masks also use dotted-decimal format. The default or standard mask for a Class B address looks like this: 255.255.0.0.

The standard or default masks for Classes A, B, and C follow:

- A—255.0.0.0
- B—255.255.0.0
- C—255.255.255.0

These default masks have all the network bits set to one.

Addressing Without Subnets

The outside world sees the organization as a single network, and no detailed knowledge of the internal structure is required. For example, in Figure 9–5, all datagrams addressed to 172.16 are treated the same way, regardless of the third and fourth octet of the address. A benefit of this configuration is the relatively short routing tables that routers can use.

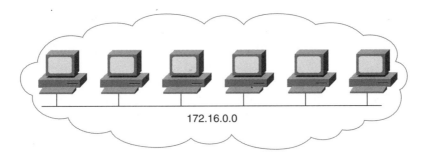

172.16.0.0

Figure 9–5
Straightforward addressing does not use subnets.

Network addressing with the scheme set up so far has no way of distinguishing individual segments (wires) within the network. Inside the cloud having no subnets, you have a single large broadcast domain; all systems on the network encounter all the broadcasts on the network. This type of configuration can result in relatively poor network performance.

By default, this Class B address space defines one wire with 65,000 workstations on it. Subnets enable you to divide this wire into segments.

Addressing with Subnets

With subnets, the network address use is more efficient. There is no change in how the outside world sees the network, but within the organization, there is additional structure.

In the example in Figure 9–6, the network 172.16.0.0 is subdivided or broken into four subnets: 172.16.1.0, 172.16.2.0, 172.16.3.0, and 172.16.4.0. Routers determine the destination network using the subnet address, thus limiting the amount of traffic on the other network segments.

Figure 9–6
From the outside world, no one can tell your network is subnetted.

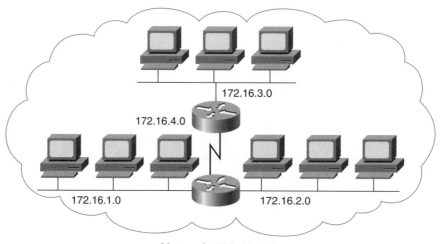

Network 172.16.0.0

A network device uses a subnet mask to determine what part of the IP address is used for the network, the subnet, and the host address or IP. A *subnet mask* is a 32-bit value containing a number of one bits that correspond to the network and subnet portions of the address, and a number of zero bits that correspond to the host portion.

Given its own IP address and subnet mask, a device can determine if an IP packet is destined for the following:

- A device on its own subnet

- A device on a different subnet on its own network

- A device on a different network

A device can determine what class of address the device has been assigned from its own IP address. The subnet mask then tells the device where the boundary is between the subnet ID and the host ID. Subnet masks are discussed in greater detail in the following pages.

Subnetting Addressing

From the addressing standpoint, subnets are an extension of the network number. Network administrators decide the size of subnets based on organization and growth needs.

Network devices use subnet masks to identify which part of the address is considered network and which part is used for host addressing, as shown in Figure 9–7.

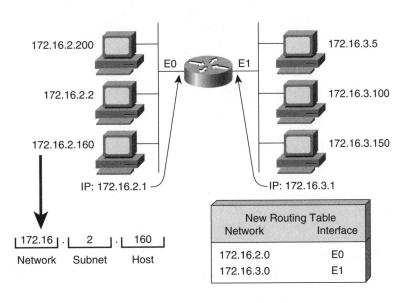

Figure 9–7
Subnet masking uses part of the host address as a subnet portion.

Subnet Mask

As noted earlier, an IP address is 32 bits in size, written as four octets. A subnet mask is also 32 bits in size, written as four octets, and consists of a series of contiguous ones followed by contiguous zeros. Like IP addresses, subnet masks can be expressed in dotted-decimal as well as binary notation.

Subnet masks indicate which of the bits in the host field of the IP address are used to specify different parts (subnets) of a particular network. Figure 9–8 shows an IP address and two relevant subnet masks. The first mask is the default mask. It reserves the first two bytes for the network portion and the last two bytes as the host portion. The second mask shown is a subnet mask that borrows bits from the host portion to increase the number of networks possible.

Figure 9–8
The subnet mask identifies sections (subnets) of a larger network.

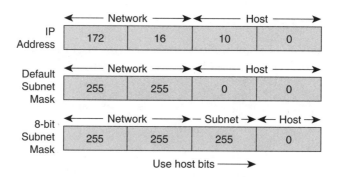

The layout of the subnet mask field is as follows:

- Binary 1 for the network bits
- Binary 1 for the subnet bits
- Binary 0 for the host bits

Decimal Equivalents of Bit Patterns

Subnet bits come from the high-order bits of the host field. To determine a subnet mask for an address, add the decimal values of each position that has a 1 in it. For example, in Figure 9–9, the values of each bit position are shown. As you'll notice, you perform binary to decimal conversion by simply adding up the bit equivalents represented by a 1 in their bit position. For example:

224 = 128 + 64 + 32

Because the subnet mask is not defined by the octet boundary but by bits, you need to convert dotted-decimal addresses to binary and back into dotted-decimal. (Refer to Appendix E for help with binary conversions.)

128	64	32	16	8	4	2	1		
1	0	0	0	0	0	0	0	=	128
1	1	0	0	0	0	0	0	=	192
1	1	1	0	0	0	0	0	=	224
1	1	1	1	0	0	0	0	=	240
1	1	1	1	1	0	0	0	=	248
1	1	1	1	1	1	0	0	=	252
1	1	1	1	1	1	1	0	=	254
1	1	1	1	1	1	1	0	=	255

Figure 9–9
Each bit position has a decimal value.

Subnet Mask Without Subnets

A default subnet mask, also known as a standard or internal subnet mask, is associated with an IP address when no subnetting is required. The first subnet mask in Figure 9–8 is a default mask. In binary, a default mask has all ones for the network portion of the IP address and all zeros for the host portion.

The router extracts the IP destination address from the packet and retrieves the internal subnet mask. The router examines the bits that have been masked off for the network portion to obtain the network number. During the process of determining the network address, the host portion of the destination address is removed. Routing decisions are then based on the network number only.

Using the example from Figure 9–7:

> Packet IP address: 172. 16.2.160

> Default subnet mask: 255.255.0.0

> Network: 172. 16.0.0

With no subnetting, the network number that is extracted is 172.16.0.0.

Subnet Mask with Subnets

When subnetting of an IP address is required, the binary subnet mask consists of all ones for the network and subnet portions of the address, and all zeros for the host portion. The second mask in Figure 9–8 is an 8-bit subnet mask. That is, compared to the default mask, eight additional bits have been "turned on" (made binary ones).

Continuing the example from Figure 9–7, with eight bits of subnetting (255.255.255.0), the extracted network (subnet) number from the address 172.16.2.160 is 172.16.2.0. You arrive at this number by masking off the first three bytes of the total address as defined by the mask 255.255.255.0.

> Packet IP address: 172.16.2.160
>
> 8-bit subnet mask: 255.255.255.0
>
> Network: 172.16.2.0

The practical effect of the extended subnet mask is to create a subnet field by "borrowing" eight host bits. This subnet field is used to represent subnetworks inside the network.

Subnet Planning

In Figure 9–10, the network has been assigned a Class C address of 201.222.5.0. Assume that 20 subnets are needed, with five hosts per subnet. You need to subdivide the last octet into a subnet and a host portion, and determine what the subnet mask will be.

Figure 9–10
Consider the number of networks and hosts per network when selecting a subnet mask.

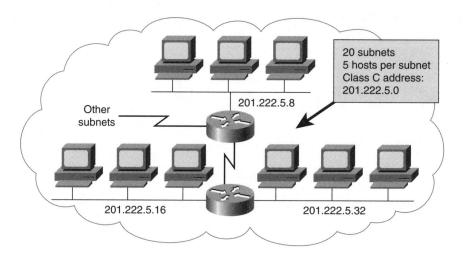

20 subnets
5 hosts per subnet
Class C address:
201.222.5.0

201.222.5.8

Other subnets

201.222.5.16   201.222.5.32

Select a subnet field size that yields enough subnetworks. In this example, choosing a 5-bit mask allows 30 ($2^5 = 32 - 2 = 30$ total subnets possible) subnets. In the example, the subnet addresses are all multiples of eight, such as 201.222.5.8, 201.222.5.16, and 201.222.5.24.

The remaining bits in the last octet are used for the host field. The three bits in the example allow enough hosts to cover the required five hosts per wire. The host numbers will be 1, 2, 3, and so forth.

The final host addresses are a combination of the network/subnet "wire" starting address plus each host value. The hosts on the 201.222.5.16 subnet would be addressed as 201.222.5.17, 201.222.5.18, 201.222.5.19, and so forth.

A host number of zero is reserved for the *wire* address, and a host value of all ones is reserved for a network broadcast because it selects all hosts.

Key Concept

Table 9–1 shows a subnet planning example for a Class B address; a routing example combines an arriving IP address with the subnet mask to derive the subnet number. The extracted subnet number should be typical of the subnets generated during this planning exercise. For an extended table of subnetting, refer to RFC 1878.

No. Bits	Subnet Mask	No. Subnets	No. Hosts
2	255.255.192.0	2	16,382
3	255.255.224.0	6	8190
4	255.255.240.0	14	4094
5	255.255.248.0	30	2046
6	255.255.252.0	62	1022
7	255.255.254.0	126	510
8	255.255.255.0	254	254
9	255.255.255.128	510	126
10	255.255.255.192	1022	62
11	255.255.255.224	2046	30
12	255.255.255.240	4094	14
13	255.255.255.248	8190	6
14	255.255.255.252	16,382	2

Table 9–1

Subnet planning example for a Class B address.

Class B Subnet Planning Example

Consider a Class B subnet based on the following information:

>   Subnet address: 172.16.2.0

>   Subnet mask: 255.255.255.0

The entire third byte of the address is available for subnetting; the entire fourth byte is available for host addresses (172.16.2.1–172.16.2.254). The network address for this subnet is 176.16.2.0. The broadcast address for this subnet is 172.16.2.255.

This network has eight bits of subnetting that provide up to 254 subnets and 254 host addresses.

Class C Subnet Planning Example

Class C addresses are much more difficult to subnet because you must split up the last byte into two portions: the subnetwork portion and the host portion. In Figure 9–11, a Class C network is subnetted to provide six host addresses and 30 subnets.

Table 9–2 should help in this process. The first column (No. Bits) indicates how many bits must be borrowed from the host portion to create a subnetwork address. The second column (Subnet Mask) provides the decimal value of the subnet mask used. Table 9–2 also indicates the number of subnetworks and hosts per subnetwork that are possible with each of these masks.

No. Bits	Subnet Mask	No. Subnets	No. Hosts
2	255.255.255.192	2	62
3	255.255.255.224	6	30
4	255.255.255.240	14	14
5	255.255.255.248	30	6
6	255.255.255.252	62	2

Table 9–2 *Class C subnet reference chart.*

Take a look at that Class C subnet shown in Figure 9–10 again. You've masked off five bits of the host ID portion to use as the subnet area (11111000). Now the subnet mask is as follows:

>   255.255.255.248

>   11111111.11111111.11111111.11111000

The subnet number is defined based on the mask, as shown in Figure 9–11.

```
IP Host Address:  201.222.5.121
    Subnet Mask:  255.255.255.248
 Subnet Address:  201.222.5.120
```

		Network		Subnet	Host
201.222.5.121:	11001001	11011110	00000101	01111	001
255.255.255.248:	11111111	11111111	11111111	11111	000
Subnet:	11001001	11011110	00000101	01111	000
IP Host Address:	201	222	5	120 +	1

Figure 9–11
In order to subnet a Class C address, you must borrow some of the host bits to use for a subnet portion.

Broadcast Addresses

Broadcasting is supported on the Internet. *Broadcast messages* are those you want every host on the network to see. The broadcast address is formed by using all ones within the IP network address.

The Cisco IOS software supports two kinds of broadcasts:

- Directed broadcasts (subnet broadcasts)
- Flooded broadcasts (local broadcasts)

Flooded broadcasts (255.255.255.255) are not propagated but are considered local broadcasts, as shown in Figure 9–12. Broadcasts directed into a specific network are allowed and are forwarded by the router. These directed broadcasts contain all ones in the host portion of the address.

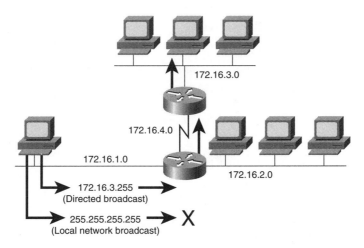

172.16.3.0

172.16.4.0

172.16.1.0

172.16.2.0

172.16.3.255 →
(Directed broadcast)

255.255.255.255 → X
(Local network broadcast)

Figure 9–12
You can broadcast locally or to a subnet.

You can also broadcast messages to all hosts within a single subnet and to all subnets within a network. To broadcast a message to all hosts within a single subnet, the host portion of the address contains all ones. The following example broadcasts messages to all hosts in network 172.16, subnet 3 (assuming a 255.255.255.0 mask):

```
All hosts on a specific subnet = 172.16.3.255
```

You can also broadcast messages to all hosts on all subnets within a single network. To broadcast a message to all hosts on all subnets within a single network, the host and subnet portions of the address contain all ones. The following example broadcasts messages to all hosts on all subnets in network 172.16:

```
All hosts on all subnets in a specific network = 172.16.255.255
```

TIPS

Broadcasts don't always end with 255. Consider a Class C network 201.222.5.0 that is being split with a router. If a company chooses a four-bit subnet (255.255.255.240) and assigns subnet address 201.222.5.160 to one side, what is the broadcast address for this subnet?

Hint: Check the binary table for a number that would put all ones in the host IP portion—201.222.5.175. Therefore, this IP address cannot be assigned to an individual station.

CONFIGURATION COMMANDS

Now that you have an understanding of the concepts behind IP addressing, this section turns to the IOS commands used in configuring IP addresses and related routing capabilities.

Configuring IP Addresses

Table 9–3 shows the address configuration commands.

Command Level	Command	Purpose	
Router(config-if)#	ip address *ip-address* *subnet-mask*	Assigns an address and subnet mask to an interface; starts IP processing	**Table 9–3** *IP address commands.*
Router#	**term ip netmask-format {bitcount|decimal| hexadecimal}**	Sets format of network mask for current session	
Router(config-if)#	**ip netmask-format {bitcount|decimal| hexadecimal}**	Sets format of network mask for a specific line	

Use the **ip address** command to establish the IP network address of this interface. The command itself is followed by an IP address and subnet mask, both in dotted decimal form.

As discussed, IP uses a 32-bit mask, called a subnet mask or netmask, that indicates which address bits belong to the network and subnetwork fields and which bits belong to the host field. The **show ip interface** command displays a summary of an interface's IP information, including its IP address and netmask. By default, the netmask is displayed in dotted-decimal notation. For example, a subnet would be displayed as 131.108.11.55 255.255.255.0.

You can also display the network mask in hexadecimal format. The hexadecimal format is commonly used on UNIX systems. An example of this format is 0xFFFFFF00 for a netmask of 255.255.255.0. The leading 0x indicates that the number is in hexadecimal formal. In this format, two characters are required to define a byte value. For example, the first byte is FF (all ones in binary or 255 in decimal format).

You can also display the netmask in a bit-count format. This format appends a slash (/) and the total number of bits in the netmask to the address itself. An example of this format is 131.108.11.55/24, which indicates that the first 24 bits are used for the network portion.

Enter the **term ip netmask-format** command at the EXEC mode prompt to specify the format of network masks for the current session. The mask format will revert to the default of bit count when you exit the current session.

To specify the network mask format for a specific line, enter the **ip netmask-format** command in line configuration mode.

IP Host Names

The Cisco IOS software maintains a table of host names and their corresponding addresses, also called *host name-to-address mapping* (discussed in more detail in the section "Name-to-Address Schemes"). Higher-layer protocols such as Telnet use host names to identify network devices (hosts). The router and other network devices must be able to associate host names with IP addresses to communicate with other IP devices.

Issue the **ip host** command from global configuration mode to manually assign host names to addresses. The complete form of the command is as follows:

```
ip host name [tcp-port-number] address [address]...
```

where the variable elements of the command have the following meanings:

- *name*—Any name you prefer to describe the destination.

- *tcp-port-number*—Optional number that identifies TCP port to use when using the host name with an EXEC connect of Telnet command. The default is port 23 for Telnet.

- *address*—IP address or addresses where the device can be reached.

Following are two examples of the **ip host** command:

```
ip host P1R1 1.0.0.5 2.0.0.8
ip host P1R2 1.0.0.4
```

The first example defines a host named P1R1 and two network addresses for reaching it. The second example defines a host named P1R2 and defines one network address for reaching it.

Name Server Configuration

The **ip name-server** command in global configuration mode defines which hosts can provide DNS name services. DNS name servers can answer name queries directly or look up answers on behalf of clients on the network. A maximum of six IP addresses can be specified as name servers in a single command. The form of the command is as follows:

```
ip name-server server-address1 [[server-address2]. . .
   server-address6]...
```

To map domain names to IP addresses, you must identify the host names and specify a name server with this command, where [*server-address*] is the address of the domain name server to use.

The Domain Name System (DNS) process is on by default. You can, however, optionally identify the default domain name with the command

```
IP domain-name [name of default domain]
```

Any time the operating system software receives a command or address it does not recognize, it refers to DNS for the IP address of that device.

Name-to-Address Schemes

Each unique IP address can have a host name associated with it. The Cisco IOS software maintains a cache of host name-to-address mappings for use by EXEC commands. This cache speeds the process of converting names to addresses.

IP defines a naming scheme that allows a device to be identified by its location in IP. A name such as ftp.cisco.com identifies the domain of the File Transfer Protocol for Cisco. To keep track of domain names, IP identifies a name server that manages the name cache.

The DNS is enabled by default with a server address of 255.255.255.255, which is a local broadcast. In case the DNS has been turned off, you can re-enable it with the command **ip domain-lookup**. The **no ip domain-lookup** command turns off name-to-address translation in the router. Doing so means that it will not broadcast to find a name server if a name is entered at the router.

Display Host Names

The **show hosts** [*host name*] command is used to display a cached list of host names and addresses.

Figure 9–13 shows output from the **show hosts** command. You can obtain the following specific information about a host name entry from the output:

- Host—Names of learned hosts
- Flags—Descriptions of how information was learned and its current status
- perm—Manually configured in a static host table
- temp—Acquired from DNS use

Figure 9–13
*Output from
the **show**
hosts com-
mand.*

```
Router# show hosts
Default domain is not set
Name/address lookup uses static mappings

Host            Flags         Age   Type   Address (es)
P1R1            (perm, OK)    5     IP     144.253.100.200  133.3.13.2
                                          133.3.5.1
P2R1            (perm, OK)    5     IP     144.253.100.201  153.50.3.2
                                          153.50.5.6
P2R2            (perm, OK)    **    IP     128.45.17.4  153.50.3.200
                                   153.50.34.17
P2R3            (perm, OK)    **    IP     172.26.40.11  153.50.5.7
                                          153.50.34.1
- - More - -
```

- OK—Entry is current

- EX—Entry has aged out; it has expired

- Age—Time measured in hours since software referred to the entry

- Type—Protocol field

- Address(es)—Logical addresses associated with the name of the host

VERIFYING ADDRESS CONFIGURATION

Addressing problems are the most common problems that occur on IP networks. It is important to recheck your address configuration before continuing with further configuration steps. Three commands allow you to verify address configuration in your internetwork, as shown in Figure 9–14.

These three commands perform the following troubleshooting functions:

- **telnet**—Verifies the application-layer software between source and destination stations. This is the most complete test mechanism available.

- **ping**—Uses the ICMP protocol to verify the hardware connection and the logical address of the network layer. This is a very basic testing mechanism. Both a simple and extended **ping** command are available.

- **trace**—Uses Time To Live (TTL) values to generate messages from each router used along the path. Using the TTL feature allows you to locate failures in the path from the source to the destination.

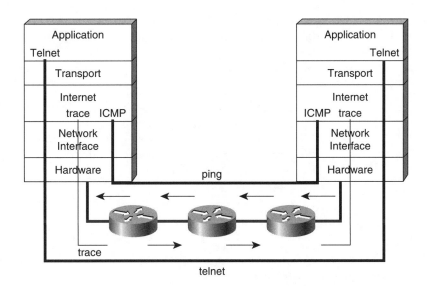

Figure 9–14
*Use the **telnet**, **ping**, and **trace** commands to verify your configuration.*

To verify that a router is properly configured for IP, follow these steps:

1. Try to Telnet to the router. This tests the application layer.
2. If you cannot Telnet, try **ping**. This lets you test end-to-end at the network layer. If the ping works, the problem is probably above the network layer.
3. If you cannot ping, try **trace**. This shows you each step along the path to the destination router and tells you the last reachable router. With this information, you can look for a misconfiguration on that router.

Telnet Command

Telnet is a simple application to see whether you can connect to the router. If you cannot Telnet to the router, but you can ping the router, you know the problem lies in the upper-layer functionality at the router. At this time, you may want to reboot the router and Telnet to it again.

Simple Ping Command

The **ping** command sends ICMP echo packets and is supported in both user and privileged EXEC mode. When an ICMP echo packet is received by a device, the receiver simply echoes back the packet to the source. In Figure 9–15, one ping timed out, as reported by the dot (.), and four were successfully received, as shown by the exclamation points (!). These are the commands that may be returned by the ping test:

Character	Definition
!	Successful receipt of an echo reply
.	Timed out waiting for datagram reply
U	Destination unreachable error
C	Congestion-experienced packet
I	Ping interrupted (for example, Ctrl-Shift-6 X)
?	Packet type unknown
&	Packet TTL exceeded

Figure 9–15
*The **ping** command tests IP network connectivity.*

```
Router> ping 172.16.101.1
Type escape sequence to abort.
Sending 5, 100-byte ICMP Echos to 172.16.101.1,
Timeout is 2 seconds:
!!!!
Success rate is 80 percent, round-trip min/avg/max =
6/6/6 ms
Router>
```

Extended Ping Command

The extended **ping** command is supported only from privileged EXEC mode. You can use the extended command mode of the **ping** command to specify the supported Internet header options, as shown in Figure 9–16. To enter the extended mode, enter **Y** (yes) at the extended commands prompt.

In Figure 9–16, the DF (Don't Fragment) bit is set, and then a ping to successive locations in the network is performed after doing a trace. The administrator in this case experiences poor performance across the network on this path and, therefore, is attempting to determine whether the cause is fragmentation.

The DF bit specifies that if the packet encounters a node in its path configured for a smaller MTU than the packet's MTU, the packet is to be dropped and an error message sent to the router at the packet's source address. A node configured with a small MTU can contribute to problems on the network. When the DF bit is set to Yes, the packet is not fragmented if it encounters a node with an MTU smaller than the packet size.

```
Router# ping
Protocol [ip] :
Target IP-address: 192.168.101.162
Repeat count [5] :
Datagram size [100] :
Timeout in seconds [2] :
Extended commands [n] : y
Source address:
Type of service [0] :
Set DF bit ip IP header? [n] : yes.
Validate reply data? [no] :
Data pattern [0xABCD] :
Loose, Strict, Record, Timestamp, Verbose [none] :
Sweep range of sizes [n] :
Type escape sequence to abort.
Sending 5, 100-byte ICMP Echos to 192.168.101.162, timeout is 2 seconds:
! ! ! !
Success rate is 100 percent (5/5), round-trip min/avg/max = 24/26/26 ms
Router#
```

Figure 9–16
*Ping sup-
ported for sev-
eral protocols.*

Trace Command

The **trace** command, often referred to as *traceroute*, enables you to see the possible end-to-end path, as shown in Figure 9–17. (Recall that using **trace** allows you to locate failures in the path from the source to the destination.) The **trace** command is supported by IP, CLNS, VINES (Banyan), as well as AppleTalk.

```
Router#trace aba.nyc.mil
Type escape sequence to abort.
Tracing the route to aba.nyc.mil (26.0.0.73)

1    debris.cisco.com (172.16.1.6) 1000 msec 8 msec 4 msec
2    barrnet-gw.cisco.com (172.16.16.2) 8 msec 8 msec 8 msec
3    external-a-gateway.stanford.edu (192.42.110.225) 8 msec 4 msec 4 msec
4    bb2.su.barrnet.net (131.119.254.6) 8 msec 8 msec 8 msec
5    su.arc.barrnet.net (131.119.3.8) 12 msec 12 msec 8 msec
6    moffett-fld-mb.in.mil (192.52.195.1) 216 msec 120 msec 132 msec
7    aba.nyc.mil (26.0.0.73) 412 msec * 664 msec
```

Figure 9–17
*The **trace**
command
shows
interface
addresses
used to reach
the destina-
tion.*

Host names are shown if the addresses are translated dynamically or via static host table entries. The times listed represent the time required for each of three probes to return.

When the trace reaches the target destination, an asterisk (*) is reported at the display. The display of one or more asterisks is normally caused by the receipt of a port-unreachable packet and the timeout in response to the probe packet.

Responses include

!H	The probe was received by the router, but not forwarded, usually due to an access list.
P	The protocol was unreachable.
N	The network was unreachable.
U	The port was unreachable.
*	Timeout.

SUMMARY

This chapter has focused on IP address classes, components, and commands. You learned about configuring IP address classes and bit patterns, as well as the first octet rule. This chapter has taken a look at subnetting, covered subnet masks, and provided additional reference material (RFCs) that deals with addressing issues.

Finally, the chapter examined three commands available to verify IP address configuration: **ping, telnet,** and **trace.** In the next chapter, you learn about IP routing configuration.

Chapter Nine Test
IP Address Configuration

Estimated Time: 15 minutes

Complete all the exercises to test your knowledge of the materials contained in this chapter. Answers are listed in Appendix A, "Chapter Test Answer Key."

Questions 9.1–9.3

In this exercise, determine the address class and calculate the subnet of a given network address. Write the address class and subnet number next to the IP address in the table that follows.

	Address	Subnet Mask	Class	Subnet
9.1	172.16.2.10	255.255.255.0	____B	172.16.2.0 _____
9.2	10.6.24.20	255.255.0.0	____A	10.6.0.0 _____
9.3	10.30.36.12	255.255.255.0	____A	10.30.36.0 _____

Questions 9.4–9.6

Correctly calculate the address class, subnet number, and the broadcast address for the subnet for each of the IP addresses and subnet masks given. Write your answers in the table next to each IP address and subnet mask.

	Address	Subnet Mask	Class	Subnet	Broadcast
9.4	201.222.10.60/29	255.255.255.248	_____	_____	_____
9.5	15.16.193.6/21	255.255.248.0	_____	_____	_____
9.6	128.16.32.13/30	255.255.255.252	_____	_____	_____

10

IP Routing Configuration

Routers learn about networks in several ways. This chapter presents static, default, and dynamic routing for IP.

This chapter also discusses how to configure IP routing (including RIP and IGRP), and examines routing configuration and transaction information.

BASIC MECHANISMS AND COMMANDS OF IP ROUTING

This section introduces the concept of IP routing and the commands required to set up routes and routing tables.

Setting Up the Initial IP Routing Table

Devices communicate with each other over routes. A *route* is a path from the sending device to the receiving device. Devices on a network learn about routes in a variety of ways. Routes can be manually configured by an administrator; devices can send out probes to discover how to get to a destination; or devices can receive updates about what routes are available. Once a device obtains information about a route, the device stores the route information in a routing table for future reference, as shown in Figure 10–1.

If the destination device is on the same network as the sending device, the sending device simply transmits the datagram directly to the destination. When a destination is not on the local network, a sending device forwards the datagram to a router. In order to forward a datagram, the sending device must first know what routers are connected to the local network.

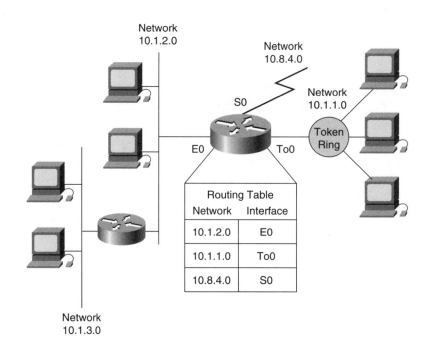

Figure 10-1
Routers maintain an address-to-port association table.

Based on the network shown in Figure 10–1, a packet destined for 10.1.3.0 would be dropped. When a router first comes up, it knows only about the networks that are directly connected to its interfaces.

A router refers to entries about networks or subnets on which the router is directly connected. Each router connection to a network is configured with an IP address and mask, which are then associated with a router interface. The Cisco IOS software learns about this IP address and mask information from configuration information input from some source, such as a network administrator.

Key Concept   **A *route* is a path from the sending device to the receiving device.**

Routers learn about nonlocal routers and the shortest path to destination devices through a variety of methods, discussed next.

Understanding How IP Routing Learns Destinations

Routers learn paths to destinations in three ways:

- *Static routes*—Manually defined by the system administrator as the only path to the destination; useful for controlling security and reducing traffic.

- *Default routes*—Manually defined by the system administrator as the path to take when no route to the destination is known.

- *Dynamically learned routes*—Router learns of paths to destinations by receiving periodic updates from other routers.

IP routing is automatically enabled in the Cisco IOS software. To disable IP routing, enter the following command in global configuration mode:

```
Router(config)#no ip routing
```

When IP routing is disabled, the router will act as an IP end host for IP packets destined for or sourced by the router. To re-enable IP routing, issue the following command in global configuration mode:

```
Router(config)#ip routing
```

Note that this book primarily focuses on dynamic routing. Refer to the Cisco Press title *Advanced Cisco Router Configuration* for more information on static and default routes.

Specifying Administrative Distance Values

An *administrative distance* is a rating of the trustworthiness of a routing information source, such as an individual router or a group of routers. An administrative distance is an integer from 0 to 255. In general, the higher the value, the lower the trust rating. An administrative distance of 255 means the routing information source cannot be trusted at all and should be ignored.

Specifying administrative distance values enables the Cisco IOS software to discriminate between sources of routing information, as shown in Figure 10–2. To get to network 128.10.0.0, Router A will choose to send the packet to Router B because Router B has a lower administrative distance than Router C.

Figure 10–2
*Administra-
tive distance
enables a
router to
select
between mul-
tiple paths.*

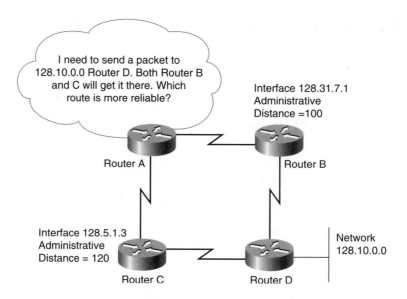

The software always picks the route whose routing protocol has the lowest administrative distance. Table 10–1 shows the default administrative distances for some routing information sources.

Table 10–1
*Comparison
of admini-
strative
distances*

Route Source	Default Distance
Connected interface	0
Static route	1
IGRP	100
RIP	120
Unknown	255

If a router has both routing protocols IGRP and RIP enabled, the Cisco IOS software uses the IGRP-derived information because the default IGRP administrative distance is lower than that for RIP and ignores the RIP-derived information. However, if you lose the source of the IGRP-derived information (for example, because of a power shutdown), the software uses the RIP-derived information until the IGRP-derived information reappears.

Configuring Static Routes

Static routes are user-defined routes that cause packets moving between a source and a destination to take a specified path. Static routes are important when the Cisco IOS software cannot build a route to a particular destination. Routers can forward packets only to known routes, and if the router cannot learn of a route dynamically, the static entry can be used to enable the router to route the incoming packet. Static routes are also useful for specifying a *gateway of last resort* to which all unroutable packets will be sent. This gateway (which is actually a router) is used as a last attempt to find some device to handle the packet.

Dynamic routing is typically preferred because static routing can be unwieldy in a large, complex, or volatile network because the administrator would have to make many manual changes. In small, simple, stable networks, however, static routing affords precision and control over the network without too much work.

To configure a static route, enter the **ip route** command in global configuration mode. A static route allows manual configuration of the routing table. No dynamic changes to this table entry will occur as long as the path is active. The complete parameters for the **ip route** command are as follows:

```
ip route network [mask] {address|interface} [distance] [permanent]
```

where the parameters have the following meanings:

- *network*—Destination network or subnet
- *mask*—Subnet mask
- *address*—IP address of next-hop router
- *interface*—Name of interface to use to get to destination network
- *distance*—The administrative distance
- **permanent** (Optional)—Specifies that the route will not be removed, even if the interface shuts down

If the mask is omitted in the **ip route** command, the router assumes it can use the default mask. Figure 10–3 provides a static route example based on the following **ip route** command:

```
Router(config)#ip route 172.16.1.0 255.255.255.0 172.16.2.1
```

In the example in Figure 10–3, the **ip route** command identifies the static route command; 172.16.1.0 specifies a static route to the destination subnetwork; 255.255.255.0

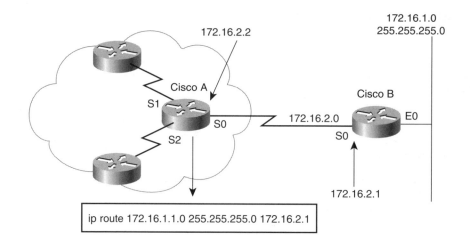

Figure 10–3
*Router A is
configured
with a static
route to
172.16.1.0.*

172.16.1.0
255.255.255.0

172.16.2.2

Cisco A

Cisco B

S1

S0   172.16.2.0   S0   E0

S2

172.16.2.1

ip route 172.16.1.1.0 255.255.255.0 172.16.2.1

indicates the subnet mask (eight bits of subnetting are in effect); and 172.16.2.1 is the IP address of next-hop router in the path to the destination.

The assignment of a static route to reach the stub network 172.16.1.0 is proper for the Cisco A router because there is only one way to reach that network. A *stub network* is one that has only one connection to another network. If the network connects to more than one network and allows traffic to cross it to get from one network to another, that network is called a *transit network*.

The assignment of a static route from Cisco B to the cloud networks is also possible. However, a static route assignment is required for each destination network, so a default route may be more appropriate.

**Key
Concept**   You can have more than one IP routing protocol operational in the same router at the same time. Each route is distinguished by administrative distance. The lower this number, the better the route is considered to be. It is basically a measurement of how good the router considers the metric of that protocol to be. For a static route, the administrative distance can be very low (for example, 0 or 1). The default administrative distance for RIP is 120 and for IGRP is 100.

Because static routes have a low default administrative distance, they are always chosen over dynamic routes. You can change this effect by overwriting the administrative distance—essentially creating a static backup route—that is effective only when the protocol is down.

Configuring Default Routers

A router might not know the routes to all other networks. To provide complete routing capability, the common practice is to use some routers as default routers and give the remaining routers default routes to those routers.

To establish a default router, issue the following command in global configuration mode:

```
Router(config)#ip default-network network-number
```

where *network-number* is equal to the IP network number or subnet number defined as the default.

When an entry for the destination network does not exist in the routing table, the packet is sent to the default network, so the default network must exist in the routing table. One benefit of default routes is that they reduce the length of routing tables.

Use the default network number when you need a route but have only partial information about the destination network. Because the router does not have complete knowledge about all destination networks, it can use a default network number to indicate the direction to take for unknown network numbers.

In addition to the normal IP network addresses, IP Routing Information Protocol (RIP) uses 0.0.0.0 as the default route.

In the example shown in Figure 10–4, the **ip default-network 192.168.17.0** global command defines the Class B network 192.168.17.0 as the destination path for packets that have no routing table entry.

To prevent unwanted updates from entering from the public network, company X could install a firewall in router A. To group those networks that will share company X's routing strategy, router A could implement an autonomous system number.

Grouping into Autonomous Systems

In Figure 10–4, you saw how company X used a default router to connect to a public network. It was mentioned that you could group routers into autonomous systems. An *autonomous system* is a set of routers and networks under the same administration. An autonomous system may consist of one router directly connected to one LAN to the Internet; or an autonomous system may be a corporate network linking several local networks through a corporate backbone. The autonomous system presents a consistent

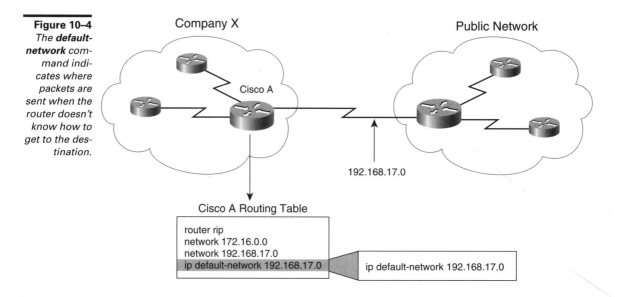

Figure 10–4
The default-network com-mand indi-cates where packets are sent when the router doesn't know how to get to the des-tination.

view of routing to the external world. For a router to belong to an autonomous system, all routers in that system must be:

- Interconnected

- Running the same routing protocol

- Assigned the same autonomous system number

The Network Information Center (InterNIC) assigns a unique autonomous system to enterprises. This autonomous system is a 16-bit number. A routing protocol such as Cisco's Interior Gateway Routing Protocol (IGRP) requires that you specify this unique, assigned autonomous system number in your configuration. A InterNIC-assigned autonomous system number is needed only if your organization plans to use an exterior router protocol, such as Border Gateway Protocol (BGP). If your company performs only interior routing, you need only ensure consistency and uniqueness of autonomous system numbers within your organization.

Using Interior or Exterior Routing Protocols

The design criteria for an interior routing protocol require it to find the best path through the network. In other words, the metric and how that metric is used is the most important element in an interior routing protocol.

Exterior protocols are used to exchange routing information between networks that do not share a common administration. IP exterior gateway protocols require the following three sets of information before routing can begin:

- A list of neighbor (or peer) routers or access servers with which to exchange routing information

- A list of networks to advertise as directly reachable

- The autonomous system number of the local router

As shown in Figure 10–5, the supported exterior gateway protocols are as follows:

- Border Gateway Protocol (BGP)

- Exterior Gateway Protocol (EGP)

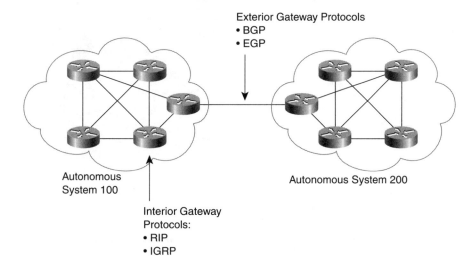

Exterior Gateway Protocols
• BGP
• EGP

Autonomous System 100

Autonomous System 200

Interior Gateway Protocols:
• RIP
• IGRP

Figure 10–5
An internetwork can use both interior and exterior routing protocols.

An exterior routing protocol must isolate autonomous systems. Basically, another autonomous system is managed by some other staff. Because you have no control over how that network is configured, you need to protect the network against errors that could arise from misconfiguration. BGP and EGP are covered in more detail in the Cisco Press title *Advanced Cisco Router Configuration*.

Routing Protocol Problems

The most common problem that could arise is a routing loop or a convergence problem. This problem could be propagated into the network as many routing updates specifying changes in metric. Hence, an exterior routing protocol usually attempts to eliminate the metric in its connection to the other network. For more information on routing loop and convergence problems, refer to Chapter 4, "Network Layer and Path Determination."

Interior IP Routing Protocols

At the Internet layer of the TCP/IP suite of protocols, as shown in Figure 10–6, a router can use the IP routing protocol to accomplish routing through the implementation of a specific routing algorithm.

Figure 10–6
*Routers use
the IP protocol
to perform
routing.*

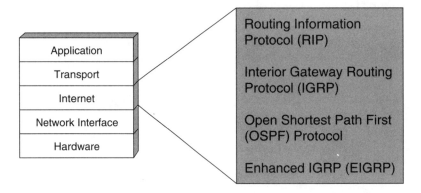

Interior protocols are used for routing networks that are under a common network administration. All IP interior gateway protocols must be specified with a list of associated networks before routing activities can begin. A routing process listens to updates from other routers on these networks and broadcasts its own routing information on those same networks. Cisco IOS software supports the following interior routing protocols:

- Routing Information Protocol (RIP)

- Internet Gateway Routing Protocol (IGRP)

- Enhanced Internet Gateway Routing Protocol (Enhanced IGRP)

- Open Shortest Path First (OSPF)

- Intermediate System-to-Intermediate System (IS-IS)

The following pages focus on how to configure the first two of these protocols: RIP and IGRP.

Completing the IP Routing Configuration Tasks

The selection of IP as a routing protocol involves the setting of both global and interface parameters. Global tasks include:

- Select a routing protocol: RIP or IGRP (see Figure 10–7).

- Assign IP network numbers without specifying subnet values.

The interface task is to assign interface-specific addresses and the appropriate subnet mask.

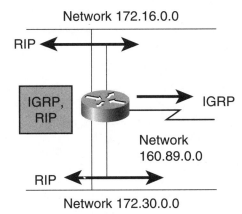

Figure 10–7
A router can use more than one routing protocol if desired.

Dynamic routing uses broadcasts and multicasts to communicate with other routers. The routing metric helps routers find the best path to each network or subnet.

Configuring Dynamic Routing

Two primary commands are used to configure dynamic routing: **router** and **network**. The **router** command starts a routing process; its form is as follows:

```
Router(config)#router protocol [keyword]
```

where the parameters specify the following:

- *protocol*—Either RIP, IGRP, OSPF, or Enhanced IGRP

- *keyword*—For example, **autonomous system number,** which is used with those protocols that require an autonomous system, such as IGRP

The **network** command is required because it allows the routing process to determine which interfaces will participate in the sending and receiving of routing updates. The **network** command starts up the routing protocol on all interfaces that the router has in the specified network. The **network** command also allows the router to advertise that network. Its form is as follows:

```
Router(config-router)#network network-number
```

where *network-number* specifies a directly connected network.

The network number must be based on the InterNIC network numbers, not subnet numbers or individual addresses. The network number also must identify a network to which the router is physically connected.

CONFIGURING RIP

IP's Routing Information Protocol (RIP) was originally specified in RFC 1058. Key characteristics of RIP include the following:

- It is a distance vector routing protocol.

- Hop count is used as the metric for path selection.

- The maximum allowable hop count is 15; 16 is considered unreachable.

- Routing updates are broadcast every 30 seconds by default.

- RIP is capable of load balancing over multiple paths.

Load balancing enables a router to use two or more equal cost paths to reach a destination (essentially turning a one-lane road into a road with two or more lanes). In Cisco routers, load balancing for RIP is enabled by defining the maximum number of parallel paths RIP will install in a routing table. If the maximum number of paths command is set to 1 (one), load balancing is disabled. Because RIP defaults to four parallel paths, load balancing is automatically enabled.

In Figure 10–8, a packet from host 1 to host 2 would cross the 19.2 kbps link because that route uses the lowest hop count.

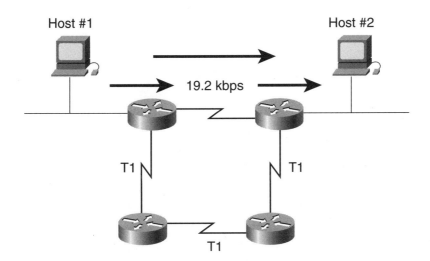

Figure 10–8
The hop count metric selects the path.

Unfortunately, in this example, the route selected is not the best route available. RIP was developed in a homogeneous network and was widely unavailable. If everything is connected via a single media type, bandwidth-based metrics reduce to hop count. But if there are different media types, RIP's hop count metric may not consistently identify the best path, as in this example.

Configuring RIP Router Commands

The two primary commands used to configure an RIP router are as follows:

```
Router(config)#router rip
Router(config-router)#network network-number
```

The **router rip** command selects RIP as the routing protocol. The **network** command assigns an IP address for the network to which the router is directly connected. The routing process will associate interfaces with the proper addresses and will begin packet processing on the specified networks.

The **network** statement contains no subnetting information. Networks are directly connected and are specified as a Class A, B, or C network number. Because of the **ip address** command with addresses and subnet masks, the routing protocol is able to determine which subnets are connected to this router. Figure 10–9 shows the configuration of the Cisco A router.

Figure 10–9
*Cisco A will
send RIP infor-
mation to net-
works 1.0.0.0
and 2.0.0.0.*

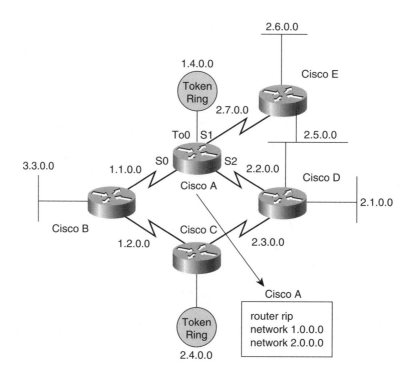

- **router rip**—Selects RIP as the routing protocol
- **network 1.0.0.0**—Specifies a directly connected network
- **network 2.0.0.0**—Specifies a directly connected network

The Cisco A router interfaces connected to networks 1.0.0.0 and 2.0.0.0 will send and receive RIP updates. These routing updates allow the router to learn the network topology.

The **network** command gives the routing protocol permission to advertise the subnets connected to the router's neighbor. Without the **network** command, nothing is advertised. With a **network** command, the router will advertise every subnet within the Class A, B, or C network specified in the configuration.

Viewing IP RIP Information

There are three primary commands used to view RIP information:

- **show ip protocol**
- **show ip route**
- **debug ip rip**

The **show ip protocol** command displays values about routing timers and network information associated with the entire router, as shown in Figure 10–10. Use this information to identify a router that is suspected of delivering bad routing information.

The router in Figure 10–10 sends updated routing table information every 30 seconds. (This interval is configurable.) It has been 17 seconds since it sent its last update, so the next one will be sent in 13 seconds. The router is also injecting routes for the networks listed following the Routing for Networks line.

```
Router> show ip protocol
Routing Protocol is "rip"
    Sending update every 30 seconds, next due in 13 seconds
    invalid after 180 seconds, hold down 180, flushed after 240
    Outgoing update filter list for all interfaces is not set
    Incoming update filter list for all interfaces is not set
Redistributing: rip
Routing for Networks
    183.8.0.0
    144.253.3.0
Routing Information Sources:
Gateway            Distance    Last Update
    183.8.128.12     120         0:00:14
    183.8.64.130     120         0:00:19
    183.8.128.130    120         0:00:03
Distance: (default is 120)
```

Figure 10–10
*Use **show ip protocol** command to observe RIP's behavior.*

The **show ip route** command displays the contents of the IP routing table, as shown in Figure 10–11.

Figure 10–11
*Use the show
ip route com-
mand to dis-
play the local
routing table.*

```
Router> show ip route
Codes: C - connected, S - static, I - IGRP, R - RIP, M - Mobile, B - BGP
       D - EIGRP, Ex - EIGRP external, O - OSPF, IA - OSPF inter area
       EI - OSPF external type 1, E2 - ISPF external type 2, E - EGP
       i - IS-IS, L1 - IS-IS level-1, L2 - IS-IS level-2, * - candidate default

Gateway of last resort is not set

        144.253.0.0 is subnetted (mask is 255.255.255.0), 1 subnet
C       144.253.100.0 is directly connected, Ethernet1

R       153.50.0.0 [120/1] via 183.8.128.12, 00:00:09, Ethernet0
        183.8.0.0 is subnetted (mask is 255.255.255.128), 4 subnets
R       183.8.0.128 [120/1] via 183.8.128.130, 00:00:17, Serial0
        [120/1] via 183.8.64.130, 00:00:17, Serial1
C       183.8.128.0 is directly connected, Ethernet0
C       183.8.64.128 is directly connected, Serial1
C       183.8.128.128 is directly connected, Serial0
```

The routing table contains entries for all known networks and subnetworks and contains a code that indicates how that information was learned. The values are defined as follows:

- C indicates a network that was configured with the networks command.

- R indicates an entry learned through RIP.

- Via refers to the router that informed you about this route.

- 00:00:09 timer value means that RIP updates are every 30 seconds.

- The administrative distance is 120.

- The hop count to 153.50.0.0 is 1.

The **debug ip rip** command displays RIP routing updates as they are sent and received.

As shown in Figure 10–12, the update is sent by 172.8.128.130. It reported on two routers, one of which is inaccessible because its hop count is greater than 15. Updates were then broadcast through 172.8.128.2.

The **no debug ip rip** command turns off the display of the RIP routing updates.

```
Router> debug ip rip
RIP protocol debugging is on
Router#
RIP: received update from 172.8.128.130 on Serial0
        172.8.0.128 in 1 hops
        172.8.64.128 in 16 hops (inaccessible)
Rip: received update from 172.8.64.130 on Serial1
        172.8.0.128 in 1 hops
        172.8.0.128.128 in 1 hops
RIP: received update from 172.8.128.130 on Serial0
        172.8.0.128 in 1 hops
        172.8.64.128 in 1 hops
RIP: sending update to 255.255.255.255 via Ethernet0 (172.8.128.2)
        subnet 172.8.0.128, metric 2
        subnet 172.8.64.128, metric 6
        subnet 172.8.128.128, metric 1
        network 10.253.0.0, metric 1
RIP: sending update it 255.255.255.255 via Ethernet 1 (10.253.100.202)
        network 10.50.0.0, metric 2
        network 172.8.0.0, metric 1
```

Figure 10–12
*Use the **debug ip rip** command to troubleshoot RIP communications.*

CONFIGURING IGRP

Internet Gateway Routing Protocol (IGRP) is an advanced distance vector routing protocol developed by Cisco in the mid-1980s. IGRP has several features that differentiate it from other distance vector routing protocols, such as RIP. These IGRP features are as follows:

- *Scalability*—Some of the largest internetworks are based on IGRP.

- *Fast response to network changes*—Unlike other distance vector routing protocols, IGRP sends updates when route topology changes occur.

- *Sophisticated metric*—IGRP uses a composite metric that provides significant route selection flexibility. Internetwork delay, bandwidth, reliability, and load are all factored into the routing decision. IGRP can be used to overcome RIP's 15-hop limit.

- *Multiple paths*—IGRP can maintain up to four nonequal paths between a network source and destination. Multiple paths can be used to increase available bandwidth or for route redundancy.

Use IGRP in IP networks that require a simple, robust, and scalable routing protocol. IGRP is also useful when trying to avoid the router processing overhead of link-state routing protocols. You can redistribute IGRP to IP RIP, OSPF, and Enhanced IGRP. Note, however, that IGRP does not support variable-length subnet masking. Variable-length subnet masking (VLSM) allows you to use some of the host address bits to define a subnet address. Subnet masking is covered in greater detail in Chapter 9, "IP Addressing." Also refer to RFC 1219 for detailed information about VLSMs and how to correctly assign addresses.

Understanding IGRP Operation

IGRP is a distance vector routing protocol. Routers using IGRP broadcast periodic routing table updates to neighbor routers at 90-second intervals. IGRP provides a number of features designed to enhance its performance and stability and at the same time reduce the possibility of routing loops. As shown in Figure 10–13, these features include:

- Flash updates
- Poison reverse
- Holddowns
- Split horizon

Flash Updates

In addition to its periodic routing updates, IGRP uses flash updates to speed up convergence of the routing algorithm. A *flash update* is sent when a network topology change is noticed.

Poison Reverse

Increases in routing metrics generally indicate routing loops. As discussed in Chapter 4, poison reverse updates are sent to remove a route and to place it in holddown. IGRP poison reverse updates are sent if a route metric has increased by a factor of 1.1 or greater.

Holddowns

IGRP has a holddown timer that prevents temporary routing loops while convergence takes place. A newly learned route is used until the holddown time expires. By default, the holddown timer is three times the update interval (90 seconds) plus 10 seconds, or

280 seconds. Holddown timers can be disabled to improve convergence time; however, removing holddown timers increases the possibility of routing loops. Use the **no metric holddown** command to disable holddowns. As a result, after a route has been removed, a new one will be accepted immediately.

Split Horizon

Recall from Chapter 4 that split horizons derive from the fact that it is not useful to send information about a route back in the direction from which it came. In Figure 10–13, for example, router B does not send route information back to router A regarding network 10.

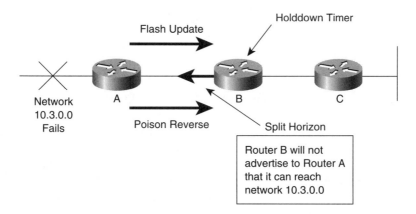

Flash Update

Holddown Timer

Network
10.3.0.0
Fails

A

Poison Reverse

B

Split Horizon

C

Router B will not
advertise to Router A
that it can reach
network 10.3.0.0

Figure 10–13
Flash updates are sent when a change in the network topology occurs.

Periodically, each router broadcasts its entire routing table (with some censoring because of the split horizon rule) to all adjacent routers. When a router gets this broadcast from another gateway, it compares the table with its existing table. Any new destinations and paths are added to the routing table. Paths in the broadcast are compared with existing paths. If a new path is better, it may replace the existing one.

In addition to the periodic updates every 90 seconds, IGRP declares a route inaccessible if it does not receive an update from the first router in the route within three update periods (270 seconds). After seven update periods (630 seconds), the route is removed from the routing table.

Using the IGRP Composite Metric

IGRP uses a composite metric to identify preferred routes. The IGRP composite metric is a 24-bit quantity that is a sum of the segment delays and the lowest segment bandwidth for a given route. This combination metric provides greater accuracy when choosing a path to a destination. In Figure 10–14, the token ring and FDDI path is preferable to the dual 19.2 kbps link. Unlike RIP's hop count, IGRP's composite metric supports the selection of the best path.

IGRP's metric does not have RIP's hop-count limitation. It includes the following components:

- *Bandwidth*—The smallest bandwidth between source and destination (expressed in kilobits per second)

- *Delay*—Cumulative interface delay along the path (expressed in tens of microseconds)

- *Reliability*—Worst between source and destination based on keepalives (expressed as an integer from 0 to 255)

- *Loading*—Worst load on a link between source and destination (based on bits per second)

- *MTU*—Smallest MTU value in path (expressed in bytes)

The path that has the smallest metric value is the best route. By default, only bandwidth and delay are used by the IGRP metric, but you can configure it to consider reliability, loading, and MTU also. In Figure 10–14, for example, you can figure that router A will send data along Path A, instead of the slower serial links. Path A includes a 16 Mbps Token Ring link, a 100 Mbps FDDI link, and some 10 Mbps Ethernet links. Even though there are more links to cross, the speed to cross Path A is much faster than crossing two 19.2 kbps links.

CAUTION

Adjusting IGRP metric values can dramatically affect network performance. Make all metric adjustment decisions carefully.

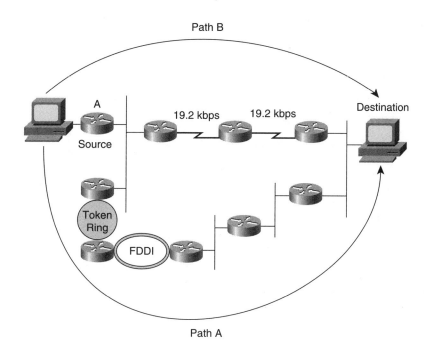

Path B

A

19.2 kbps   19.2 kbps

Source

Destination

Token Ring

FDDI

Path A

Figure 10–14
IGRP will select Path A because the metric to cross that path is lower than the metric to cross Path B.

Unequal-Cost Load Balancing Feature

The IGRP composite routing metric supports multiple paths between source and destination. This feature is known as *unequal-cost load balancing*. Unequal-cost load balancing allows traffic to be distributed among up to four unequal-cost paths to provide greater overall throughput and reliability.

The following general rules apply to IGRP unequal-cost load balancing:

- IGRP will accept up to six (four is the default) paths for a given destination network.

- The next-hop router in any of the paths must be closer to the destination than to the local router.

- The alternative path metric must be within the specified variance of the best local metric.

  For example, the alternative route may only be a specified factor worse, as measured by the IGRP metric, than the best local route. This variance can be configured.

If these conditions are met, the route is considered feasible and can be added to the routing table. In Figure 10–15, for example, a second unequal route has been added to the initial route between the source and the destination.

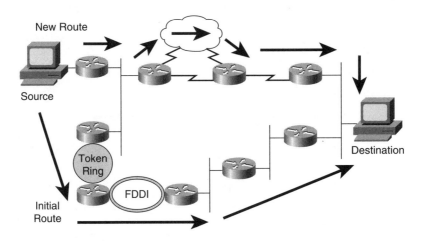

You can use the **default-metric** command to change the default metric.

Creating an IGRP Routing Process

Use the **router igrp** and **network** commands to create an IGRP routing process:

```
Router(config)#router igrp autonomous-system-number
Router(config-router)#network network-number
```

For example, in Figure 10–16, the following configuration has been set up:

router igrp 109	Enables the IGRP routing process for autonomous system 109
network 10.0.0.0	Associates network 10.0.0.0 with the IGRP routing process
network 172.31.0.0	Associates network 172.31.0.0 with the IGRP routing process

IGRP sends updates out to interfaces in networks 10.0.0.0 and 172.31.0.0 and includes information about networks 10.0.0.0 and 172.31.0.0.

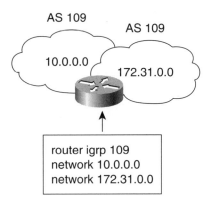

Figure 10–16
*Use **router
igrp** and **net-
work** com-
mands to
create an IGRP
router.*

Even though each IGRP routing process can provide routing information to only one autonomous system, the Cisco IOS software must run a separate IGRP process and maintain a separate routing database for each autonomous system it services.

You may want to establish different autonomous systems when your company is merging with another company, when connecting to a service provider, or when you want to isolate certain departments of the company.

To configure two IGRP routing processes, use the **router igrp** and **network** commands to define each IGRP process. For example, in Figure 10–17, network 10.0.0.0 is in autonomous system 71, and network 172.68.7.0 is in autonomous network 109.

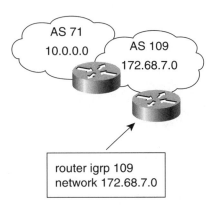

Figure 10–17
*Configuring
separate IGRP
routing pro-
cesses for
each autono-
mous system.*

Displaying IGRP Routing Information

You can use the following commands to display IGRP routing configuration and table update information:

- show ip protocols
- show ip interfaces
- show ip route
- debug ip igrp transaction
- debug ip igrp events

The following examines each of these commands and views their results.

The **show ip protocols** command displays parameters, filters, and network information about the entire router, as shown in Figure 10–18. You can see in Figure 10–18 that this router is injecting routes for 183.8.0.0 and 144.253.0.0.

Figure 10–18
The show ip protocols command indicates that IGRP is active.

```
Router> show ip protocols
Routing Protocol is "igrp 300"
Sending updated every 90 seconds, next due in 55 seconds
Invalid after 270 seconds, hold down 280, flushed after 630
Outgoing update filter list for all interfaces is not set
Incoming update filter list for all interfaces is not set
Default networks flagged in outgoing updates
Default networks accepted from incoming updates
IGRP metric weight K1=1, K2=0, K3=0, K4=0, K5=0
IGRP maximum hopcount 100
IGRP maximum metric variance 1
Redistributing: igrp 300
Routing for Networks:
183.8.0.0
144.253.0.0
Routing Information Sources:
    Gateway         Distance    Last Update
    144.253.100.1   100         0:00:52
    183.8.128.12    100         0:00:43
    183.8.64.130    100         0:01:02
Distance: (default is 100)
--More--
```

The **show ip interfaces** command displays the status and global parameters associated with an interface, as shown in Figure 10–19.

```
Router> show ip interfaces
Ethernet0 is up, line protocol is up
    Internet address is 183.8.128.2 subnet mask is 255.255.255.128
    Broadcast address is 255.255.255.255
    Address determined by non-volatile memory
    MTU is 1500 bytes
    Helper address is not set
    Directed broadcast forwarding is enabled
    Outgoing access list is not set
    Inbound access list is not set
    Proxy ARP is enabled
    Security level is default
    Split horizon is enabled
    ICMP redirects are always sent
    ICMP unreachables are always sent
    ICMP mask replies are never sent
    IP fast switching is enabled
    IP fast switching on the same interface is disabled
    IP SSE switching is disabled
    Router Discovery is disabled
    IP output packet accounting is disabled
```

Figure 10–19
The show ip interfaces command indicates that the line is up.

The Cisco IOS software automatically enters a directly connected route in the routing table if the interface is one through which software can send and receive packets. Such an interface is marked "up." If the interface is unusable, it is removed from the routing table. Removing the entry allows implementation of backup routes, if they exist.

The **show ip route** command displays the contents of an IP routing table. The table contains a list of all known networks and subnets and the metrics associated with each entry.

Note in Figure 10–20 that the information was derived from IGRP or from direct connections.

You can use two commands to display routing table updates:

```
Router#debug ip igrp transaction [ip-address]
Router#debug ip igrp events [ip-address]
```

The **debug ip igrp transaction** command displays transaction information on IGRP routing transactions. If the IP address of an IGRP neighbor is specified, the resulting output includes messages describing updates from that neighbor and updates that the router broadcasts toward that neighbor.

Figure 10–20
*The **show ip**
route com-
mand is used
to read the
routing table.*

```
Router> show ip route
Codes: C - connected, S - static, I - IGRP, R - RIP, M - Mobile, B - BGP
    D - EIGRP, EX - EIGRP external, O - OSPF, IA - OSPF inter area
    E1 - OSPF external type 1, E2 - OSPF external type 2, E- EGP
    i - IS-IS, L1 - IS-IS level-1, L2 - IS-IS level-2, * - candidate default

Gateway of last resort is not set

       144.253.0.0 is subneted (mask is 255.255.255.0) 1 subnets
C      144.253.100.0 is directly connected, Ethernet1
I      133.3.0.0 [100/1200] via 144.253.100.200, 00:00:57, Ethernet1
I      153.50.0.0 [100/1200] via183.8.128.12, 00:00:05, Ethernet0
       183.8.0.0 is subnetted (mask is 255.255.255.128), 4 subnets
I      183.8.0.128 [100/18067] via 183.8.64.130, 00:00:27, Serial1
       [100/18067] via 183.8.128.130, 00:00:27, Serial0
C      183.8.64.0 is directly connected, Ethernet0
C      183.8.64.128 is directly connected, Serial1
C      183.8.128.128 is directly connected, Serial0
I      172.16.0.0 [100/1200] via 144.253.100.1, 00:00:55, Ethernet1
I      192.3.63.0 [100/1300] via 144.253.100.200, 00:00:58, Ethernet1
```

Use the **no debug ip igrp transaction** command to disable the debugging output. When there are many networks in your routing table, displaying every update for every route can flood the console and make the router unusable. In this case, the **debug ip igrp events** command is used to display a summary of the IGRP routing information. This command indicates the source and destination of each update as well as the number of routes in each update. Messages are not generated for each route.

If the IP address of an IGRP neighbor is specified when issuing the **debug ip igrp events** command, the resulting output includes messages describing updates from that neighbor and updates that the router broadcasts toward that neighbor.

Use the **no debug ip igrp events** command to disable the debugging output.

SUMMARY

In this chapter, you've learned that routers can be configured to use one or more IP routing protocols. (This chapter focused on RIP and IGRP.) You've also examined the commands used to enable, configure, and examine those protocols on a Cisco router, including the commands required to view and debug your routing configuration. Although TCP/IP is the more popular worldwide internetworking protocol, it is not the only protocol used on large networks. Novell's IPX protocol is widely used, as well. Chapter 11, "Configuring Novell IPX," defines how to configure a router to support Novell's IPX protocol.

Chapter Ten Test
IP Routing Configuration

Estimated Time: 15 minutes

Complete all the exercises to test your knowledge of the materials contained in this chapter. Answers are listed in Appendix A, "Chapter Test Answer Key."

Question 10.1

T F To verify that IP routing is enabled, you can enter the **show protocols** command.

Question 10.2

T F You issue the **router rip** command to enable the RIP routing protocol.

Question 10.3

T F You issue the **network** *network-number* subnet mask command to associate a network with a routing process.

Question 10.4

T F You issue the **show ip protocol** command to verify that the RIP routing protocol is enabled.

Question 10.5

T F You issue the **show ip rip route** command to display the current status of the RIP routing table.

Question 10.6

In the RIP routing table display, how do you determine which networks are discovered by the RIP routing protocol?

Question 10.7

What command displays the RIP routing updates sent from and received by your router?

Question 10.8

What command disables the display of the RIP routing updates sent from and received by your router?

Question 10.9

What command do you issue to enable the IGRP routing protocol?

Question 10.10

Can you add multiple autonomous groups in a single **enable IGRP** command?

Question 10.11

What command do you issue to verify that the IGRP routing protocol is enabled?

Question 10.12

What command do you issue to display the current state of the IGRP routing table?

_____

Question 10.13

What command displays the IGRP routing updates events sent from the router?

_____

Question 10.14

What type of routing metric does IGRP use, and what are the five components of that metric?

_____

a. _____

b. _____

c. _____

d. _____

e. _____

Configuring Novell IPX

This chapter presents an introduction to the Novell IPX protocol suite and how it operates using Cisco IOS software configurations. The chapter also explains how to plan for IPX parameters, how RIP, SAP, and GNS work, and how to verify IPX routing.

IPX ROUTING OVERVIEW

This section presents an overview of IPX routing. It includes a discussion of how Cisco routers fit in NetWare networks, the IPX protocol stack, IPX addressing, IPX encapsulations, and IPX protocols.

Cisco Routers in NetWare Networks

In today's networking environment, no one manufacturer can provide all the hardware and software required to support the computing needs of a business. As a result, most networks include a variety of vendor products, each one chosen for the powerful features it provides. For that reason, Cisco routers are often found in NetWare networks even though Novell offers routing products.

Cisco's routers offer the following features in Novell network environments:

- Support for a wide range of interfaces, including native ISDN and ATM.

- Access lists and filters for IPX, RIP, SAP, and NetBIOS.

- Scalable routing protocols, including Enhanced IGRP and NLSP. Cisco uses generic routing encapsulation (GRE) to allow IPX datagram transmission across IP networks.

- Configurable RIP and SAP updates and packet sizes.

- Serverless LAN support.

- Dial-on-demand routing and spoofing for IPX and SPX. Traffic is routed across lines only when necessary, limiting the amount of time each line is used.

- Rich diagnostics, management, and troubleshooting features. The **ping IPX** command and many **show** commands provide complete information about IPX performance.

Novell NetWare Protocol Suite

Novell IPX is a proprietary suite of protocols derived from the Xerox Network Systems (XNS) protocol suite. IPX is a datagram, connectionless protocol that does not require an acknowledgment for each packet. It is a Layer 3 protocol that defines the internetwork (network and node) addresses. In the NetWare environment, a user's workstation is most often referred to as a *node*.

Novell NetWare uses the following:

- IPX Routing Information Protocol (RIP) to facilitate the exchange of routing information

- Proprietary Service Advertisement Protocol (SAP) to advertise network services

- NetWare Core Protocol (NCP) to provide client-to-server connections and applications

- Sequenced Packet Exchange (SPX) service for Layer 4 connection-oriented services

As an alternative to RIP and SAP, Novell has a link-state routing protocol called NetWare Link Services Protocol (NLSP). Because it is a link-state routing protocol, NLSP offers more reliable and effective routing processes than IPX RIP. Novell also has a directory service called Novell Directory Service (NDS).

The NetWare protocol stack supports all common media access protocols, as shown in Figure 11–1. The data link and physical layers are accessed through the Open Data-Link (ODI) Interface.

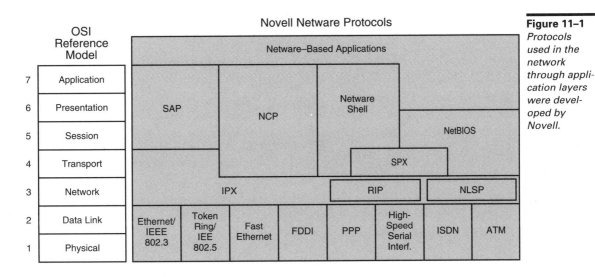

Figure 11–1
Protocols used in the network through application layers were developed by Novell.

Layers 3 through 7 are unique to Novell.

- Layer 3 encompasses IPX, a datagram service.

- The Service Advertising Protocol (SAP) provides services for part of Layer 3 and all services from Layers 4 through 7.

- Layer 4 is characterized by SPX, which provides a reliable connection-oriented service. NCP and the NetWare shell also provide Layer 4 services.

NetBIOS (Network Basic Input/Output System) emulation performs tasks applicable to the ISO/OSI model transport and session layers. Novell NetWare defines a special IPX packet called Type 20 (flooded packets) for NetBIOS applications.

Key Novell NetWare Features

As shown in Figure 11–2, a Novell IPX address has 80 bits: 32 bits for the network number and 48 bits for the node number. The node number contains the MAC address of an interface.

Novell IPX supports multiple logical networks on an individual interface; each network requires a single encapsulation type.

Novell RIP is the default routing protocol on older NetWare products; NLSP is the default routing protocol on NetWare 4.11 and higher.

Figure 11–2
*Each IPX inter-
face has a
unique 10-
byte address.*

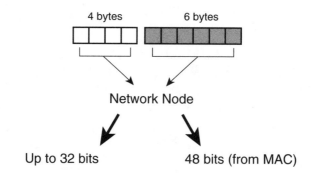

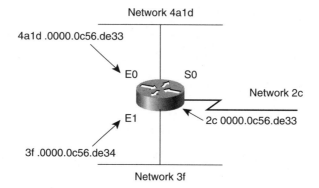

NetWare clients automatically discover available network services because Novell servers and routers announce the services using SAP broadcasts. The filtering of service
advertisements is a critical issue in Novell networks. SAP traffic can become excessive
and can severely impact bandwidth available for user data traffic.

One type of SAP advertisement is Get Nearest Server (GNS), which enables a client to
locate the nearest server for login. These features are discussed in detail later in this
chapter.

Novell IPX Addressing

Novell IPX addressing uses a two-part address: the network number and the node number.

- The IPX network number can be up to 4 bytes (8 hexadecimal digits) in length. Usually, only the significant digits are listed. This number is assigned by the network administrator. Figure 11–2 features the IPX network 4a1d. Other IPX networks shown are 2c and 3f.

- The IPX node number is 6 bytes (12 hexadecimal digits) in length. This number is usually the MAC address obtained from a network interface card. Figure 11–2 features the IPX node 0000.0c56.de33. Another node address is 0000.0c56.de34.

Notice in Figure 11–2 that the same node number appears for both E0 and S0. Serial interfaces do not have MAC addresses, so the router created this node number for S0 by using the MAC address from E0.

Each interface retains its own address. The use of the MAC address in the logical IPX address eliminates the need for an Address Resolution Protocol (ARP).

How to Determine the IPX Address

You must use a valid IPX network address when you configure the Cisco router. Because Novell NetWare networks are likely to be established already with IPX addresses, you can determine the existing IPX address from these already established networks. The IPX network address refers to the logical wire; all routers on the same wire must share the same IPX network address.

The first and recommended way to find out what address to use is to ask the NetWare administrator. Make sure that the NetWare administrator specifies the IPX network address for the same network where you want to enable IPX on your Cisco router. The Cisco router must use the same network as the NetWare file server (or other source of the address) specified by the NetWare administrator for that cabling system.

If you cannot obtain an IPX address to use from the NetWare administrator, you can get the IPX address directly from a neighbor router. Pick the most appropriate of the several methods available to do this.

- If the neighbor router is another Cisco router, you can use a Cisco IOS command to show cdp neighbors detail.

- You can Telnet to the neighbor router, enter the appropriate mode, and then display the running configuration on the neighbor.

- If the neighbor router is not a Cisco router (for example, a NetWare PC-based router or a NetWare file server), you might be able to attach or log in and use the NetWare config utility to determine the address.

On the Cisco router, you must use the same IPX network address as the address that already exists on that network.

Alternately, if you have access to the server console, you can use the NetWare **config** command. The **config** command displays a window with the IPX address of the segment that the file server shares with the Cisco router.

Multiple Novell Encapsulations

NetWare allows multiple Layer 2 frame structures for Novell IPX packets. Cisco routers support all of the framing variations. For example, there are four different Ethernet framing types, as shown in Figure 11–3. Each encapsulation type is appropriate in specific situations:

- *Ethernet 802.3*—Also called raw Ethernet; the default for NetWare versions 2 through 3.11

- *Ethernet 802.2*—The default for NetWare 3.12, NetWare 4, and NetWare 5, and also used for OSI routing

- *Ethernet II*—Used with TCP/IP and DECnet

- *Ethernet SNAP*—Used with TCP/IP and AppleTalk

Multiple encapsulations can be specified on an interface, but only if multiple network numbers have also been assigned. Although several encapsulation types can share the same interface, clients and servers with different encapsulation types cannot communicate directly with each other. The default encapsulation on Cisco routers is novell-ether (Novell Ethernet_802.3).

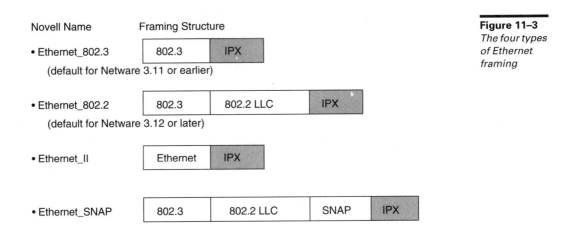

Figure 11–3
*The four types
of Ethernet
framing*

In addition to the four encapsulation types discussed here, you can specify a fifth type, HDLC, for serial connections. HDLC is covered in more detail in Chapter 14, "Introduction to WAN Connections."

Novell and Cisco Encapsulation Names

Novell and Cisco use different names for each frame type, as shown in Figure 11–4.

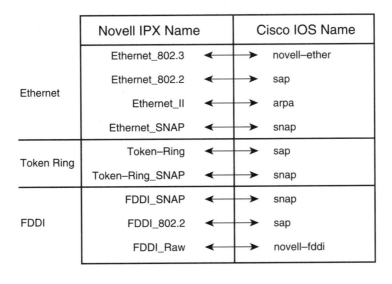

Figure 11–4
*Specify the
encapsulation
type when you
configure IPX
networks.*

When you configure an IPX network, you may need to specify a nondefault encapsulation type on either the Novell servers and clients or on the Cisco router. To help you specify the appropriate encapsulation type, use the table in Figure 11–4. The table matches the Novell term to the equivalent Cisco IOS term for the same framing types.

When you configure Cisco IOS software for Novell IPX, use the Cisco name for the appropriate encapsulation. Make sure the encapsulations on the clients, servers, and routers all match. Devices that use different encapsulation methods cannot communicate with each other.

If you do not specify an encapsulation type when you configure the router for IPX, the router will use the default encapsulation type on its interfaces.

CAUTION

The default Ethernet encapsulation type on Cisco routers does not match the default Ethernet encapsulation type on Novell servers after NetWare 3.11. This means that you have to change or add the 'sap' frame type to a Cisco router on an Ethernet network that supports more current versions of NetWare.

The default encapsulation types on Cisco router interfaces and their keywords are as follows:

- Ethernet—**novell-ether**
- Token Ring—**sap**
- FDDI—**snap**
- Serial—**hdlc**

Novell Uses RIP for Routing

Novell RIP is a distance vector routing protocol. RIP uses two metrics to make routing decisions: ticks (a time measure) and hop count (a count of each router traversed).

RIP checks its two distance vector metrics by first comparing the ticks for path alternatives. If two or more paths have the same tick value, RIP compares the hop count. If two or more paths have the same hop count, the router will use the age of the entry as the tiebreaker; the most recent entry in the tables will be preferred over the older entry.

Each IPX router periodically broadcasts copies of its RIP routing table to its directly connected networks, as shown in Figure 11–5.

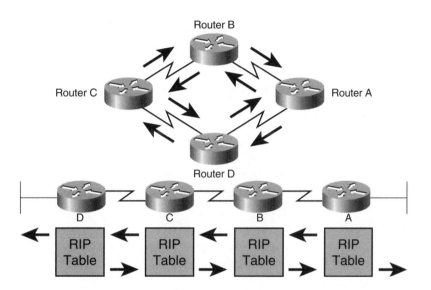

Figure 11–5
RIP routers periodically broadcast updates of their routing tables.

Upon receipt of these broadcasts, the neighbor IPX routers add distance vectors as required before broadcasting copies of their RIP tables to their other attached networks.

A split-horizon algorithm prevents the neighbor from broadcasting RIP tables about IPX information back to the networks from which it received that information.

RIP also uses an information aging mechanism to handle conditions where an IPX router goes down without any explicit message to its neighbors. Periodic updates reset the aging timer.

Routing table updates are sent at 60-second intervals. This update frequency can cause excessive overhead traffic on some internetworks.

TIPS

Enhanced IGRP for IPX is an alternative to RIP as a routing protocol and is supported on Cisco-to-Cisco connections. Enhanced IGRP can provide faster convergence, reduced broadcast traffic, and better routes.

NLSP, Novell's link-state routing protocol, is another alternative to RIP. NLSP is supported for Cisco-to-Novell multiprotocol PC-based router (MPR) connections and NetWare 4.11 and later versions.

NLSP is derived from the OSI Intermediate System-to-Intermediate System (IS-IS) protocol. NLSP will interoperate with RIP and SAP to ease the transition and provide backward compatibility with RIP internetworks that have no need for link-state routing. NLSP (and IPXWAN) is covered in more detail in the Cisco Press book *Advanced Cisco Router Configuration*.

Many Novell customers want to reduce the excessive distance vector overhead packet traffic in RIP and SAP. Link-state routing requires less ongoing bandwidth, but link-state updates can also have problems, especially in large networks using a single (Level 1) area.

SAP: Supporting Service Advertisements

All the servers on NetWare internetworks can advertise their services and addresses. All versions of NetWare support SAP broadcasts to announce and locate registered network services, as shown in Figure 11–6. Adding, finding, and removing services on the internetwork are dynamic because of SAP advertisements.

Each SAP service is an object type identified by a hexadecimal number. Examples:

4	NetWare file server
7	Print server
278	Directory server

All servers and routers keep a complete list of the services available throughout the network in server information tables. Like RIP, SAP also uses an aging mechanism to identify and remove table entries that become invalid.

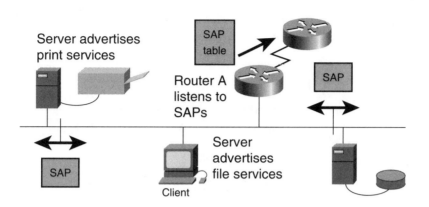

Figure 11–6
SAP packets advertise all NetWare network services.

By default, service advertisements occur at 60-second intervals. However, although service advertisements might work well on a LAN, broadcasting services can require too much bandwidth to be acceptable on large internetworks or in internetworks linked on WAN serial connections.

Routers do not forward SAP broadcasts. Instead, each router builds its own SAP table and forwards the SAP table to other routers. By default, this occurs every 60 seconds.

SAP advertisements can be filtered on input or output, or from a specific router:

- An IPX input—SAP filter allows the administrator to control services that are added to the router's SAP table from a specified interface.

- The IPX output—SAP filter allows the administrator to specify services included in SAP updates sent out to a specified interface.

- The IPX router—SAP filter statement is used to filter SAP messages received from a specified router on a specified interface.

All these filters must refer to an earlier IPX access list numbered from 1000 to 1099.

GNS: The Get Nearest Server Process

The NetWare client/server interaction begins when the client powers up and runs its client startup programs. These programs use the client's network adapter on the LAN and initiate the connection sequence for the NetWare client software to use.

GNS is a broadcast that comes from a client using SAP. NetWare file servers respond with a SAP reply (Give Nearest Server), as shown in Figure 11–7. From that point on, the client can log in to the target server, make a connection, set the packet size, and proceed to use server resources, provided the client is an authorized user of those resources.

Figure 11–7
*GNS is a
broadcast
from a client
needing a
server.*

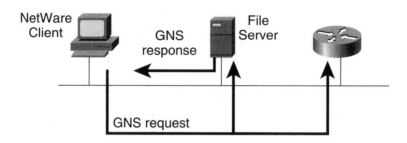

If a NetWare server is located on the segment, it will respond to the client request. If there are no NetWare servers on the local network, the Cisco router will respond to the GNS query with the address of the nearest server (or service) specified by the client.

An administrator might want to filter the extent of GNS responses. To filter GNS responses, the administrator uses a GNS output filter to limit the SAP table listing of nearest or preferred servers that respond to the GNS broadcast. This process of filtering is covered in more detail in Chapter 13, "Basic Traffic Management with Access Lists."

CONFIGURING IPX ROUTING

Configuration of Novell IPX as a routing protocol involves both global and interface parameters.

- Global tasks include:

  Start the IPX routing process.

  Enable load sharing if appropriate for your network. Load sharing is the process of dividing routing tasks evenly among multiple routers to balance the work and improve network performance.

- Interface tasks include:

  Assign unique network numbers to each interface, as shown in Figure 11–8. Multiple network numbers can be assigned to an interface, allowing support of different encapsulation types.

  Set the optional encapsulation type if it is different from the default.

Network 9e encap Novell-ether

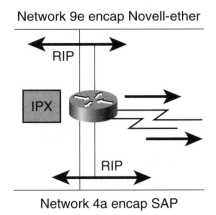

Figure 11–8
*Each interface
is assigned a
unique net-
work address.*

Network 4a encap SAP

Novell IPX Global Configuration Commands

You can use three commands to set up the IPX global configuration, if desired:

- **ipx routing** [*node*]
- **ipx maximum-paths** [*paths*]
- **ipx route destination-net next-hop** [*floating-static*]

Each of these commands is discussed in more detail in the following sections.

ipx routing Command

The **ipx routing** [*node*] command enables Novell IPX routing. If no node address is specified, the Cisco router uses the MAC address of the interface.

If a Cisco router has only serial interfaces, an address must be specified.

ipx maximum-paths Command

The **ipx maximum-paths** [*number of paths*] command enables load sharing. Load sharing occurs when parallel metric paths are available between the source and directly connected networks leading to the destination. The maximum-paths parameter indicates how many identical paths can be considered when load balancing decisions are being made. The default value of *number of paths* is 1, which means load balancing is disabled by default.

ipx route destination-net next-hop Command

This variation of the **ipx routing** command is available in Release 10.3 and later and uses the following syntax:

```
ipx route {network [network-mask] ¦ default} {network.node ¦ interface}
    [floating-static]
```

This command extension allows static routes to be overridden by a dynamically learned route. These alternate routes are learned from IPX RIP, Enhanced IGRP, and the new NLSP.

Use a floating static route so that if a static route goes down, another dynamic route is available for use. The IPX floating static routes allow you to switch to this other dynamic path any time the static route to a destination is lost.

Novell IPX Interface Configuration Commands

Two commands are used to set up the IPX configuration:

- interface
- ipx network

Both of these commands are discussed in more detail in the following sections.

interface Command

To assign network numbers to interfaces that support multiple networks, you normally use subinterfaces. A subinterface is a mechanism that allows a single physical interface to support multiple logical interfaces or networks. That is, several logical interfaces or networks can be associated with a single hardware interface.

Each subinterface must use a distinct encapsulation, and the encapsulation must match that of the clients and servers using the same network number. To run the NLSP on multiple networks on the same physical LAN interface, you must configure subinterfaces.

ipx network Command

The syntax of this command is:

```
ipx network network [encapsulation encapsulation-type [secondary]]
```

When assigning network numbers to interfaces that support multiple networks, you can also configure primary and secondary networks. However, in future Cisco IOS software releases, primary and secondary networks will not be supported.

The first logical network you configure on an interface is considered the primary network. Any additional networks are considered secondary networks. Again, each network on an interface must use a distinct encapsulation, and it should match that of the clients and servers using the same network number. Assigning the second network number is necessary if an additional encapsulation type is linked to an individual network.

There are two commands that affect individual interfaces:

```
Router(config-if)#
Router (config)#interface type number.subinterface-number
Router (config-subif)#ipx network network [encapsulation encapsulation
type]
ipx network network [encapsulation encapsulation-type] [secondary]
```

Novell IPX Configuration Example

Consider the IPX configuration example shown in Figure 11–9.

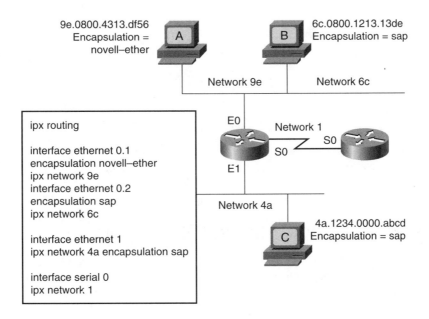

Figure 11–9
Networks 9e and 6c reside on the same physical media.

The following list defines the functions configured for the local router in Figure 11–9:

ipx routing	Selects IPX as a routing protocol and starts the routing process.
ipx maximum-paths 2	Allows load sharing over parallel metric paths to the destination. The number of parallel paths used is limited to two.
interface ethernet 0.1	Indicates the first subinterface on interface E0.
encapsulation novell-ether	Specifies that Novell's unique frame format is used on this network segment. Cisco's keyword is **novell-ether**; Novell's terminology is Ethernet_802.3.
ipx network 9e	Network number assigned to subinterface E0.1.
interface ethernet 0.2	Indicates the second subinterface on interface E0.
encapsulation sap	Specifies that Ethernet 802.2 frame format is used on this network segment. Cisco's keyword is **sap**.
ipx network 6c	Network number assigned to subinterface E0.2.
interface ethernet 1	Indicates the first (and only) interface on interface E1.
ipx network 4a encapsulation sap	Specifies that the new default frame format is used on this network, which has network address 4a.
interface serial 0	Indicates the first interface on serial interface S0.
ipx network 1	Sets the serial link's IPX network address to 1.

VERIFYING AND MONITORING IPX ROUTING

It's always important to verify your configuration once it is complete. Use the commands defined in this section to ensure that your router is set up properly. Then focus on the monitoring section to learn how to check and troubleshoot SAP, RIP, and IPX traffic through the router.

Once IPX routing is configured, you can monitor and troubleshoot it using the following commands:

Monitoring Command	Displays
show ipx interface	IPX status and parameters
show ipx route	Routing table contents
show ipx servers	IPX server list
show ipx traffic	Number and type of packets

Troubleshooting Command	Displays
debug ipx routing activity	Information about RIP update packets
debug ipx sap	Information about SAP update packets

Each of these commands is discussed in detail in the following sections.

Monitoring the Status of an IPX Interface

The **show ipx interface** command shows the status of IPX interface and IPX parameters configured on each interface, as shown in Figure 11–10. The first highlighted line shows the IPX address, the type of encapsulation, and the status of the interface. The second highlighted area shows that the SAP filters are not set. The last highlighted line shows that fast switching is enabled.

You can manually set the tick metric. Use the command **ipx delay** *number* where *number* is the ticks to associate with an interface. This command manually overrides the following defaults on the Cisco router:

- For LAN interfaces, one tick

- For WAN interfaces, six ticks

Some of the display fields shown include:

- *IPX address...*—Network and node address of the local router interface, followed by the type of encapsulation configured on the interface and the interface's status. Refer to the **ipx network** command for a list of possible values.

- *SAP Input filter list*—Number of the input SAP filter applied to the interface with the **ipx input-sap-filter** command

- *SAP Output filter list*—Number of the output SAP filter applied to the interface with the **ipx output-sap-filter** command

Figure 11–10
Use the show ipx interface command to gather interface configuration details.

```
Router#show ipx interface ethernet 0
Ethernet 0 is up, line protocol is up
   IPX address is 3010.aa00.0400.0284, NOVELL-ETHER [up] line-up, RIPPQ: 0, SAPPQ: 0
   Delay of this Novell network, in ticks is 1
   IPXWAN processing not enabled on this interface.
   IPX SAP update interval is 1 minute(s)
   IPX type 20 propagation packet forwarding is disabled
   Outgoing access list is not set
   IPX Helper access list is not set
   SAP Input filter list is not set
   SAP Output filter list is not set
   SAP Router filter list is not set
   SAP GNS output filter list is not set
   Input filter list is not set
   Output filter list is not set
   Router filter list is not set
   Netbios Input host access list is not set
   Netbios Input bytes access list is not set
   Netbios Output host access list is not set
   Netbios Output bytes access list is not set
   Update time is 60 seconds
   IPX accounting is disabled
   IPX fast switching is configured (enabled)
   IPX SSE switching is disabled
   RIP packets received 1, RIP packets sent 10006
   SAP packets recieved 1, SAP packets sent 6
--More--
```

- *SAP Router filter list*—Number of the router SAP filter applied to the interface with the **ipx router-sap-filter** command

- *IPX fast switching*—Indicates whether IPX fast switching is enabled (default) or disabled for this interface, as configured with the **ipx route-cache** command

Monitoring IPX Routing Tables

The **show ipx route** command displays the contents of the IPX routing table, as shown in Figure 11–11.

The first highlighted line provides routing information for a remote network:

- The information was learned from a RIP update (indicated by an R).

- The network is number 3030.

```
Router#show ipx route
Codes: C  -  Connected primary network, c - Connected secondary network
       R  -  RIP, E  -  EIGRP, S - static, W  -  IPXWAN connected
5 Total IPX routes

Up to 2 parallel paths allowed Novell routing algorithm variant in use

R  Net  3030  [6/1] via 3021.0000.0c03.13d3, 23 sec,  Serial1
                    via 3020.0000.0c03.13d3, 23 sec,  Serial0
C  Net  3020  (x25),  Serial0
C  Net  3021  (HDLC),  Serial1
C  Net  3010  (NOVELL-ETHER),  Ethernet0
C  Net  3000  (NOVELL-ETHER),  Ethernet1
```

Figure 11-11
*Use the **show ipx route** command to view an IPX routing table.*

- It is located six ticks or one hop away. This information is used to determine best routes. If there is a tie between ticks, hops are used to break the tie.

- The next hop in the path is router 3021.0000.0c03.13d3.

- The information was updated 23 seconds ago.

- The updates will be sent through the interface named Serial1.

The second line of highlighting provides information about a direct connection (indicated by a C):

- The network number is 3010.

- The encapsulation type is NOVELL-ETHER.

Of particular interest are the delay and metric values. The delay factor is specified in ticks ([1/18]th of a second). Ticks are not implemented by all networking equipment; therefore, path decisions can be based on inaccurate information. The metric values specified as 6/1 refer to ticks and hop count.

TIPS

With Cisco IOS Release 10.0 and later, you can use Novell's new protocol IPXWAN to test the WAN link when establishing an IPX connection. The router uses a timer request and response exchange to learn the link delay. Then the router can use the learned link delay to set the WAN interface tick value. Enter the **ipx ipxwan** command as you configure for the serial interface.

Monitoring the Novell IPX Servers

The **show ipx servers** command lists the IPX services discovered through SAP advertisements, as shown in Figure 11–12.

Figure 11–12
*Use the **show ipx servers** command to view the Server Information Table.*

```
Router> show ipx servers
Codes: P - Periodic, I - Incremental, H - Holdown, S - static
1 Total IPX Servers

Table ordering is based on routing and server info

Type    Name    Net      Address         Port    Route      Hops    Itf
P4      Maxine  AD33000.0000.1b04.0288:0451  332800/1   2       Et3
```

This example provides the following information:

- The service learned about the server from a SAP update
- The server name, network location, device address, and source socket number
- The ticks and hops for the route (taken from the routing table)
- The number of hops (taken from the SAP protocol)
- The interface through which to reach the server

Monitoring IPX Traffic

The **show ipx traffic** command displays information about the number and type of IPX packets received and transmitted by the router, as shown in Figure 11–13.

```
Router#show ipx traffic
System Traffic for 2018.0000.0000.0001 System-Name: dtp-18
Rcvd:    23916 total, 13795 format errors, 0 checksum errors, 0 bad
hop count,
         0 packets pitched, 23916 local destination, 0 multicast
Bcast:   17111 received, 9486 sent
Sent:    16707 generated, 0 forwarded
         0 encapsulation failed, 0 no route
SAP:     6 SAP requests, 6 SAP replies, 2309 servers
         0 SAP Nearest Name requests, 0 replies
         0 SAP General Name requests, 0 replies
         1521 SAP advertisements received, 2212 sent
         0 SAP flash updated sent, 0 CAP format errors
RIP:     6 Rip request, 6 RIP replies, 2979 routes
         8033 RIP advertisements received, 4300 sent
         154 RIP flash updates sent, 0 RIP format errors
Echo:    Rcvd 0 requests, 0 replies
         Sent 0 requests, 0 replies
         0 unknown: 0 no socket, 0 filtered, 0 no helper
         0 SAPs throttled, freed NDB len 0
Watchdog:
         0 packets received, 0 replies spoofed
Queue Lengths:
         IPX input: 0. SAP 0, RIP 0, GNS 0
         SAP throttling length: 0/(no limit), 0 nets pending lost route
reply
         Delayed process creation: 0
```

Figure 11–13
*The **show ipx traffic** command tracks packets that have arrived at the router and that were sent by the router.*

Notice in Figure 11–13 that a high percentage of the total number of packets received and sent were RIP advertisements, because this sample was taken from a lab network with essentially no user traffic on it. This screen shows how much overhead traffic SAP and RIP generate.

Troubleshooting IPX Routing

The **debug ipx routing activity** command displays information about IPX routing update packets that are transmitted or received, as shown in Figure 11–14.

A router sends an update every 60 seconds. Each update packet can contain up to 50 entries. If there are more than 50 entries in the routing table, the update will include more than one packet.

In Figure 11–14, the router is sending updates but not receiving them. Updates received from other routers would also appear in this listing.

Figure 11–14
If IPX routing problems occur, run the **debug ipx routing activity** *command.*

```
Router#debug ipx routing activity
IPX routing debugging is on
Router#
IPXRIP: posting full update to 3010.ffff.ffff.ffff via Ethernet 0 (broadcast)
IPXRIP: posting full update to 3000.ffff.ffff.ffff via Ethernet 1 (broadcast)
IPXRIP: posting full update to 3020.ffff.ffff.ffff via Serial0 (broadcast)
IPXRIP: posting full update to 3021.ffff.ffff.ffff via Serial1 (broadcast)
IPXRIP: sending update to 3020.ffff.ffff.ffff via Serial0
IPXRIP: src=3020.0000.0c03.14d8, dat=3020.ffff.ffff.ffff, packet sent
        network 3021, hops 1, delay 6
        network 3010, hops 1, delay 6
        network 3000, hops 1, delay 6
IPXRIP: sending update to 3021.ffff.ffff.ffff via Serial1
IPXRIP: src=3021.0000.0c03.14d8, dat=3021.ffff.ffff.ffff, packet sent
        network 3020, hops 1, delay 6
        network 3010, hops 1, delay 6
        network 3000, hops 1, delay 6
IPXRIP: sending update to 3010.ffff.ffff.ffff via Ethernet0
IPXRIP: src=3010.aa00.0400.0284, dat=3010.ffff.ffff.ffff, packet sent
        network 3030, hops 2, delay 7
        network 3020, hops 1, delay 1
        network 3021, hops 1, delay 1
        network 3000, hops 1, delay 1
IPXRIP: sending update to 3000.ffff.ffff.ffff via Ethernet1
```

Troubleshooting IPX SAP

The **debug ipx sap** [activity | events] command displays information about IPX SAP packets that are transmitted or received.

Like RIP updates, these SAP updates are sent every 60 seconds and may contain multiple packets. Each SAP packet appears as multiple lines in the output and includes a packet summary message and a service detail message.

SAP packets may be one of these types:

- 0×1—General query
- 0×2—General response

- 0×3—Get Nearest Server request

- 0×4—Get Nearest Server response

In each line, the address and distance of the responding or target router is listed.

Each update takes multiple lines, one summary, and the rest detail. Figure 11–15 shows three SAPs:

- An Input SAP (indicated by the "I")

- A SAP update sent to IPX network 160

- An Output SAP (indicated by the "O") with information about the file server named Magnolia

```
Router#debug ipx sap events
IPX service events debugging is on
Router#
NovellSAP: at 0023F778
I SAP Response type 0x2 len 160 src:160.0000.0c00.070d dest:160.ffff.ffff.ffff(452)
  type 0x4,  "HELL02", 199.0002.0004.0006 (451), 2 hops
  type 0x4,  "HELL02", 199.0002.0004.0008 (451), 2 hops
NovellSAP: sending update to 160
NovellSAP: at 00169080
o SAP Update type 0x2 len 96 ssoc:0x452 dest:160.ffff.ffff.ffff(452)
Novell: type 0x4 "Magnolia", 42.0000.0000.0001 (451), 2 hops
```

Figure 11–15
*The **debug ipx** sap [activity | events]* command can be used to check out any service availability problems.

Activity and events are not really options because one or the other is required. The debug ipx **activity** command provides more details, and the debug ipx **events** option provides fewer details by focusing on SAP packets that contain interesting events. For the most useful information, use these two commands together.

SUMMARY

In this chapter, you learned about Novell's 10-byte IPX address elements and encapsulation options. You also learned how Novell's service discovery mechanism and routing information exchange methods work for clients and server. You viewed the IPX global and interface configuration commands used by Cisco. In Chapter 12, "Configuring AppleTalk," you will learn how AppleTalk addressing works and how to configure a Cisco router to connect AppleTalk networks.

Chapter Eleven Test
Configuring Novell IPX

Estimated Time: 15 minutes

Complete all the exercises to test your knowledge of the materials contained in this chapter. Answers are listed in Appendix A, "Chapter Test Answer Key."

Question 11.1

Determine the required IPX address and encapsulation type for a given router port. In this exercise, four routers are configured to run Novell IPX, as shown in Figure 11–16.

Figure 11–16
These four routers are configured to run Novell IPX.

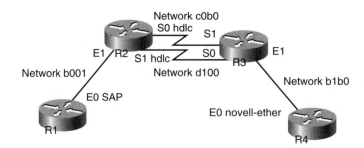

You must determine the IPX network addresses and encapsulation types to use when you configure the R3 router. Given the crucial IPX details necessary for configuration for only three of the four routers, you must solve for router R3. This information is summarized in the following table:

Router Name	Interface Name	IPX Network Address	Encapsulation Type
R1	E0	b001	sap
R2	S1	d100	hdlc
R2	S0	c0b0	hdlc
R4	E0	b1b0	novell-ether

Write your answers in the following table.

R3 Interface Name	Network Address	Encapsulation
S0		
S1		
E1		

Question 11.2

What command do you issue to enable IPX routing on a router?

Question 11.3

What command mode must the router be in before you can issue the **ipx routing** command?

Question 11.4

What command do you issue to assign IPX network numbers on a router?

Question 11.5

What command do you issue to verify IPX address assignment on a router?

Question 11.6

What command do you issue to verify entries in the routing table?

Configuring AppleTalk

This chapter presents an introduction to the AppleTalk protocol suite and how it operates using Cisco IOS software configurations. The chapter includes information on AppleTalk addressing, logical zones, and locating services, as well as details on configuring and verifying AppleTalk routing.

APPLETALK OVERVIEW

AppleTalk was designed by Apple Computer to provide communication and resource sharing among its Macintosh computers and peripherals. In this section, you learn about the AppleTalk protocol stack, the addressing system, and the service discovery mechanism.

AppleTalk Protocol Stack

Figure 12–1 compares the AppleTalk protocol architecture to the OSI reference model.

At the hardware layers, most standard media types are supported using AppleTalk Phase 2 (an extended network). Many Apple products contain a LocalTalk interface that operates over twisted-pair cabling at 230 kbps. The LocalTalk interface is not available on Cisco products; therefore, LocalTalk devices can be adapted to Ethernet or other LAN environments.

At Layer 3 in the AppleTalk architecture, the Datagram Delivery Protocol (DDP) provides a connectionless datagram service.

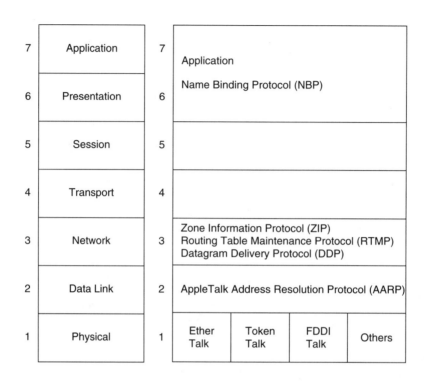

Figure 12–1
*The AppleTalk
protocol
architecture.*

At Layer 4 in the AppleTalk architecture, the Name Binding Protocol (NBP) provides name-to-address association. Routing table updates are provided by the Routing Table Maintenance Protocol (RTMP).

Other important AppleTalk protocols at Layer 4 are as follows:

- AppleTalk Transaction Protocol (ATP)—Uses transactions to ensure that DDP packets are delivered to a destination without losses.

- AppleTalk Data Stream Protocol (ADSP)—Uses sliding window transport to ensure that the flow of data from one side of a connection does not overwhelm the buffer space of a slower node on the other side of the connection.

At Layer 5 in the AppleTalk architecture, the *Zone Information Protocol (ZIP)* maps logical zones to network numbers and coordinates name lookup within zones. ZIP is used by NBP to find out which networks contain nodes that belong to a specific zone. A good additional resource for this chapter is *Inside AppleTalk, Second Edition*, from Addison-Wesley. Chapter 1 contains the AppleTalk reference model and shows the relationship between the various protocols.

AppleTalk Features

AppleTalk was designed as a *client-distributed network system*. This means that users share network resources (such as files and printers) with other users. Computers supplying these network resources are called *servers*; computers using a server's network resources are called *clients*. Interaction with servers is essentially transparent to the user because the computer determines the location of the requested materials and accesses it without further information from the user.

Clients use broadcasts to learn about available services. The AppleTalk environment allows propagation of lookups by the router, ensuring that all available services will be located by the user.

AppleTalk addresses are composed of a 16-bit network number and an 8-bit node number:

- The network portion of the address is manually configured by the administrator.

- The node identifier portion is dynamically acquired during device startup. The node identifier can also be manually configured on the Cisco router. This process is useful when configuring AppleTalk for multipoint WANs and for remote access using dialer maps.

Random selection of node numbers makes troubleshooting with a network analyzer somewhat more difficult, but a Macintosh or PowerPC usually saves its address in NVRAM so it can reuse the same address on its network environment.

RTMP (Routing Table Maintenance Protocol) provides routing information updates at Layer 3. RTMP is a Routing Information Protocol (RIP) derivative, using hop count as its metric for routing decisions. RTMP routing protocol updates occur at 10-second intervals (thus, the reputation of AppleTalk as a "chatty" protocol). Use of *zones* (which are discussed later in this chapter) groups networks and services to help reduce this chattiness. Hosts listen to RTMP updates to learn the router's address.

Nonextended or Extended Networks

Early releases of AppleTalk (pre–1988) used an addressing scheme referred to as Phase 1. This scheme did not allow large numbers of hosts on a single wire. An equal number of servers and hosts was allocated, at most 127 of each. Any Macintosh can be a host or server. Nonextended networks could have only a single network on the wire. These characteristics define a *nonextended network*, as shown at the top of Figure 12–2.

Figure 12–2
*Extended
AppleTalk net-
works are
more flexible
than nonex-
tended
networks.*

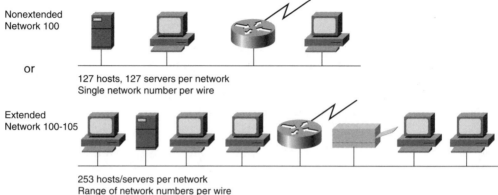

Nonextended
Network 100

or

127 hosts, 127 servers per network
Single network number per wire

Extended
Network 100-105

253 hosts/servers per network
Range of network numbers per wire

Later releases of AppleTalk use an extended form of addressing known as Phase 2. In an *extended network*, multiple network numbers can exist on the same wire. The number of devices on a single logical network is limited to 253, but these devices can be any combination of hosts and services. The network and node addresses are considered in combination, greatly enlarging the available address space. Extended network details are summarized in the lower portion of Figure 12–2.

Note that most AppleTalk networks have been upgraded to Phase 2. Phase 1 AppleTalk may be all but extinct on internetworks that use Cisco routers. Phase 2 addressing is the standard (since AppleTalk version 6.3B).

AppleTalk Addressing

In a Phase 1 network, only one network number is allowed for each wire. Node numbers are assigned dynamically when clients start up. In a Phase 2 network, multiple network numbers are available for each wire, as shown in Figure 12–3.

The range of logical network numbers on a single physical wire is called a *cable range* or *cable group*. The range of numbers is assigned by the administrator. Node numbers from 1 to 253 are assigned dynamically for both hosts and routers.

Both addressing schemes require that a unique network.node address be applied to each router interface. Notice serial 0 in Figure 12–3, which has the same number for both parts of the range (1000-1000). This means a range of one number.

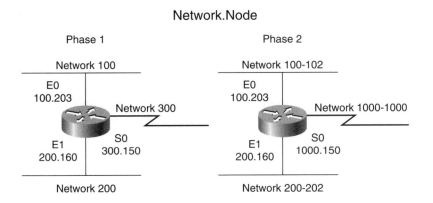

Network.Node

Phase 1   Phase 2

Network 100   Network 100-102

E0   E0
100.203   100.203

　　　Network 300   　　　Network 1000-1000

E1   S0   E1   S0
200.160   300.150   200.160   1000.150

Network 200   Network 200-202

Figure 12–3
*Phase 2 inter-
faces must
use unique
network.node
numbers.*

Extended Addressing

In an extended network, the network numbers of the nodes can be different, as shown in Figure 12–4. There may be a wide network range on a single logical network:

- Network number—16 bits

  A cable range states the span of network numbers available on this media.

  Narrow range networks (networks with a single network number) are supported.

  A network number of 0 is reserved by the protocol for a newly attached node to use when it does not yet know the network number to use on its attached cable.

- Node number—8 bits

  Numbers in the range 1 to 253 represent any node (user, printer, and other devices).

  The numbers 0, 254, and 255 are reserved on extended networks.

Node numbers are dynamically assigned.

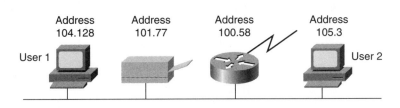

Address   Address   Address   Address
104.128   101.77   100.58   105.3

User 1   User 2

Figure 12–4
*Extended
addressing
permits a
range of net-
work numbers
per wire.*

AppleTalk Address Acquisition

When user 2's system is powered on, it has no address stored in its permanent memory (RAM). User 2's software selects a provisional network address from the FF00-FFE0 range and a random node number. The new node sends 10 AppleTalk Address Resolution Protocol (AARP) probes to verify the node ID availability, as shown in Figure 12–5. AARP is the protocol that reconciles protocol addresses to hardware address, much in the same manner as IP ARP does. In this case, the new node sends out a probe to see if anyone has the same node number. If the packet goes unanswered, the new node believes the address is available.

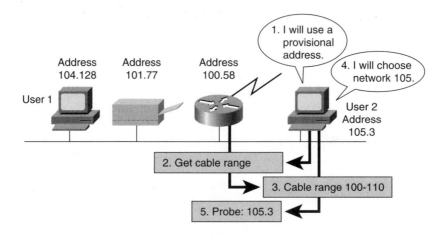

Figure 12–5
The AppleTalk host uses ZIP and AARP to obtain an address.

A "get cable range" ZIP request is issued by user 2. The router's response indicates the range of network numbers available on the wire. If user 2's provisional network address is invalid, user 2 selects a network number from the cable range.

User 2 issues 10 more AARP probes to verify the uniqueness of the chosen node ID:

- If there is a response that the node ID is in use, user 2 tries another node ID.
- If there is no response to the probe, user 2 uses this ID.

User 2's address becomes 105.3. After an address is acquired, it is saved in RAM. The stored address is probed for at the next power-up sequence, and if it is in use, dynamic assignment is initiated.

Limiting Requests for Services

One method for controlling broadcast traffic is to allocate nodes to zones, as shown in Figure 12–6. A node can be in only one zone. A *zone* is a logical group of networks and services that helps reduce AppleTalk's chattiness by enabling devices to perform discovery and communication within a smaller area.

Each interface in the router must be assigned to a zone as part of its configuration. Many devices, including the Cisco router interfaces, are visible in the default zone for a cable range.

In Figure 12–6, the LAN in Bldg. 1 has been separated from the WAN that includes Bldg. 2 and 3.

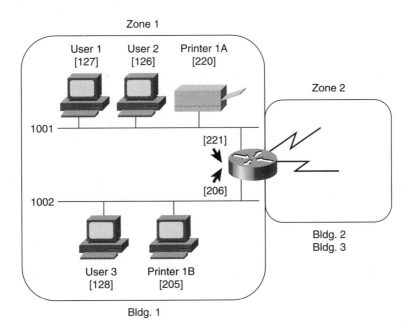

Figure 12–6
Zones are used to control broadcast traffic.

AppleTalk Services

Consider Figure 12–7, where nodes are assembled into zones. (Recall that each device can be in only one zone.) When a Macintosh user requires a service (such as an Appleshare file server or printer), the following takes place:

- The Chooser sends a request to the router for a list of zones.

- The NBP (Name Binding Protocol) looks for the services in the zone that the Macintosh user specifies.

- The router forwards the request to each cable grouped in the selected zone.

- A multicast (one-to-many) goes to all devices that match the device type requested.

- Available matching services reply to the address of the Macintosh that originated the NBP process.

Figure 12–7
The Chooser initiates service discovery on behalf of the client.

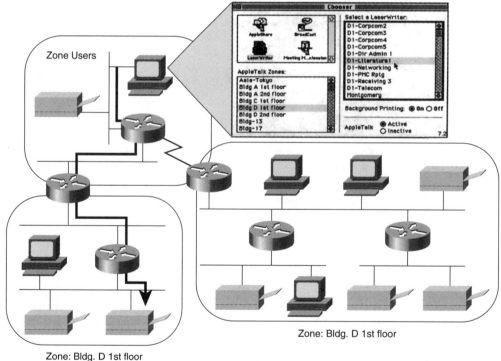

- Routers in the path forward these replies until they reach the originating router.

- The originating router sends the reply to the end user. The user selects the preferred service.

A logical link for that service is retained in the Macintosh for future reference, and a list of services and zones is maintained within the router for local reference.

Remember that each device can be in only one zone.

Key Concept

Locating AppleTalk Services

As mentioned earlier, users on the AppleTalk network locate specific services using NBP requests.

In Figure 12–8, user 2 looks for printers in the zone named Users. The router checks a local table, the zone information table (ZIT), which contains a complete network-to-zone-name mapping of the Internet. The router then creates a packet to forward the requests to each segment of the selected zone. The router will create one request to send out cable 1001-1001, and another request to send out cable 1002-1002. Responses that the router forwards to user 2 inform user 2 about printer 1A and printer 1B.

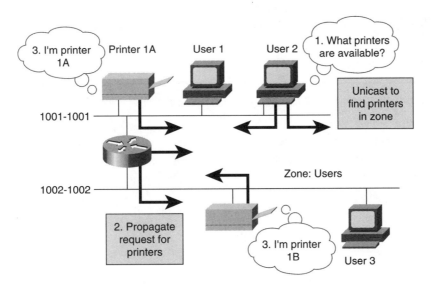

Figure 12–8
User 2 searches the zone named "Users."

CONFIGURING APPLETALK

Now that you know the basic elements of the AppleTalk protocol, including addressing and service discovery, the following looks at how you can configure a Cisco router to handle AppleTalk communications.

AppleTalk Configuration Tasks

Configuration of AppleTalk as a routing protocol requires setting both global and interface parameters.

- Global task:

  Select AppleTalk routing to start the routing process.

- Interface tasks:

  Assign a range of network numbers to each interface. A narrow range can be an appropriate assignment.

  Assign each interface to a zone. Phase 2 allows multiple zones per segment. Zone assignment is a mandatory configuration statement; it is necessary to establish a zone for each interface to become enabled.

TIPS

Cisco engineering recommends that you use one network number per 50 nodes.

After an address and zone name are assigned, the interface is enabled for packet processing. All routers in a network or data link must agree on the cable range, default zone, and zone list.

AppleTalk Configuration Commands

There are four basic commands used to configure an AppleTalk network:

- Global config tasks/commands: **appletalk routing** (Required)
- Interface commands: **appletalk protocol** (Optional)
- **appletalk cable-range** (Required)
- **appletalk zones** (Required)

Each of these commands is discussed in more detail in the following sections.

appletalk routing Command

```
Router(config)#appletalk routing
```

The **appletalk routing** command starts the AppleTalk routing process.

appletalk protocol Command

```
Router(config-if)#appletalk protocol {rtmp ¦ eigrp ¦ aurp}
```

The **appletalk protocol** command selects one or more routing protocols for use on this interface.

appletalk protocol Command	Description
rtmp	The routing protocol is RTMP, which is the default.
eigrp	This specifies that the routing protocol to use is Enhanced IGRP.
aurp	This specifies that the routing protocol to use is AppleTalk Updated-Based Routing Protocol (AURP). This is the method of encapsulating AppleTalk traffic in the header of a foreign protocol, allowing the connection of two or more discontiguous AppleTalk internetworks through a foreign network (such as TCP/IP) to form an AppleTalk WAN.

Note that if the **appletalk protocol** command is omitted in the interface specification, RTMP is selected by default.

appletalk cable-range Command

```
Router(config-if)#appletalk cable-range cable-range [network.node]
```

The **appletalk cable-range** command specifies a range of network numbers to the interface. This command has two parameters, described in the following table.

appletalk cable-range Command	Description
cable-range	This argument defines the value of the cable range and specifies the start and end of the cable range, separated by a hyphen. These values are a decimal number from 0 to 65279. The starting network number must be less than or equal to the ending network number and numbers must be contiguous.
network.node	This argument is optional and defines the suggested AppleTalk address for the interface. The argument network is the 16-bit network number, and the argument node is the 8-bit node number. Both numbers are decimal. The suggested network number must fall within the specified range of network numbers.

The optional *network.node* argument allows the network administrator to specify a unique address. Use hardcoded addresses for AppleTalk over Frame Relay and Switched Multimegabit Data Service (SMDS), as well as when using dialers.

appletalk zone Command

```
Router(config-if)#appletalk zone zone-name
```

The **appletalk zone** command assigns the zone name to the data link. Multiple zones can be assigned to one interface in a Phase 2 installation. The first zone name is the default zone name.

AppleTalk Configuration Example

Figure 12–9 provides a sample AppleTalk configuration. The following list defines the commands shown.

Command	Description
appletalk routing	Starts the AppleTalk routing process
interface ethernet 0	Defines the interface being configured

Command	Description
appletalk cable-range 100–105	Establishes a range of six network numbers available to devices on E0
appletalk zone engineering	Places interface E0 into a zone named engineering. This zone is the default because it is specified first.
interface ethernet 1	Defines the next interface being configured
appletalk cable-range 200–205	Establishes a range of six network numbers available to devices on E1
appletalk zone engineering	Places interface E1 into the engineering zone
appletalk zone headquarters	Places interface E1 into the headquarters zone
interface serial 0	
appletalk cable-range 1000–1000 1000.128	Assigns a narrow cable range of 1000 to interface serial 0 and specifies the network.node address of 1000.128
appletalk zone engineering	Places interface S0 into the engineering zone

All interfaces are using RTMP as the default routing protocol because **no appletalk protocol** commands are specified.

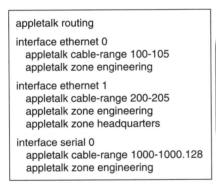

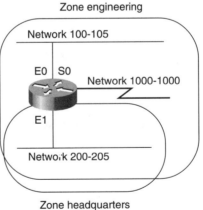

Figure 12–9
This AppleTalk network supports two zones.

After AppleTalk routing is enabled, interface E0 dynamically acquires a node number on one of six available network numbers. Serial 0 has a hard-coded address of 1000.128. All interfaces in the router are part of the zone engineering, and E1 is also part of zone headquarters.

Discovery Mode

AppleTalk routers can dynamically discover network number ranges and zones by using a technique called *discovery*. A seed router is a router that contains configuration information, whereas a nonseed router is not privy to this information.

Seed routers seed the AppleTalk Internet with configuration information such as network number ranges and zones. The network administrator sets up a router as a seed router to provide configuration information to other nonseed routers.

Placing a nonseed router interface in discovery mode allows the interface to dynamically learn its cable range and zone information from a seed router. There are two ways to place a nonseed interface into discovery mode:

- In Phase 2 only—Assign the cable range as 0–0 like so:

  ```
  Router(config-if)#appletalk cable-range 0-0
  ```

- Assign an address to the interface using normal configuration steps and then allow dynamic learning from other routers by using the default **appletalk discovery** command. AppleTalk discovery mode does not work over serial interfaces.

  ```
  Router(config-if)#appletalk cable-range cable-range
  Router(config-if)#appletalk discovery
  ```

TIPS

The second command shown (**appletalk discovery**) is on by default. To disable it, use the command **no appletalk discovery**.

Once a nonseed router obtains configuration information, it participates in routing updates. In effect, it acts like a seed router for other routers coming up on the network.

Discovery Mode Example

Figure 12–10 provides an example of how to configure one router as a seed router and another router to use discovery mode. The following list explains the configuration commands used:

Command	Description
Interface E0 appletalk cable-range 0-0	Places E0 into discovery mode

Interface E1
  appletalk cable-range
  3000-3002
	Assigns a network range to E1

appletalk discovery	Places E1 into discovery mode

Both E0 and E1 dynamically learn their addresses and zones.

In the live configuration file, after discovery, for E0:

Command	Description
appletalk cable-range 100-105	The acquired network range
appletalk Zone Bldg-17	The acquired zone name

In the live configuration file, for E1:

Command	Description
appletalk cable-range 200-205	The acquired network range
appletalk Zone Bldg-13	The acquired zone name

Figure 12–10
*You must have
a seed router
on the net-
work to use
discovery
mode.*

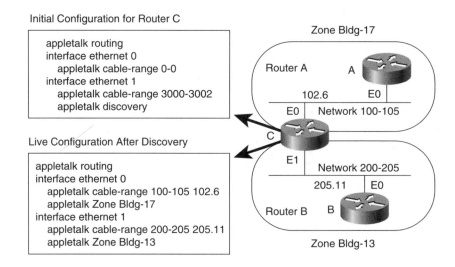

Initial Configuration for Router C

```
appletalk routing
interface ethernet 0
    appletalk cable-range 0-0
interface ethernet 1
    appletalk cable-range 3000-3002
    appletalk discovery
```

Live Configuration After Discovery

```
appletalk routing
interface ethernet 0
    appletalk cable-range 100-105 102.6
    appletalk Zone Bldg-17
interface ethernet 1
    appletalk cable-range 200-205 205.11
    appletalk Zone Bldg-13
```

Verifying Your AppleTalk Configuration

Use the **show appletalk interface** command to display status about all AppleTalk interfaces, including individual addressing, line status, timers, access lists assigned, and other details, as shown in Figure 12–11.

Figure 12–11
*Use the **show
appletalk
interface** com-
mand to check
the protocol
configuration.*

```
Router#show appletalk interface ethernet 0
Ethernet0 is up, line protocol is up
    AppleTalk cable range 3010-3019
    AppleTalk address is 3012.93, Valid
    AppleTalk zone is "1d-e0"
    AppleTalk port configuration verified by 3017.170
    AppleTalk address gleaning is enabled
    AppleTalk route cache is enabled
```

The **show appletalk interface** command is particularly useful when you first enable AppleTalk on a router interface. The display in Figure 12–11 shows you this information:

- The interface is Ethernet 0.

- The cable range contains an address value from which an address was selected. The address is marked as valid, which means it does not conflict with another node on that segment.

- The zone name is listed.

- AppleTalk address gleaning is enabled, which means the router can add addresses to its address resolution table by examining incoming packets.

- AppleTalk route cache is enabled, which means that fast switching is enabled on this interface.

MONITORING APPLETALK

Use the **show appletalk route** command to display the contents of the AppleTalk routing table, as shown in Figure 12–12.

Figure 12–12 shows the zones assigned to each cable range. The highlighted line shows an example of a wide cable range in the entry derived from RTMP.

The information indicates C for directly connected interfaces and R for routes derived from the RTMP routing protocol updates.

```
Router#show appletalk route
Codes:  R - RTMP derived, E - EIGRP derived, C - connected, A - AURP
        S - static, P - proxy
5 routes in internet

The first zone listed for each entry is its default (primary) zone.

C Net 3000-3005 directly connected, Ethernet1, zone ozone
C Net 3010-3019 directly connected, Ethernet0, zone ld-e0
C Net 3020-3020 directly connected, Serial0, zone dc-s0
C Net 3021-3021 directly connected, Serial1, zone dc-s1
R Net 3030-3039 [1/G] via 3020.259 4 sec, Serial0, zone cfeo
```

Figure 12–12
*Use the **show appletalk route** command to check the AppleTalk routing table.*

The **show appletalk zone** command displays entries in the AppleTalk zone information table, as shown in Figure 12–13.

Notice that the wide range of networks, 3000–3005, occur in zone ld-e0 as well as in zone ozone. The NBP lookup process is limited to the zone specified by the Macintosh user's zone selection in the Chooser.

The **show appletalk globals** command displays information and settings about the router's global AppleTalk configuration parameters, as shown in Figure 12–14. The highlighted line indicates Phase 1 compatibility through the use of unary cable ranges and single zones per interface.

Two other commands you may be interested in are the **show appletalk arp** and **debug apple events** commands.

Figure 12–13
*Use **show**
appletalk zone
command to
view the
router's zone
information
table.*

```
Router#show appletalk route
Codes: R - RTMP derived, E - EIGRP derived, C - connected, A - AURP
          S - static, P - proxy
5 routes in internet

The first zone listed for each entry is its default (primary) zone.

C Net 3000-3005 directly connected, Ethernet1, zone ozone
C Net 3010-3019 directly connected, Ethernet0, zone 1d-e0
C Net 3020-3020 directly connected, Serial0, zone dc-s0
C Net 3021-3021 directly connected, Serial1, zone dc-s1
R Net 3030-3039 [1/G] via 3020.25, 4 sec, Serial0, zone cf-e0
```

Figure 12–14
*Use the **show**
appletalk glo-
bals com-
mand for an
overview of
Phase 1 to
Phase 2
compatibility.*

```
Router#show appletalk route
Name                        Network(s)
1d-e0                       3010-3019 3000-3005
ozone                       3000-3005
cf-e0                       3030-3039
dc-s0                       3020-3020
dc-s1                       3021-3021
```

The **show appletalk arp** command displays the entries in the AARP cache. It is a privileged EXEC command. Use the **debug apple errors** EXEC command to display errors occurring in the AppleTalk network. The **no** form of this command disables debugging output.

The **debug apple routing** command displays output from the RTMP routines, as shown in Figure 12–15. This command is used to monitor acquisition, aging, and advertisement of routes. It also reports conflicting network numbers on the same network.

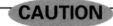

CAUTION

The **debug apple routing** command can generate many messages. It should be used only when CPU use is less than 50 percent.

```
Router#debug apple routing
AppleTalk RTMP routing debugging is on
AppleTalk EIGRP routing debugging is on
Router#
AT:  RTMP from 3002.5 (new 0, old 0, bad 0, ign 0, dwn 0)
AT:  RTMP from 3017.170 (new 0, old 0, bad 0, ign 0, dwn 0)
AT:  src=Ethernet0:3012.93, dst=3010-3019, size=34, 4 rtes, RTMP pkt sent
AT:  Route ager starting on Main AT RoutingTable (5 active nodes)
AT:  Route ager finished on Main AT RoutingTable (5 active nodes)
AT:  RTMP from 3020.25 (new 0, old 1, bad 0, ign 1, dwn 0)
AT:  RTMP from 3021.193 (new 0, old 1, bad 0, ign 3, dwn 0)
AT:  RTMP from 3020.25 (new 0, old 1, bad 0, ign 1, dwn 0)
AT:  RTMP from 3002.5 (new 0, old 0, bad 0, ign 0, dwn 0)
AT:  RTMP from 3017.170 (new 0, old 0, bad 0, ign 0 dwn 0)
AT:  src=Ethernet0:3012.93, dst=3010-3019, size=34, 4 rtes, RTMP pkt sent
AT:  src=Ethernet0:3000.175, dst=3000-3005, size=34, 4 rtes, RTMP pkt sent
AT:  src=Serial0:3020.26, dst=3020-3020, size=28, 3 rtes, RTMP pkt sent
AT:  src=Serial1:3021.144, dst=3021-3021, size=34, 4 rtes, RTMP pkt sent
```

Figure 12–15
*The **debug apple routing** command displays all RTMP update information.*

SUMMARY

In this chapter, you learned how AppleTalk addressing is designed, as well as the difference between Phase 1 and Phase 2 AppleTalk networks. Recall that AppleTalk was designed as a client-distributed network system, which means that users share network resources with other users. You learned about the AppleTalk protocol stack, examined the AppleTalk configuration and monitoring commands, and learned how clients perform service discovery. Chapter 13, "Basic Traffic Management with Access Lists," covers traffic management techniques using Cisco access lists.

Chapter Twelve Test
Configuring AppleTalk

Estimated Time: 15 minutes

Complete all the exercises to test your knowledge of the materials contained in this chapter. Answers are listed in Appendix A, "Chapter Test Answer Key."

Question 12.1

T   F   AppleTalk addressing is composed of *network.node.*

Question 12.2

T   F   Node numbers are dynamically acquired.

Question 12.3

T   F   Multiple network numbers can exist on one wire.

Question 12.4

T   F   Servers use broadcasts to learn about available clients.

Question 12.5

What command do you issue to enable AppleTalk routing on a router?

Question 12.6

What command do you issue to assign cable ranges to each interface on a router?

Question 12.7

What command do you issue to assign zones to each link on a router?

_____

Question 12.8

What command do you issue to verify the address assignment on a router?

_____

Question 12.9

What command do you issue to verify entries in the routing table?

_____

13

Basic Traffic Management with Access Lists

This chapter presents standard and extended access lists as a means to control network traffic. It explains general concepts about access lists and defines how to configure IP, IPX, and AppleTalk access lists.

Cisco IOS software features access lists for most protocols. Refer to Appendixes B and C, respectively, for more information on DECnet and VINES.

ACCESS LISTS OVERVIEW

This section presents an overview of access lists, including where and when they should be used and how they work. It also includes a discussion of wildcards in access lists and shows sample access list configurations.

An *access list* is used to define the type of traffic that should be allowed or restricted from crossing a router.

Why Use Access Lists?

The earliest routed networks connected a modest scale of LANs and hosts. Next, the network administrator enlarged router connections to legacy and outside partners' networks. Increased use of the Internet brought new challenges to access control. Newer technology—from optical backbones to broadband services and high-speed LAN switches—increased control challenges again.

Network administrators face the following dilemma: how to deny unwanted traffic while allowing appropriate access. Although other tools such as passwords, callback equipment, and physical security devices are helpful, they often lack the flexible expression and specific controls most administrators prefer.

Access lists offer another powerful tool for network control. These lists add the flexibility to filter the packet flow in or out router interfaces. Such control can help limit network traffic and restrict network use by certain users or devices. Access lists differentiate packet traffic into categories that permit or deny other features. You can also use access lists to:

- Identify packets for priority or custom queuing. Prioritization enables you to designate certain packets to be processed by a router before other traffic, on the basis of protocol. Custom queuing is used to balance traffic based on protocol, type, and overall purpose.

- Restrict or reduce the contents of routing updates. These restrictions are used to limit information about specific networks from propagating through the network.

- Identify which packets will bring up a dial-on-demand routing connection. This ensures that packets that are not vital to the communications processes do not bring up a WAN link.

Access lists also process packets for other security features to:

- Provide IP traffic dynamic access control with enhanced user authentication using the lock-and-key feature

- Identify packets for encryption

- Identify Telnet access allowed to the router virtual terminals

As just noted, access lists can be used to define the type of traffic that can bring up a WAN link (see Figure 13–1). Compared to LAN or campus-based networking, the traffic that uses dial-on-demand routing (DDR) is typically low volume and periodic. DDR initiates a WAN call to a remote site only when there is traffic to transmit. To identify this traffic, you specify the packets that the DDR processes on the router will interpret as interesting traffic. For example, the destination network address, source network address, and service or network information could be defined as interesting and therefore be restricted or permitted on the WAN link.

When you configure for DDR, you must enter configuration commands that indicate what protocol packets constitute interesting traffic to initiate the call. To configure for

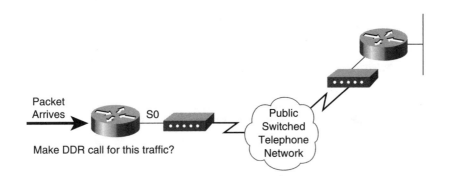

Figure 13–1
Access lists can be used to specify packet traffic for dialing remote sites using dial-on-demand routing.

Packet Arrives

S0

Public Switched Telephone Network

Make DDR call for this traffic?

call initiation, enter access list statements to identify the source and destination addresses and choose specific protocol selection criteria for initiating the call.

Then you establish the interfaces where the DDR call initiates. This step designates a dialer group. The *dialer group* associates the results of the access list specification of interesting packets to the router's interfaces for dialing a WAN call.

For DDR as well as other applications, access lists express the set of rules that give added control for packets that enter inbound interfaces, packets that relay through the router, and packets that exit outbound interfaces of the router. Access lists do not act on packets that originate in the router itself. Instead, they are statements that specify conditions for how the router will handle the traffic flow through specified interfaces. Access lists give added control for processing the specific packets in a unique way.

The two main types of access lists are standard and extended access lists.

Standard Access Lists

Standard access lists for IP check the source address of packets that could be routed. The result permits or denies output for the entire IP protocol suite, based on the network/subnet/host address.

Extended Access Lists

Extended access lists check for both source and destination packet addresses. They also can check for specific protocols, port numbers, and other parameters, which allows administrators more flexibility to describe what checking the access list will do. Packets can be permitted or denied output based on where the packet originated and on its destination.

For example, in Figure 13–2, packets coming in E0 are checked for source address, destination address, and protocol. If the packets belong to a permitted protocol and source and destination addresses, they are output through S0, which is grouped to the access list. If not, they are dropped.

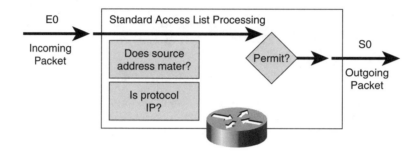

The extended access list also permits or denies with more granularity. For example, it can allow electronic mail traffic from E0 to specific S0 destinations, while denying remote logins or file transfers.

The extended access list can also match other packet header fields, for example, IP protocol and TCP port number.

How Access Lists Work

The beginning of the process is the same regardless of whether access lists are used: As a packet enters an interface, the router checks its routing table to see if the packet is routable or bridgeable, as shown in Figure 13–3. If it is neither, the packet will be dropped. If the packet is routable, a routing table entry indicates a destination network, some routing metric or state, and the interface to use for forwarding the packet.

Next, the router checks to see if the destination interface is grouped to an access list. If it is not, the packet can be sent to the output buffer. For example, if the packet will use To0 as a destination interface, and To0 has no access lists in effect, the packet uses To0 directly.

If the destination interface of the packet is grouped to an access list, the packet must be filtered through that list. For example, suppose that an interface, E0, has been grouped to an extended access list. The administrator used precise, logical expressions to set the access list. Before a packet can proceed to that interface, it is tested by a combination of access list statements associated with that interface.

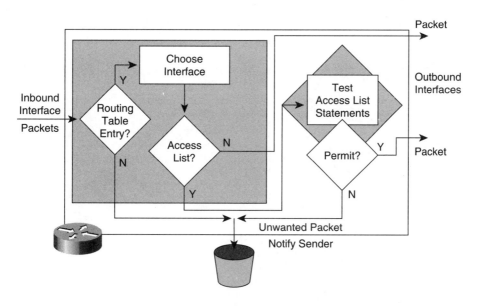

Figure 13–3
The access list process.

Based on the extended access list tests, the packet can be permitted or denied. To deny permission means to discard the packet. When discarding packets, some protocols return a special packet that notifies the sender of the unreachable destination. For example, an ICMP packet may be sent to indicate that the destination is unreachable. To permit the packet means to grant it access through interface E0. The router's access list provides effective control in denying the use of the E0 interface.

Note that E0 in this case is an outbound access list. For outbound lists, the test process determines whether or not the packet is permitted access to the outbound interface. For inbound lists, the test process determines whether or not the router continues processing the packet after receiving it on an inbound interface.

A List of Tests: Deny or Permit

Access list statements operate in sequential, logical order. They evaluate packets from the top down. If a packet header and access list statement match, the packet skips the rest of the statements. If a condition match is true, the packet is permitted or denied. There can be only one access list per protocol per interface.

In Figure 13–4, for instance, by matching the first test, a packet is denied access to destination interfaces. It will be discarded and dropped into the bit bucket. The packet is not exposed to any access list tests that follow.

Figure 13-4
*The access list
statements
are a sequen-
tial collection
of permit
or deny
conditions.*

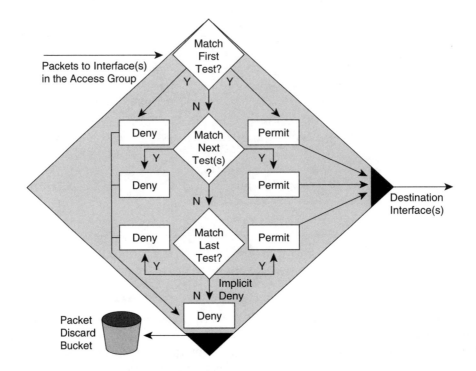

Only if the packet does not match conditions of the first test will it drop to the next access list statement. Assume a different packet's parameters match the next test, a permit statement. The permitted packet proceeds to the destination interface.

Another packet does not match the conditions of the first or second test but does match conditions of the next access list statement; again, a permit results.

For logical completeness, an access list must have conditions that test true for all packets using the access list. A final implied statement covers all packets for which preceding conditions did not test true. All such packets match the final test condition. The final test condition is a deny, often called an *implicit deny* because it does not actually appear as a line of configuration code. Instead of proceeding in or out an interface, all packets that reach the implicit deny are dropped.

Access List Command Overview

There are two general elements of access list configuration commands:

1. The access list process contains global statements:

```
Router(config)#access-list access-list-number {permit | deny}
   {test conditions}
```

- This global statement identifies the access list, usually an access list number. This number refers to the type of access list this will be. In Cisco IOS Release 11.2 or later, access lists for IP may also use an access list name rather than a number.

- The permit or deny term in the global access list statement indicates how packets that meet the test conditions will be handled by Cisco IOS software. Permit usually means the packet will be allowed to use one or more interfaces you will specify later.

- The final term or terms specifies the test conditions used by this access list statement. The test can be as simple as checking for a single source address, but usually test conditions are extended to include several test conditions. Use several global access list statements with the same identifier to stack several test conditions into a logical sequence or list of tests.

2. The **access-group** command activates access lists on an interface.

```
Router(config)#protocol access-group access-list-number
```

How to Identify Access Lists

Access lists can control most protocols on a Cisco router. Table 13–1 shows the protocols and number ranges of the access list types covered in this chapter.

Access List Type		Number Range/Identifier
IP	Standard	1–99
	Extended	100–199
		Named (Cisco IOS 11.2 and later)
IPX	Standard	800–899
	Extended	900–999
	SAP filters	1000–1099
		Named (Cisco IOS 11.2F and later)
AppleTalk		600–699

Table 13-1
Access list protocols and number ranges.

An administrator enters a number in the protocol number range as the first argument of the global access list statement. The router identifies which access list software to use based on this numbered entry. Access list test conditions follow as arguments. These arguments specify tests according to the rules of the given protocol suite. The meaning or validity of the standard and extended identification scheme for access lists varies by protocol.

CAUTION

Exceptions to the numbering classification scheme include AppleTalk and DECnet, where the same number range can identify various access list types. For the most part, number ranges do not overlap between different protocols.

Banyan VINES is the notable exception. Its number ranges 1 to 100 and 101 to 200 work with the single server and clients node group in each logical network. Numbers in these ranges will not conflict with IP access list numbers because the administrator uses a different command: **vines access-list**. More on this point appears in Appendix C.

Many access lists are possible for a protocol. Select a different number from the protocol number range for each new access list; however, the administrator can specify only one access list per protocol per interface.

Number ranges generally allow 100 different access lists per type of protocol. When a given 100-number range designates a standard access list, the rule is that the next 100-number range is for extended access lists for that protocol.

With Cisco IOS Release 11.2 and later, you can also identify a standard or extended IP access list with an alphanumeric string (name) instead of the current numeric (1 to 199) representation. This can be an easier identification method to administer. Named IP access lists provide other advantages, covered later in this chapter.

TCP/IP ACCESS LISTS

This section focuses on TCP/IP standard, extended, and named access lists. It concludes with three access list examples.

Testing Packets with IP Access Lists

For TCP/IP packet filters, Cisco IOS access lists check the packet and upper-layer headers, as shown in Figure 13–5.

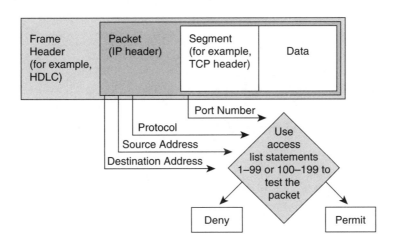

Figure 13–5
Access lists can check the packet's source address, destination address, or upper-layer port numbers.

For example, access lists can check the packet for:

- Source IP addresses using standard access lists; identify these with a number in the range 1 to 99

- Destination and source IP addresses or specific protocols using extended access lists; identify these with a number in the range 100 to 199

- Upper-level TCP or UDP port numbers in addition to the other tests in extended access lists; also identify these with a number in the range 100 to 199

For all these TCP/IP access lists, after a packet is checked for a match with the access list statement, it can be denied or permitted to use an interface in the access group.

Key Concepts for IP Access Lists

Create access lists using the normal global router configuration process.

Specifying an access list number from 1 to 99 instructs the router to accept standard IP access list statements. Specifying an access list number from 100 to 199 instructs the router to accept extended IP access list statements.

The administrator must carefully decide specific access controls logically and order the statements to achieve intended controls. Permitted protocols must be specified. All other TCP/IP protocols are denied.

Select which IP protocols to check. Any other IP protocols are not checked. Later in the procedure, the administrator can also specify an optional destination port for more granularity.

Address filtering occurs using access list address wildcard masking to identify how to check or ignore corresponding IP address bits (0=check, 1=ignore). The next section examines how to use wildcard mask bits in access lists.

How to Use Wildcard Mask Bits

IP access lists use wildcard masking to identify single or multiple IP addresses for permit or deny tests. A wildcard mask is paired with an IP address and uses the numbers 1 and 0 to identify how to treat the corresponding IP address bits (see Figure 13–6):

- A wildcard mask bit 0 means "check the corresponding bit value in the IP address."

- A wildcard mask bit 1 means "do not check (ignore) the corresponding bit value in the IP address."

Figure 13–6
Wildcard masking uses 1s and 0s to identify how to treat IP address bits.

128	64	32	16	8	4	2	1	Octet bit position and address value for bit
								Examples
0	0	0	0	0	0	0	0	= check all address bits (match all)
0	0	1	1	1	1	1	1	= ignore last 6 address bits
0	0	0	0	0	1	1	1	= ignore last 4 address bits
1	1	1	1	1	1	0	0	= check last 2 address bits
1	1	1	1	1	1	1	1	= do not check address (ignore bits in octet)

Although both are 32-bit quantities, wildcard masks and IP subnet masks operate differently. Recall that the zeros and ones in a subnet mask determine the network, subnet, and host portions of the corresponding IP address. The zeros and ones in a wildcard mask, as just noted, determine if the corresponding bits in the IP address should be checked or ignored for access list purposes.

The term *wildcard masking* is a nickname for the access list mask-bit-matching process. This nickname comes from an analogy of a wildcard that matches any other card in a poker game.

You've seen how the zero and one bits in an access list wildcard mask cause the access list to either check or ignore the corresponding bit in the IP address. In Figure 13–7, this wildcard masking process is applied.

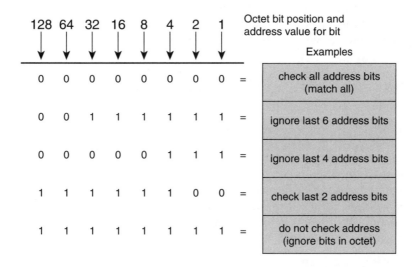

Figure 13–7
*Using wild-
card mask
0.0.15.255.*

An administrator wants to test an IP address for subnets that will be permitted or denied. Assume the IP address is Class B (the first two octets are the network number) with eight bits of subnetting (the third octet is for subnets). The administrator wants to use IP wildcard masking bits to match subnets 172.30.16.0 to 172.30.31.0. Here is how to use the wildcard mask to do this:

- To begin, the wildcard mask will check the first two octets (172.30) using corresponding zero bits in the wildcard mask.

- Because there is no interest in individual host addresses (a host ID will not be .00 at the end of the address), the wildcard mask will ignore the final octet using corresponding one bits in the wildcard mask.

- In the third octet, where the subnet address occurs, the wildcard mask will check that the bit position for the binary 16 is on and all the higher bits are off using corresponding zero bits in the wildcard mask.

  For the final (low end) four bits in this octet, the wildcard mask will ignore the value. In these positions, the address value can be binary 0 or binary 1, and the corresponding wildcard bits will be ones.

In this example, the address 172.30.16.0 with the wildcard mask 0.0.15.255 matches subnets 172.30.16.0 to 172.30.31.0. The wildcard mask will not match any other subnets.

Matching Any IP Address

Working with decimal representations of binary wildcard mask bits can be tedious. For the most common uses of wildcard masking, you can use abbreviation words. These abbreviation words reduce how many numbers an administrator will be required to enter while configuring address test conditions. One example you can use is an abbreviation instead of a long wildcard mask string when you want to match any address.

Consider a network administrator who wants to specify that any destination address will be permitted in an access list test, as shown in Figure 13–8. To indicate any IP address, the administrator would enter 0.0.0.0; then to indicate that the access list should ignore (allow without checking) any value, the corresponding wildcard mask bits for this address would be all ones (that is, 255.255.255.255).

The administrator can use the abbreviation **any** to communicate this same test condition to Cisco IOS access list software. Instead of typing 0.0.0.0 255.255.255.255, the administrator can use the word **any** by itself as the keyword.

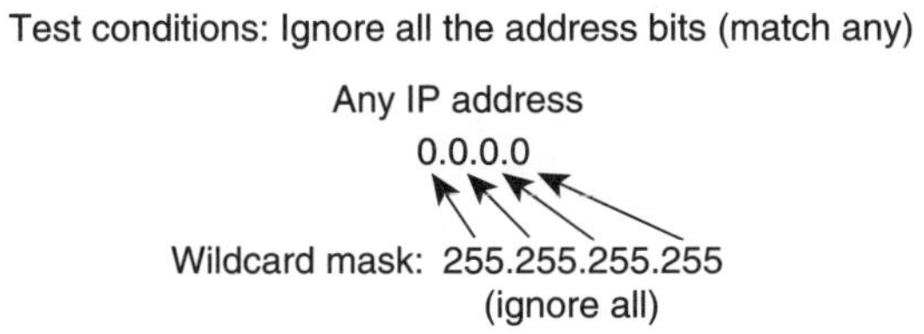

Test conditions: Ignore all the address bits (match any)

Any IP address
0.0.0.0

Wildcard mask: 255.255.255.255
(ignore all)

Figure 13–8
This abbrevia-tion means ignore any bit value in all bit positions, which has the effect of matching any-thing in all bit positions.

Matching a Specific IP Host Address

A second common condition in which Cisco IOS software will permit an abbreviation term in the extended access list wildcard mask is when the administrator wants to match all the bits of an entire IP host address, as shown in Figure 13–9.

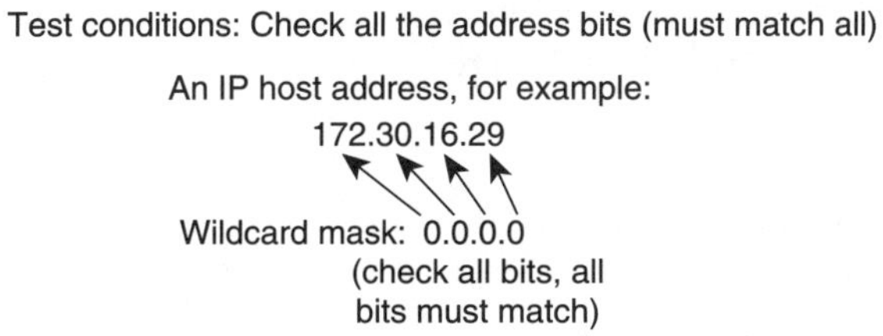

Test conditions: Check all the address bits (must match all)

An IP host address, for example:
172.30.16.29

Wildcard mask: 0.0.0.0
(check all bits, all
bits must match)

Figure 13–9
This abbrevia-tion means check the bit value in all bit positions, which has the effect of matching only the specified IP host address in all bit positions.

Consider a network administrator who wants to specify that a specific IP host address will be denied in an access list test. To indicate a host IP address, the administrator would enter the full address, for example, **172.30.16.29**. Then, to indicate that the access list should check all the bits in the address, the corresponding wildcard mask bits for this address would be all zeros; that is, 0.0.0.0.

The administrator can use the abbreviation **host** to communicate this same test condi-tion to Cisco IOS access list software. In the example, instead of typing **172.30.16.29 0.0.0.0**, the administrator can use the string **host 172.30.16.29.**

IP Standard Access Configuration

The **access-list** command creates an entry in a standard traffic filter list. The format of the command is:

```
Router(config)#access-list access-list-number {permit | deny} source
    [source-mask]
```

where the parameters and keywords have the following meanings:

- *access-list-number*—Identifies the list to which the entry belongs; a number from 1 to 99.

- **permit | deny**—Indicates if this entry allows or blocks traffic from the specified address.

- *source*—Identifies source IP address. Use the keyword **any** as an abbreviation for a source and source-wildcard of 0.0.0.0 255.255.255.255.

- *source-mask*—Identifies which bits in the address field are matched. It has a 1 in positions indicating "don't care" bits and a 0 in any position that is to be strictly followed.

The **ip access-group** command links an existing access list to an outbound interface. Only one access list per port per protocol per direction is allowed. The form of the command is:

```
Router(config)#ip access-group access-list-number {in | out}
```

where the parameters and key words have the following meanings:

- *access-list-number*—Indicates the number of the access list to be linked to this interface.

- **in | out**—Selects whether the access list is applied to the incoming or outgoing interface. If **in** or **out** is not specified, **out** is the default.

Example 1: Permit My Network Only

Figure 13–10 illustrates a situation in which the access list allows only traffic from source network 172.16.0.0 to be forwarded. Non-172.16.0.0 network traffic is blocked.

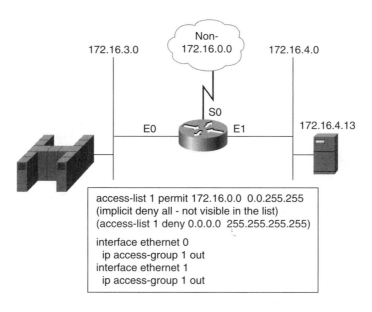

Figure 13–10
Permitting traffic from 172.16.0.0 to pass.

In the figure, the parameters of the **access-list** command are as follows:

- **1**—Access list number; indicates this is a standard list.
- **permit**—Traffic that matches selected parameters will be forwarded.
- **172.16.0.0**—IP address that will be used with the wildcard mask to identify the source network.
- **0.0.255.255**—Wildcard mask; 0s indicate positions that must match, and 1s indicate "don't care" positions.

Also in Figure 13–10, the command **ip access-group 1 out** links the access list 1 to an outgoing interface.

To remove an access list, first enter the **no access-group** command with all of its set parameters, and then enter the **no access-list** command with all of its set parameters.

Example 2: Deny a Specific Host

Figure 13–11 depicts a sample access list to block traffic from a specific address.

This access list is designed to block traffic from a specific address, **172.16.4.13**, and to allow all other traffic to be forwarded on interface Ethernet 0. The first **access-list**

Figure 13–11
*Denying a
specific host.*

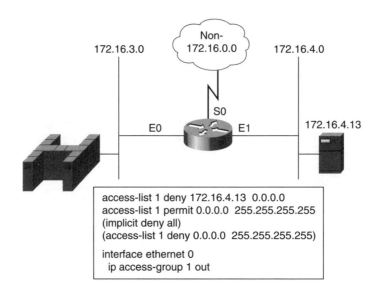

access-list 1 deny 172.16.4.13 0.0.0.0
access-list 1 permit 0.0.0.0 255.255.255.255
(implicit deny all)
(access-list 1 deny 0.0.0.0 255.255.255.255)

interface ethernet 0
  ip access-group 1 out

command uses the **deny** parameter to deny traffic from the identified host. The address mask **0.0.0.0** in this line requires the test to match all bits. If this mask is omitted, the router assumes an implicit mask.

In the second **access-list** command, the **0.0.0.0 255.255.255.255** IP address/wildcard mask combination identifies traffic from any source. This combination can also be written using the keyword **any**. All 0s in the address indicate a placeholder, and all 1s in the wildcard mask indicate that all 32 bits will not be checked in the source address.

Any packet that does not match the first line of the access list will match the second one and be forwarded.

Example 3: Deny a Specific Subnet

Figure 13–12 depicts an access list to block traffic from a specific subnet.

This access list is designed to block traffic from a specific subnet, **172.16.4.0**, and to allow all other traffic to be forwarded. Note the wildcard mask, **0.0.0.255**: 0s in the first three octets indicate those positions must match; the 255 in the last octet indicates a "don't care" condition. Note also that the **any** abbreviation has been used for the IP address of the source.

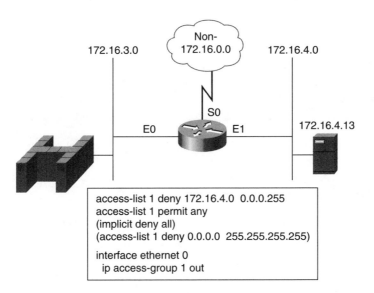

Figure 13–12
An access list blocking traffic from subnet 172.16.4.0.

172.16.3.0

Non-172.16.0.0

172.16.4.0

S0

E0   E1

172.16.4.13

```
access-list 1 deny 172.16.4.0  0.0.0.255
access-list 1 permit any
(implicit deny all)
(access-list 1 deny 0.0.0.0  255.255.255.255)

interface ethernet 0
  ip access-group 1 out
```

Extended IP Access Lists

The standard access list (numbered 1 to 99) may not provide the traffic-filtering control you need. Standard access lists filter based on a source address and mask; they permit or deny the entire IP protocol suite. You may need a more precise way to control traffic and access.

For more precise traffic-filtering control, use extended IP access lists. Extended IP access list statements check for source address and for destination address. In addition, at the end of the extended access list statement, you gain additional precision from a field that specifies the optional TCP or UDP protocol port number. These port numbers can be the well-known port numbers for TCP/IP. (See Table 2–1 to review some of the most common port numbers.)

You can specify the logical operation that the extended access list will perform on specific protocols. Extended access lists use a number from 100 to 199.

> ### TIPS
>
> For both standard and extended IP access lists, enter an address mask that identifies which bits in the address field you want the access list to match. For both types of access lists, the ip **access-group** command allows packet filtering into or out of the router.

Extended Access List Configuration

The **access-list** command creates an entry to express a condition statement in a complex filter. The complete format is:

```
Router(config)#access-list access-list-number {permit | deny} protocol
    source source-mask destination destination-mask [operator operand]
    [established]
```

where the parameters and keywords have the following meanings:

- *access-list-number*—Identifies the list using a number in the range 100 to 199

- **permit | deny**—Indicates whether this entry allows or blocks the specified address

- *protocol*—IP, TCP, UDP, ICMP, GRE, IGRP

- *source* and *destination*—Identifies source and destination IP addresses

- *source-mask* and *destination-mask*—Wildcard mask

- *operator* and *operand*—lt, gt, eq, neq (less than, greater than, equal, not equal), and a port number or name

- **established**—Allows TCP traffic to pass if packet uses an established connection (for example, has ACK bits set)

The **ip access-group** command links an existing extended access list to an outbound interface. The format of the command is:

```
Router(config)#ip access-group access-list-number {in | out}
```

where the parameters and keywords have the following meanings:

- *access-list-number*—Indicates the number of the access list to be linked to this interface.

- **in | out**—Selects whether the access list is applied to the incoming or outgoing interface. If **in** or **out** is not specified, **out** is the default.

Remember that only one access list per port per protocol is allowed.

Key Concept

Example 1: Deny FTP for E0

Figure 13–13 shows an extended access list that blocks FTP traffic.

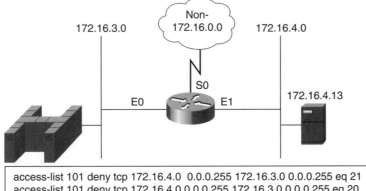

Figure 13–13
An access list for a specific upper-layer application.

```
access-list 101 deny tcp 172.16.4.0  0.0.0.255 172.16.3.0 0.0.0.255 eq 21
access-list 101 deny tcp 172.16.4.0 0.0.0.255 172.16.3.0 0.0.0.255 eq 20
access-list 101 permit ip 172.16.4.0 0.0.0.255 0.0.0.0 255.255.255.255
(implicit deny all)
(access-list 101 deny ip 0.0.0.0  255.255.255.255 0.0.0.0 255.255.255.255)

interface ethernet 0
 ip access-group 101 out
```

The **permit** statement allows traffic from subnet 172.16.4.0 to be forwarded to all other networks or subnetworks via interface E0. The rest of the parameters of the access list in this example have the following meanings:

- **101**—This access list number indicates an extended IP access list.
- **deny**—Traffic that matches selected parameters will be blocked.
- **tcp**—Transport-layer protocol.
- **172.16.4.0** and **0.0.0.255**—Source IP address and mask; the first three octets must match, but the last octet consists of "do not care" bits.
- **172.16.3.0** and **0.0.0.255**—Destination IP address and mask; the first three octets must match, but the last octet consists of "do not care" bits.
- **eq 21**—Specifies the well-known port number for FTP.
- **eq 20**—Specifies the well-known port number for FTP data.

The effect of the **interface E0 access-group 101** command is to link access list 101 to outgoing port interface E0.

Example 2: Deny Only Telnet out of E0; Permit All Other Traffic

Figure 13–14 shows another extended access list.

Figure 13–14
An extended access list can deny specific protocol traffic from one source.

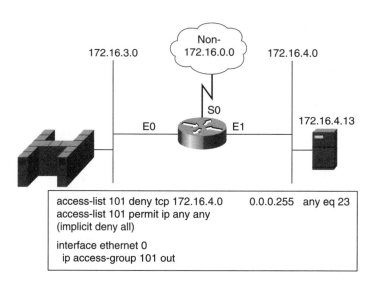

```
access-list 101 deny tcp 172.16.4.0     0.0.0.255  any eq 23
access-list 101 permit ip any any
(implicit deny all)

interface ethernet 0
  ip access-group 101 out
```

This example denies Telnet traffic (**eq 23**) from 172.16.4.0 being sent out interface E0. All IP traffic from any other source to any destination is permitted, as indicated by the keywords **any any**. Interface E0 is configured with the **ip access-group 101 out** command; that is, access list 101 is linked to outgoing port interface E0.

Named Access Lists

Named access lists allow IP standard and extended access lists to be identified with an alphanumeric string (name) instead of the current numeric (1 to 199) representation.

With prior numbered IP access list statements, an administrator wanting to alter an access list first would be required to delete all the statements in the numbered access list. This deletion uses the word **no** preceding each statement.

Named IP access lists can be used to delete individual entries from a specific access list. Deleting individual entries enables you to modify your access lists without deleting and then reconfiguring them. Use named IP access lists when:

- You want to identify access lists intuitively using an alphanumeric name.

- You have more than 99 simple and 100 extended access control lists to be configured in a router for a given protocol.

Consider the following before implementing named IP access lists:

- Named IP access lists are not compatible with Cisco IOS releases prior to Release 11.2.

- You cannot use the same name for multiple access lists. In addition, access lists of different types cannot have the same name. For example, it is illegal to specify a standard access control list named "George" and an extended access control list with the same name.

Most of the commonly used IP access list commands accept named IP access lists.

The following commands can be used to define named IP address lists. To name the access list, use the following command:

```
Router(config)#ip access-list {standard ¦ extended} name
```

In access-list configuration mode, specify one or more conditions allowed or denied. This determines if the packet is passed or dropped:

```
Router(config (std- ¦ ext-)nacl)#
deny {source [source-wildcard] ¦ any}
```

or

```
permit {source [source-wildcard] ¦ any}
```

The following configuration creates a standard access list named Internet_filter and an extended access list named marketing_group:

```
interface Ethernet0/5
ip address 2.0.5.1 255.255.255.0
ip access-group Internet_filter out
ip access-group marketing_group in
...
ip access-list standard Internet_filter
permit 1.2.3.4
deny any
ip access-list extended marketing_group
permit tcp any 171.69.0.0 0.0.255.255 eq telnet
deny tcp any any
permit icmp any any
deny udp any 171.69.0.0 0.0.255.255 lt 1024
deny ip any any log
```

To activate the named access list on an interface, use the following command:

```
Router(config-if)#ip access-group {name ¦ 1-199 {in ¦ out}}
```

The following is a configuration output example:

```
ip access-list extended come_on
permit tcp any 171.69.0.0 0.0.255.255 eq telnet
deny tcp any any
permit icmp any any
deny udp any 171.69.0.0 0.0.255.255 lt 1024
deny ip any any
interface Ethernet0/5
ip address 2.0.5.1 255.255.255.0
ip access-group over_and out
ip access-group come_on in
ip access-list standard over_and
permit 1.2.3.4
deny any
```

Where to Place IP Access Lists

Access lists are used to control traffic by filtering and eliminating unwanted packets. Where the administrator places an access list statement can reduce unnecessary traffic. Traffic that will be denied at a remote destination should not use network resources along the route to that destination.

Suppose an enterprise's policy aims at denying Token Ring traffic on router A to the switched Ethernet LAN on router D's E1 port, as shown in Figure 13–15. At the same time, other traffic must be permitted. Several approaches can accomplish this policy.

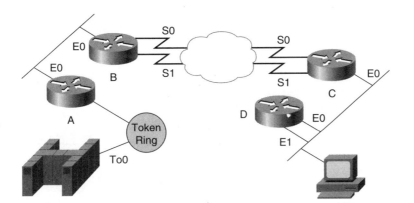

Figure 13–15
Place standard access lists close to the destination; place extended access lists close to the source.

The recommended approach uses an extended access list. It specifies both source and destination addresses. Place this extended access list in router A. Then packets do not cross router A's Ethernet, do not cross the serial interfaces of routers B and C, and do not enter router D. Traffic with different source and destination addresses can still be permitted.

The rule, with extended access lists, is to put the extended access list as close as possible to the source of the traffic denied.

Standard access lists do not specify destination addresses. The administrator would have to put the standard access list as near the destination as possible. For example, in Figure 13–15, place an access list on E0 of router D to prevent traffic from router A.

Verifying Access Lists

The **show ip interface** command displays IP interface information and indicates if any access lists are set. Figure 13–16 shows the result of the **show ip interface** command.

Figure 13–16
*This **show ip interface** output indicates that only an outgoing access list is set.*

```
Router#show ip interface

Ethernet 0 is up, line protocol is up
        Internet address is 192.54.222.2, subnet mask is 255.255.255.0
        Broadcast address is 255.255.255.255
        Address determined by non-volatile memory
        MTU is 1500 bytes
        Helper address is 192.52.71.4
        Secondary address 131.192.115.2, subnet mask 255.255.255.0
        Outgoing access list 10 is set
        Inbound access list is not set
        Proxy APR is enabled
        Security level is default
        Split horizon is enabled
        ICMP redirects are always sent
        ICMP unreachables are always sent
        ICMP mask replies are never sent
        IP fast switching is enabled
        Gateway Discovery is disabled
        IP accounting is disabled
        TCP/IP header compression is disabled
        Probe proxy name replies are disabled
Router#
```

Monitoring Access List Statements

The **show access-lists** command displays the contents of all access lists, as shown in Figure 13–17. This Cisco IOS command provides more details about the access list statements. By entering the access list name or number as an option for this command, you can see a specific list.

The highlighted line points out a deny statement. If you explicitly type a deny statement, it is not dynamic and will appear in the access list. However, if you use the "implicit deny" feature, the deny statement does not show up in the access list.

You can also see access lists in the Cisco IOS commands **show running-config** and **show startup-config** (Cisco IOS Release 10.3 and later). However, with these other commands, the access list details are interspersed among all the other statements of the configuration file.

NOVELL IPX ACCESS LISTS

In this section, you learn how you can block or permit specific IPX traffic through a Cisco router. Start with testing packets using access lists.

```
Router#> show access-list
Standard IP access list 19
      permit  172.16.19.0
      deny   0.0.0.0. wildcard bits 255.255.255.255
Standard IP access list 49
      permit 172.16.31.0  wildcard bits 0.0.0.255
      permit 172.16.194.0  wildcard bits 0.0.0.255
      permit 172.16.195.0  wildcard bits 0.0.0.255
      permit 172.16.195.0  wildcard bits 0.0.0.255
      permit 172.16.196.0  wildcard bits 0.0.0.255
      permit 172.16.197.0  wildcard bits 0.0.0.255
Extended IP access list 101
      permit tcp 0.0.0.0 255.255.255.255 0.0.0.0 255.255.255.255  eq 23
Type code access list 201
      permit 0x6001 0x0000
Type code access list 202
      permit 0x6004 0x0000
      deny   0x0000 0xFFFF
Router>
```

Figure 13–17
*Use the **show access-lists** command to view all blocked or permitted traffic.*

Testing Packets Using Novell Access Lists

For the Novell IPX packet filters covered in this chapter, Cisco IOS access lists check the packet header for:

- Destination and source IPX addresses using standard access lists; identify these lists with a number in the range 800 to 899, as shown in Figure 13–18.

- Service advertisement numbers in addition to the other tests in SAP filter access lists; identify these lists with a number in the range 1000 to 1099.

For all these Novell IPX access lists, after a packet is checked for a match with the access list statement, it can be denied or permitted use of an interface in the access group. (Cisco IOS software offers several other forms of access lists for Novell IPX packets. Refer to www.cisco.com for more information.)

Key Concepts for IPX Access Lists

Novell addressing is based on the *network.node.socket* format. The network number is assigned by the administrator; the node portion is derived from the MAC address of the individual interface. Serial lines adopt the MAC address of another interface in the creation of their logical addresses. The socket number refers to a process or application (somewhat like the TCP segment).

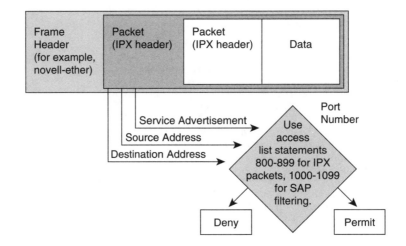

For example, AABB0001.00001B03AC33.0452 can be interpreted as follows:

- AABB0001 is the network address.
- 00001B03AC33 is the node address.
- 0452 is the socket number.

Every NetWare file server has an internal IPX network number and performs IPX routing. External IPX networks attach to router interfaces. The IPX network number assigned on a Cisco router's interface must be unique and consistent with the network numbers known to the file server.

IPX standard access lists use numbers in the range 800 to 899. These access lists check for either source address or both source and destination addresses. To identify parts of the address to check or ignore, IPX standard access lists use a wildcard mask that operates like the mask used with IP addresses.

To control the traffic from the SAP, use SAP filters that use numbers in the range 1000 to 1099. Several other packet and route filters can help manage IPX overhead traffic. For example, access lists can control Get Nearest Server (GNS) from clients to servers, Routing Information Protocol (RIP), and NetWare Link Services Protocol (NLSP).

Extended IPX access lists, on the other hand, filter on specific protocols or designated sockets. Filtering IPX sockets use the access lists numbered in the range 900 to 999.

Controlling IPX Overhead

IPX routing and advertising processes were developed to run on LANs. As LANs interconnect with slower, more costly WAN links, as in Figure 13–19, overhead from IPX control packets can reduce the bandwidth available for user applications traffic. IPX servers broadcast service advertising (SAPs) details every 60 seconds.

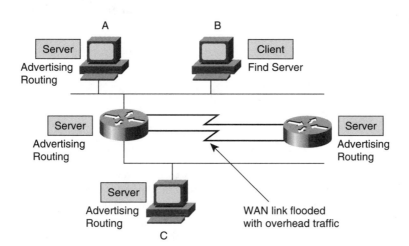

Figure 13–19
Frequent updates reduce the bandwidth available for user traffic.

Routers broadcast routing information and metrics to other IPX routers. Figure 13–19 shows four IPX networks and multiple servers that advertise routes and services.

When a client workstation starts up, it sends its own SAP broadcast to find a server; then from the nearest server, the client can log in to a target server and run network applications from network drives.

When packets from these protocols are unwanted, a network administrator can set up IPX access lists. With the standard access lists in this chapter, the permit/deny filtering acts on all IPX packets for the interface addresses.

IPX Standard Access List Configuration

Use the **access-list** command to filter traffic in an IPX network. Using filters on the out-going router interface allows or restricts different protocols and applications on individual networks. The complete form of the **access-list** command is:

```
Router(config)#access-list access-list-number {deny | permit} source-
network [.source-node][source-node-mask][destination-network]
[.destination-node][destination-node-mask]
```

where the parameters and keywords have the following meanings:

- *access-list-number*—Number of the access list. This is a decimal number from 800 to 899.

- **deny**—Denies access if the conditions are matched.

- **permit**—Permits access if the conditions are matched.

- *source-network*—Number of the network from which the packet is being sent. This is an eight-digit hexadecimal number that uniquely identifies a network cable segment. It can be a number in the range 1 to FFFFFFFE. A network number of 0 matches the local network. A network number of –1 matches all networks. You need not specify leading zeros in the network number; for example, for the network number 000000AA, you can enter just AA.

- *source-node* (Optional)—Node on source network from which the packet is being sent. This is a 48-bit value represented by a dotted triplet of four-digit hexadecimal numbers (xxxx.xxxx.xxxx).

- *source-node-mask* (Optional)—Mask to be applied to source node. This is a 48-bit value represented as a dotted triplet of four-digit hexadecimal numbers (xxxx.xxxx.xxxx). Place 1s in the bit positions you want to mask.

- *destination-network* (Optional)—Number of the network to which the packet is being sent. This is an eight-digit hexadecimal number that uniquely identifies a network cable segment. It can be a number in the range 1 to FFFFFFFE. A network number of 0 matches the local network. A network number of –1 matches all networks. You need not specify leading zeros in the network number; for example, for the network number 000000AA, you can enter just AA.

- *destination-node* (Optional)—Node on destination network to which the packet is being sent. This is a 48-bit value represented by a dotted triplet of four-digit hexadecimal numbers (xxxx.xxxx.xxxx).

- *destination-node-mask* (Optional)—Mask to be applied to destination node. This is a 48-bit value represented as a dotted triplet of four-digit hexadecimal numbers (xxxx.xxxx.xxxx). Place 1s in the bit positions you want to mask.

Use the **ipx access-group** command to link an IPX traffic filter to an interface, as follows:

```
Router(config-if)#ipx access-group access-list-number [in ¦ out]
```

where *access-list-number* is a specific IPX filter list from 800 to 899.

Standard IPX Access List Example

Figure 13–20 shows an access list permitting IPX traffic from network 2b destined for network 4d to be forwarded out Ethernet0.

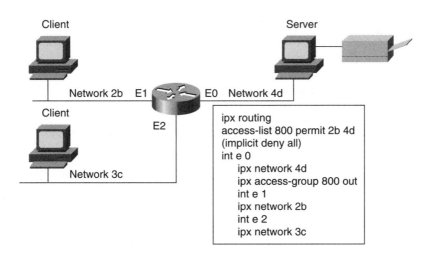

Figure 13–20
A sample standard access list for IPX.

Details of the **access-list 800** command in Figure 13–21 are as follows:

- **800**—Specifies a Novell IPX standard access list
- **permit**—Traffic matching the selected parameters will be forwarded
- **2b**—Source network number

- **4d**—Destination network number

- **(implicit deny all)**—Not a valid configuration command, just a reminder that access lists filter traffic not specified to be forwarded

Interface E0 is configured with the **ipx access-group 800 out** command; that is, access list 800 is linked to outgoing interface E0.

The access list is applied to an outgoing interface and filters outbound packets. Notice that the other interfaces, E1 and E2, are not subject to the access list; they lack the access group statement to link them to the access list 800.

IPX Extended Access List Configuration

To define an extended Novell IPX access list, use the extended version of the access list global configuration command. Extended IPX access lists filter on protocol type with all other parameters being optional. For some versions of NetWare, the protocol type field is not a reliable indicator of the type of packet encapsulated by the IPX header. In these cases, use the source and destination socket fields to make this determination.

The complete form of the IPX extended access list command is:

```
Router(config)#access-list access-list-number {deny | permit} novell-
    protocol source-network.[source-address [source-mask]] source-
    socket] destination-network.[destination-address [destination-mask]]
    destination-socket][log]
```

Some of the parameters shown in this access list were covered earlier in the discussion of standard access lists. Following are the meanings of the new parameters and keywords:

- *access-list-number*—Number of the access list. This is a decimal number from 900 to 999.

- *protocol*—Name or number (decimal) of an IPX protocol type. This is sometimes referred to as the packet type.

- *source-socket*—Socket name or number (hexadecimal) from which the packet is being sent

- *destination-socket* (optional)—Socket name or number (hexadecimal) to which the packet is being sent.

- **log** (optional)—Logs IPX access control list violations whenever a packet matches a particular access list entry. The information logged includes source address, destination address, source socket, destination socket, protocol type, and action taken (permit/deny).

Use the following command to activate the IPX extended access list on an interface:

```
Router(config-if)#ipx access-group access-list-number [in ¦ out]
```

Normal IPX SAP Operation

SAP broadcasts synchronize the list of available services. The NetWare file server acts like an IPX router. The Cisco router acts like a SAP server.

If the router passed a SAP every time it received one, the WAN link would be flooded with SAP traffic. The router will not forward SAP broadcasts.

Instead, both file servers and routers listen to SAP messages and build a SAP table, as shown in Figure 13–21. All devices that build SAP tables advertise this information every 60 seconds.

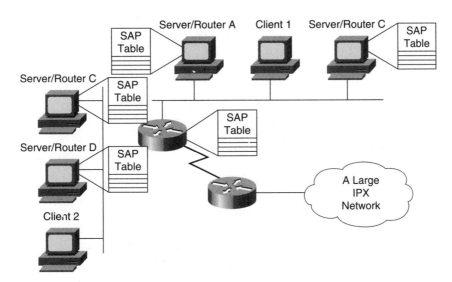

Figure 13–21
In normal IPX SAP operations, the router does not forward SAP broadcasts; instead, devices build SAP tables and advertise the information.

Advertising SAP tables can still result in considerable overhead, because all these servers and routers send their own complete SAP table every 60 seconds. Over a WAN link (Cisco to Cisco), the 60-second SAP interval can be changed. However, other IPX devices may not tolerate the interval change. For interfacing with these other devices, you may need to leave the SAP interval at 60 seconds.

To change the SAP update interval, use the command:

```
ipx sap-interval interval
```

where the *interval* between SAP updates sent by the router is specified in minutes. The default value is 1 minute. If interval is 0, periodic updates are never sent.

How to Use SAP Filters

You must carefully plan for SAP filtering before configuring it. Make sure that all clients will see advertisements necessary for their application processing. You will need to enter the SAP filters in all routers in which you want them to operate. Table 13–2 lists the most common SAP numbers.

SAP Number	Server Type
4	NetWare file server
7	Print server
278	NetWare directory server
Ø	All Services

Table 13–2 *Commonly used SAP numbers.*

Place SAP filters close to the source. Proper placement of SAP filters conserves critical bandwidth, especially on serial links.

The following list provides an example of SAP filter goals:

> deny type 7 (print server) SAP from 2a
> deny type 98 (access server) SAP from 5b
> deny type 24 (router) SAP to 7c
> deny type 4 (file server) SAP from 4a
> deny type 26a (NMS)
> deny type 7a (NetWare for VMS) from *8

permit the remaining SAPs

When a SAP advertisement arrives at the router interface, the contents are placed in the SAP table portion of main memory. The contents of the table are propagated during the next SAP update.

Figure 13–22 shows the two types of access list filters that control SAP traffic, listed here:

- IPX input SAP filter. When a SAP input filter is in place, the number of services entered into the SAP table are reduced. The propagated SAP updates represent the entire table but contain only a subset of all services. Use this type of filter when you want to reduce the size of the SAP table.

- IPX output SAP filter. When a SAP output filter is in place, the number of services propagated from the table is reduced. The propagated SAP updates represent a portion of the table contents and are a subset of all the known services. When you use this filter, your router still contains a list of all the known services.

Input filter: Do not add filtered SAPs to SAP table

Output filter: Do not add filtered SAPs to the SAP table sent

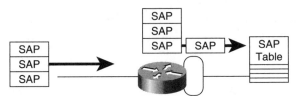

Figure 13–22
You can apply the access list to the interface as an input or output SAP filter.

SAP Filter Configuration Commands

Use the **access-list** command to control propagation of the SAP messages, as follows:

```
Router(config)#access-list access-list-number {deny ¦ permit} network
    [.node] [network-mask node-mask][service-type [server-name]]
```

where the parameters and keywords have the following meanings:

- *access-list-number*—Number from 1000 to 1099. Indicates a SAP filter list.
- *network* [.node]—Novell source internal network number with optional node number; −1 is all networks.
- *network-mask node-mask*—Mask to be applied to the network and node. Place 1s in the positions to be masked.
- *service-type*—SAP service type to filter. Each SAP service type is identified by a hexadecimal number. See Table 13–2 for some commonly used service types.
- *server-name*—Name of the server providing the specified service type.

The **ipx input-sap-filter** and **ipx output-sap-filter** commands place a SAP filter on an interface. The use of **input** or **output** determines if SAPs are filtered before entry into the SAP table or if the SAP table contents are filtered during the next update:

```
Router(config-if)#ipx output-sap-filter access-list-number
Router(config-if)#ipx input-sap-filter access-list-number
```

Why would you use one type of filter rather than the other? Consider, for example, a router that connects to two Ethernet networks (E0 and E1) and two serial links (S0 and S1). Suppose you want the devices on the LAN ports (E0 and E1) to know about local services, but you do not want these same services advertised over the serial links. In this case, you would use a SAP output filter on S0 and S1.

SAP table content can be filtered on input by using the **ipx router-sap-filter** command, which identifies from which router SAP advertisements can be received. Note that only one of the three kinds of SAP filters can be active on an interface at any one time.

Example 1: SAP Filtering on Output

In Figure 13–23, the network administrator wants to prevent file server advertisements from Server C from being forwarded on interface Serial 0 (S0). All other SAP services from any source should be forwarded on interface S0. The configuration to accomplish this is shown in Figure 13–23.

Details of the **access-list 1000 deny** command in Figure 13–23 are as follows:

- **1000**—An access list number in the Novell SAP filter range
- **deny**—SAP services matching selected parameters will be blocked

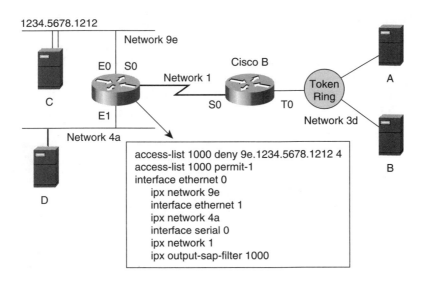

Figure 13–23
*SAP advertise-
ments are fil-
tered through
the access list.*

- **9e.1234.5678.1212**—Source network address of SAP advertisement

- **4**—Type of SAP service; advertises file service

Details of the **access-list 1000 permit –1** command are as follows:

- **1000**—Access list number

- **permit**—SAP services matching parameters will be forwarded

- **–1**—Source network number; –1 means all networks

The **ipx output-sap-filter 1000** command places list 1000 on interface S0 as an output
SAP filter. By applying the list as an outgoing interface, you can restrict the advertisements onto the serial link. It is preferable to restrict traffic at the point closest to the
source.

Server C in Figure 13–23 is in the upper-left corner. Specifically, network 9e is the source
with a MAC address ending in 1212. The complete identifier for the server is
9e1234.5678.1212. The value 4 at the end of the first line of the access list identifies the
type of service. This value indicates a file server SAP advertisement. So the complete line
denies file service advertisements, from server 9e.1234.5678.1212, out of the router.

The **access-list 1000 permit –1** command allows all other advertisements to propagate
onto the serial line.

Example 2: SAP Filtering on Input

In Figure 13–24, print server advertisements from Servers A and B will not be entered into the SAP table. All other SAP services from any source will be added into the SAP table. Note that the first line of the access list specifies service type 7, print services. Note also the **ipx input-sap-filter 1001** command, which places list 1001 on interface serial 0 as an input SAP filter.

Figure 13–24
This access list will not allow print server information from servers A and B into the local SAP table.

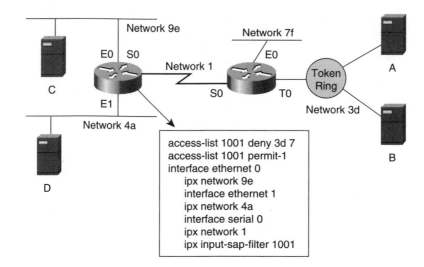

Verifying IPX Access Lists

The **show ipx interface** command displays information about the configuration of the interface. Figure 13–25 shows that the input filter list is 800, and the output filter list is 801.

Also in Figure 13–25, the **show access-lists** command displays the contents of lists 800 and 801.

APPLETALK ACCESS LISTS

At this point, you have learned about TCP/IP and Novell access lists. This section covers AppleTalk access lists and how they differ from the previous two. Start with testing packets using AppleTalk access lists.

```
Router#show ipx interface1/1

Ethernet 1/1 is up, line protocol is up
   IPX addess is 10.0000.0c0d.724f, NOVELL-ETHER [up]
   Delay of this IPX network, in ticks is 1 through 0 link delay 0
   IPXWAN processing not enabled on this interface.
   IPX SAP update interval is 1 minute(s)
   IPX type 20 propagation packet forwarding is disabled
   Incoming access list is not set
   Outgoing access list is not set
   IPX helper access list is not set
   SAP GNS processing enabled, delay 0 ms, output filter list is not set
   SAP Input filter list is not set
   Sap Output filter list is not set
   SAP Router filter list is not set
   Input filter list is 800
   Output filter list is 801                    •
                                                •
                                                •
```

```
dtp-19#show access-lists
IPX access list 800
   deny 8000
IPX access list 801
   deny all
```

Figure 13–25
*Always verify
IPX access
lists after you
configure
them.*

Testing Packets Using AppleTalk Access Lists

As shown in Figure 13–26, Cisco IOS access lists check the AppleTalk packet header for:

- Cable range or network numbers; identify these with a number in the range 600 to 699.

- Zone Information Protocol (ZIP) replies; identify these also with a number in the range 600 to 699. (This type of access list is referred to more specifically as a zip-reply-filter access list.)

After a packet is checked for a match with the access list statement, it can be denied or permitted use of an interface in the access group. Note that Cisco IOS software offers several other forms of access lists for AppleTalk packets that will not be covered in this chapter.

Figure 13–26
Filtering AppleTalk traffic.

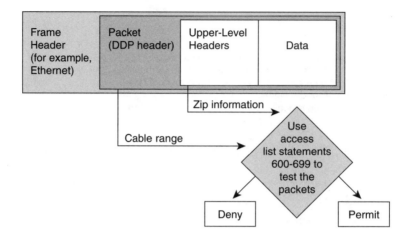

AppleTalk Network Structures

It has become commonplace for routed AppleTalk networks to evolve into complex internetworks. As growth extends across LANs and serial lines, access list controls involve several AppleTalk network structures.

The first network structure is the grouping of networks and their resources into zones. These *zones* are arbitrary subsets of nodes within the AppleTalk internetwork. In Figure 13–27, one zone, called Users, contains a separate group of resources from those in zones Bldg-D 1st floor and Bldg-13.

Current AppleTalk internetworks use extended network addresses. For example, an Ethernet transmission medium in zone Users can contain networks in the contiguous network number range of 200 to 205. Older internetworks continue using nonextended network addressing such as 130 in Bldg-D 1st floor. An extended AppleTalk network is a physical-network segment that can be assigned multiple network numbers. A nonextended AppleTalk network can contain only one network number per wire.

The user's application sends output to the print manager. For network access, the routing tables and zone information helps direct the user's output from its source to the destination zone containing the selected printer. The administrator can use access lists to control traffic based on network and cable-range selections.

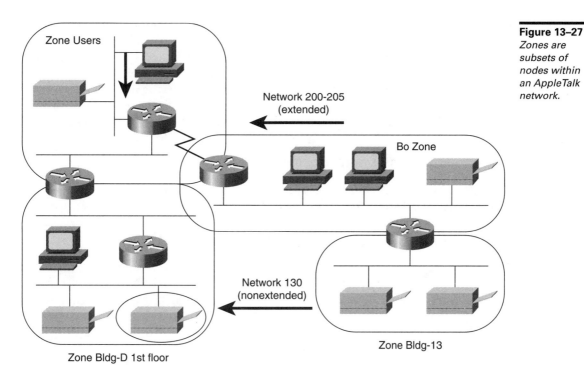

Figure 13–27
Zones are subsets of nodes within an AppleTalk network.

Key Concepts for AppleTalk Access Lists

A key AppleTalk concept hides network numbering from end users. End users may see zones and resources, but numeric configuration is a hidden issue for the network administrator.

Administrators can use AppleTalk filters to control traffic by referring to the 16-bit network number portion of a full 24-bit address. Because the node portion is dynamically assigned as AppleTalk nodes come up, these node numbers are not predictable for access list entries.

Although earlier AppleTalk networks offered a single nonextended network on a single medium, current AppleTalk uses extended addressing, which means that more than one AppleTalk network can occupy the same physical media. Express one or more AppleTalk networks on the medium as the cable range.

An administrator can filter an entire cable range. ZIP filters are one method for reducing AppleTalk zone information update distribution traffic.

Using the **access-list within cable range** command allows an administrator to select portions of a cable range for access list testing. One use of a partial cable range is when an AppleTalk administrator establishes a broad cable range for an interface to a location (say, a remote regional office), and then wants to identify subsets of the cable range for the various departments at the regional office. The administrator specifies access list filters appropriate to the different departments. Then access to the interface can be permitted or denied within the cable range subsets appropriate for each of the departments.

AppleTalk Access List Procedures

To configure for AppleTalk number access lists, select a unique access list number from within the range 600 to 699. As with the other protocols, the AppleTalk access list statement requires a permit or deny in each statement to specify traffic controls to potential outgoing interfaces.

With phase 1 addressing, specify a single network number such as network 130. More commonly, specify phase 2 addresses by entering a cable range such as 100 to 105.

As another alternative, the administrator can specify AppleTalk networks from within a partial cable range. For example, consider Figure 13–28. An access list statement can target AppleTalk networks 201 to 204 from within the complete cable range of 200 to 205 with configuration such as the following:

```
appletalk routing
access-list 601 deny within cable-range 201-204
access-list 601 permit within cable-range 204-205
interface ethernet 0
appletalk access-group 601
```

Figure 13–28
Packets can be filtered on a "per-interface" basis.

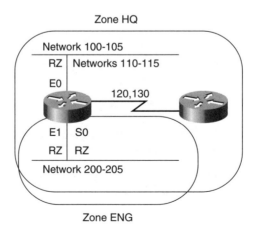

Zone HQ

Network 100-105
RZ | Networks 110-115
E0
120,130
E1 | S0
RZ | RZ
Network 200-205

Zone ENG

As with other access lists, an implicit deny performs the last test of the access list. In AppleTalk, the default is to deny all other network access. The **appletalk access-group** command is used to apply the AppleTalk access lists to one or more interfaces. In addition to the normal implicit deny, the AppleTalk-specific **access-list other-access** command may define checks for networks or cable ranges not explicitly tested in previous statements.

By filtering networks, the administrator can permit or prevent data packets and routing update packets on the specified interface. Routing updates use the AppleTalk Layer 3 protocol Routing Table Maintenance Protocol (RTMP).

Zone filtering is a process used in large AppleTalk networks to keep ZIP traffic to a minimum. These access lists focus on GetZoneList (GZL) packets. GZL packets are sent by a node to obtain a list of all the zones in the internetwork. The administrator must use separate access list statements for zone filtering.

AppleTalk Access List Commands

The **access-list** command permits or denies an entire cable range, as follows:

```
Router(config)#access-list number {permit | deny} cable-range cable-
    range
```

The **access-list other-access** command defines the default action (permit or deny) to take for other networks or cable ranges, as follows:

```
Router(config)#access-list number {permit | deny} other-access
```

The **appletalk access-group** command links the access list to one or more specified interfaces, as follows:

```
Router(config)#appletalk access-group access-list-number
```

For nonextended AppleTalk networks, use the following **access-list** command:

```
Router(config)#access-list number {permit | deny} network network-
    number
```

In Figure 13–29, interface E1 connects to a cable that supports networks 100–105. The network administrator wants to allow some but not all of these networks to access interface E0. The access list shows configuration statements in router A.

The number of the access list (601) indicates that it is an AppleTalk access list. The first line of the access list denies traffic within cable range 100–102 from being forwarded. The second line of the access list permits traffic within the cable range 103–105 to be forwarded.

The command **appletalk access-group 601** applies list 601 to interface E0 as a cable-range filter for AppleTalk networks.

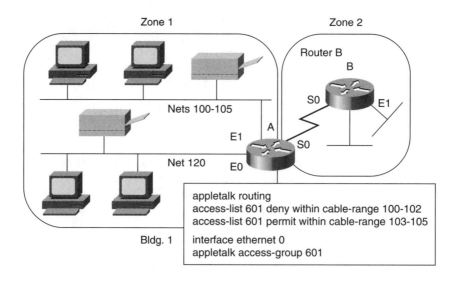

Figure 13–29
*Router A's
access list per-
mits or denies
access to E0
for traffic from
cable range
100–105.*

ZIP Reply Filter Configuration

Zone Information Protocol (ZIP) maintains network-number-to-zone-name mappings in zone information tables (ZITs). Routers learn about zones by exchanging their zone information tables. As noted earlier, ZIP filtering can be done to keep ZIP traffic to a minimum.

Use the **access-list zone** command to create an entry in the zone filter. It must use an access list number in the number range 600 to 699:

```
Router(config)#access-list number {permit ¦ deny} zone zone-name
Router(config)#access-list number {permit ¦ deny} additional-zones
```

where *zone-name* is the name assigned to the zone being filtered, and *additional-zones* specifies the action to take for all other zones not specified in the **access-list zone** command.

Use the **appletalk zip-reply-filter** command to assign the access list to an incoming interface:

```
Router(config)#appletalk zip-reply-filter access-list number
```

The *zip-reply-filter* parameter limits the zones that are visible from the router by other AppleTalk routers.

Verifying AppleTalk Access Lists

Use the **show appletalk access-lists** command to display the access lists that are set up for AppleTalk. For example, the output in Figure 13–30 verifies the access lists that permit zone information from Zone A to Zone B.

```
Router>show appletalk access-lists

AppleTalk access list 601:
    permit zone ZoneA
    permit zone ZoneB
    deny additional-zones
    permit network 55
    permit network 500
    permit cable-range 900-950
    deny includes 970-990
    permit within 991-995
    deny other-access
```

Figure 13–30
Use the **show appletalk access-lists** *command to verify your access lists after configuring them.*

SUMMARY

In review, the following access lists were covered in this chapter:

- IP standard access lists (1–99)

- IP extended access lists (100–199)

- New for Cisco IOS Release 11.2, named IP access lists

- Novell IPX standard access lists (800–899)

- Novell SAP filter access lists (1000–1099)

- AppleTalk standard access lists (600–699)

- AppleTalk zone reply filter access lists (also 600–699)

In this chapter, you've focused on how access lists are used to implement security and reduce unnecessary network traffic. You have looked primarily at TCP/IP, IPX/SPX, and AppleTalk access lists. You learned how to test, configure, and verify these three kinds of access lists. Refer to Appendixes B and C for DECnet and VINES information. Chapter 14, "Introduction to WAN Connections," covers WAN services using serial links and the Point-to-Point Protocol (PPP).

Chapter Thirteen Test
Basic Traffic Management with Access Lists

Estimated Time: 15 minutes

Complete all the exercises to test your knowledge of the materials contained in this chapter. Answers are listed in Appendix A, "Chapter Test Answer Key."

Question 13.1

What command do you issue to verify that no prior access lists exist on a router?

_____

Question 13.2

What command activates the IP extended access list on a router interface?

_____

Question 13.3

What command mode must the router be in before you can issue the command in Question 13.2?

_____

Questions 13.4

What command do you issue to verify that a new IP access list is active on an interface?

_____

Question 13.5

What command do you issue to display the contents of an access list?

_____

Question 13.6

What range of list numbers must you use to define a standard IPX access list?

_____

Question 13.7

What command do you issue to verify that a new IPX access list is now active on the interface?

_____

Question 13.8

What command do you issue to display the contents of an IPX access list?

_____

Question 13.9

What range of list numbers must you use to define an AppleTalk access list?

_____

Question 13.10

What command do you issue to display the contents of an AppleTalk access list?

_____

PART 3

Wide-Area Networking

Introduction to WAN Connections

This chapter discusses how WANs are set up, how a user subscribes to phone services for the network, and what a WAN frame looks like. It also presents the Point-to-Point Protocol (PPP).

WAN S ERVICES

A WAN is different from a LAN. With a WAN, you must subscribe to an outside WAN provider to use network resources that your organization does not own. Basic telephone service is the most commonly used WAN service. Telephone service and data service routed from the customer premises interface with the service provider's cloud at a central office (CO), as shown in Figure 14–1.

An overview of the WAN cloud organizes WAN provider services into three main types:

- Call setup service
- Time-division multiplexing (TDM)
- X.25 or Frame Relay service

Call setup service sets up and clears connections between endpoints. For example, when a branch office needs to exchange data with the corporate headquarters office, the call setup process establishes a connection between the routers on each end of the WAN

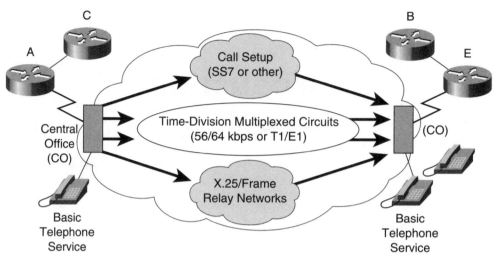

link. Once the connection has been established, the link can be used to transfer data across.

In TDM, information from multiple sources has bandwidth allocation on a single media. Circuit switching uses signaling to determine the call route, which is a dedicated path between the sender and the receiver. By multiplexing traffic into fixed time slots, TDM avoids congested facilities and variable delays. Basic telephone service and Integrated Services Digital Network (ISDN) use TDM circuits.

TDM for most WANs uses statistical multiplexing schemes to allocate bandwidth to the voice and data channels that carry subscriber traffic.

In X.25 and Frame Relay service, information contained in packets or frames shares nondedicated bandwidth with other data traveling along the WAN. X.25 packet switching uses Layer 3 routing with sender and receiver addressing contained in the packet. Frame Relay uses Layer 2 identifiers and permanent virtual circuits (PVCs).

Interfacing WAN Service Providers

When your organization subscribes to an outside WAN provider for network resources, the provider assigns your organization the parameters for connecting WAN calls. Your organization makes connections to destinations as point-to-point calls, as shown in Figure 14–2.

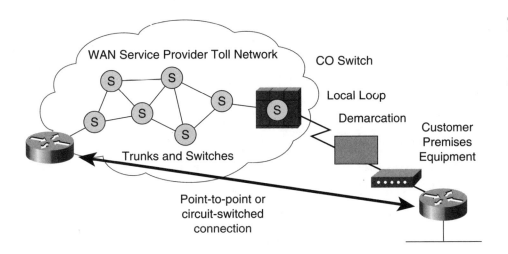

The most commonly used terms for the main parts of a user-provider WAN are:

- Customer premises equipment
- Demarcation
- Local loop
- Central office switch
- Toll network

Customer Premises Equipment (CPE)

CPE devices are physically located on the subscriber's premises. They include both devices owned by the subscriber and devices leased to the subscriber by the service provider. For example, terminals, telephones, and modems that connect to the provider service are considered customer premises equipment. The WAN subscriber should know how to interface the CPE elements to the provider service.

Demarcation

The demarcation point (or demarc) is the juncture at which the CPE ends and the local loop portion of the service begins. The demarc often occurs at a telecommunication closet (a room containing a punch-down block of provider wiring).

Local Loop

The local loop consists of cabling (usually copper wiring) that extends from the demarc into the WAN service provider's central office. The local loop is sometimes referred to as the "last mile." Usually the local loop extends for a relatively short distance to the nearest telephone company premises.

Central Office (CO) Switch

The CO switch is a switching facility that provides the nearest point of presence for the provider's WAN service.

The central office acts as:

- An entry point to the WAN cloud for calling
- An exit point from the WAN for called devices
- A switching point for calls that traverse the facility

Inside the long-distance toll network are several types of central offices. For example, a calling subscriber's connection on a local loop can enter an *end central office* switch and access an interoffice trunk to a *toll central office*. In most U.S. locations, AT&T, Sprint, and MCI offer toll offices to handle their subscribers' calls.

A called subscriber can receive a call that has traversed the trunks and switches of a similar hierarchy of central offices. The called subscriber receives the call over the local loop from the called subscriber's end central office.

Toll Network

The collective switches and facilities (called trunks) inside the WAN provider's cloud make up the toll network. The caller's traffic may cross a trunk to a primary center, and then go to a sectional center, and then to a regional- or international-carrier center as the call goes the long distance to its destination. Switches operate in provider offices with toll charges based on tariffs or authorized rates.

Often, for point-to-point circuits spanning regional or national boundaries, several providers handle a connection in the toll network.

Subscriber to Provider Interface

A key interface in the customer premises occurs between the data terminal equipment (DTE) and the data circuit-terminating equipment (DCE), as shown in Figure 14–3. The lower portion of Figure 14–3 depicts the DTE/DCE at each endpoint of a switched WAN.

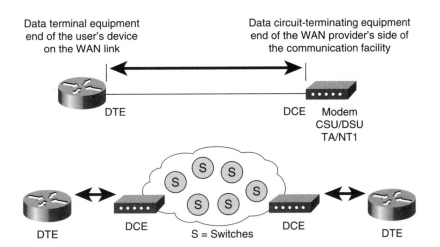

Data terminal equipment
end of the user's device
on the WAN link

Data circuit-terminating equipment
end of the WAN provider's side of
the communication facility

DTE   DCE   Modem
                              CSU/DSU
                              TA/NT1

DTE   DCE   S = Switches   DCE   DTE

Figure 14–3
The DTE/DCE is the point where responsibility passes from the calling subscriber to the provider and then back to the called subscriber on the other end of the connection.

Typically, DTE is the router where the packet switching application resides. The DCE is the device used to convert the user data from the DTE into a form acceptable to the WAN service's facility. As shown in Figure 14–3, the DCE can be a modem, channel service unit/data service unit (CSU/DSU), or Terminal Adapter/Network Termination 1 (TA/NT1).

Data communication over WANs interconnects DTEs so they can share resources over a wide area. The WAN path between the DTEs is called the *link, circuit, channel,* or *line.* The DCE primarily provides the interface of the DTE into the communication link in the WAN cloud and may provide the clocking. The DTE/DCE interface acts as a boundary where responsibility for the traffic passes between the WAN subscriber and the WAN provider.

The DTE/DCE interface uses one of various protocols available, such as EIA/TRA 232 or X.25. These protocols establish the codes that the devices use to communicate with each other. This communication determines how call setup operates and how user traffic crosses the WAN.

Data switching equipment (DSE) is an additional term sometimes used to describe the switch components that appear inside the WAN cloud. The DSE adds and removes channels assigned inside the WAN. The DSE connects traffic from various sources to their final destinations through other switches.

Using WAN Services with Routers

You can access three forms of WAN services with Cisco routers:

- The first form uses switched or relayed services. Examples of this form of WAN include X.25, Frame Relay, and ISDN. Chapters on X.25 and Frame Relay follow this chapter. ISDN is outside the scope of this book, but is covered in the book *Advanced Cisco Router Configuration,* also published by Cisco Press.

- The second form of WAN service provides an interface front end to the IBM enterprise data center computers. This form of WAN uses Synchronous Data Link Control (SDLC) for the point-to-point or point-to-multipoint connection of remote devices to the central mainframe. This topic is outside the scope of this book.

- With the third form, you can access the services of WAN providers using protocols that connect peer devices. This form uses High-Level Data Link Control (HDLC) or PPP encapsulation on the peer devices. The section "An Overview of PPP," later in this chapter, provides an introduction to PPP.

  This third form of WAN access can use Dial-on-Demand Routing (DDR) as a trigger for the Cisco router to make a WAN call. For example, a router uses DDR statements when local user traffic needs to set up an ISDN call over a WAN so it can access a remote network. DDR routing leaves the link idle until data needs to be transmitted from one side or the other. When data is queued to be sent, the routers establish a dial-up connection. After the data has been sent, the routers tear down the connection again.

Figure 14–4 depicts these three forms of WAN service.

WAN Frame Format Summary

The different forms of WAN services available with Cisco routers use different frame types. If the default frame type is not used, you must specify the framing needed, because there are field differences among the different frame types.

Figure 14–5 shows the differences between commonly used WAN frame formats.

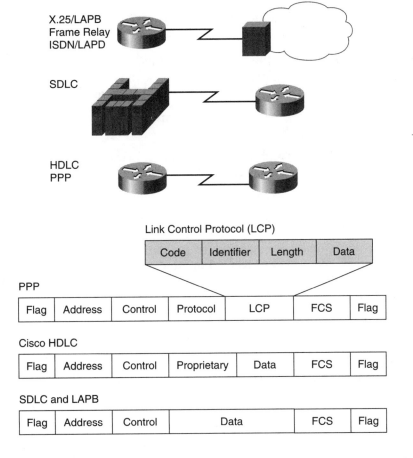

Figure 14–4
The three
forms of WAN
service.

Figure 14–5
These for-
mats assume
framing on
dedicated
WAN facilities.

LAPB, used by X.25, is derived from HDLC. HDLC is the popular ISO-standard bit-oriented data-link protocol that encapsulates data on synchronous serial data links. Frame Relay also uses a variation of HDLC.

HDLC does not inherently support multiprotocols on a single link because it does not have a standard way to indicate which protocol it is carrying. The Cisco HDLC frame uses a proprietary type field that acts as a protocol field, which makes it possible for multiple network-layer protocols to share the same serial link.

PPP extends the basic HDLC frame by incorporating a protocol field. The protocol field identifies the protocol encapsulated in the information field of the frame.

The PPP fields have inherited some attributes that are not typically used. For example, the address field is assigned the binary sequence 11111111, the standard broadcast address, because PPP has no need to assign individual station addresses.

The Link Control Protocol (LCP) used by PPP provides a method of establishing, configuring, maintaining, and terminating the point-to-point connection. LCP serves much the same function as the 802.2 logical link control (LLC) in the LAN protocols.

This book deals mainly with the WAN frame formats for PPP and HDLC. Serial connections use WAN framing that is similar.

It is important to understand and recognize the frame formats used by WAN routers to effectively troubleshoot WAN connections.

AN OVERVIEW OF PPP

Developers on the Internet designed PPP to make the connection for point-to-point links. PPP, originally described in RFCs 1661 and 1332, encapsulates network-layer protocol information over point-to-point links. RFC 1661 is updated by RFC 2153, *PPP Vendor Extensions*.

You can configure PPP on the following types of physical interfaces:

- Asynchronous serial
- HSSI (High-Speed Serial Interface)
- ISDN
- Synchronous serial

PPP uses its Network Control Programs (NCPs) component to encapsulate multiple protocols, as shown in Figure 14–6. This use of NCPs surpasses the limits of PPP's predecessor Serial Line IP (SLIP), which could only set up transport for IP packets.

PPP uses another of its major components, the Link Control Protocol (LCP), to negotiate and set up control options on the WAN data link.

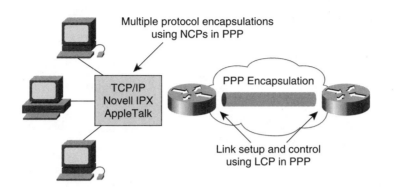

Figure 14–6
PPP can carry packets from several protocol suites using Network Control Programs.

Layering PPP Elements

PPP uses a layered architecture, as shown in Figure 14–7. With its lower-level functions, PPP can use:

- Synchronous physical media like those that connect ISDN

- Asynchronous physical media like those that use basic telephone service for modem dialup connections

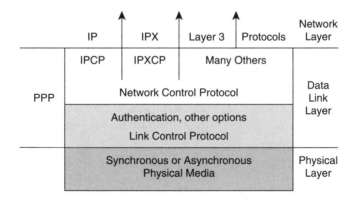

Figure 14–7
The PPP layered architecture.

PPP offers a rich set of services that control setting up a data link. These services are options in LCP and are primarily negotiation and checking frames to implement the point-to-point controls an administrator specifies for the call.

With its higher-level functions, PPP carries packets from several network-layer protocols in NCPs. These are functional fields containing standardized codes to indicate the network-layer protocol type that PPP encapsulates.

PPP LCP Configuration Options

RFC 1548 describes PPP operation and LCP configuration options. RFC 1548 is updated by RFC 1570, *PPP LCP Extensions*.

Cisco routers that use PPP encapsulation include the LCP options shown in Table 14–1.

Table 14–1

PPP LCP configuration options.

Feature	How It Operates	Protocol
Authentication	Require a password	PAP
	Perform Challenge Handshake	CHAP
Compression	Compress data at source;	Stacker or
	Reproduce data at destination	Predictor
Error Detection	Monitor data dropped on link	Quality
	Avoid frame looping	Magic Number
Multilink	Load balancing across	Multilink
	multiple links	Protocol (MP)

Authentication options require that the calling side of the link enter information to help ensure the caller has the network administrator's permission to make the call. Peer routers exchange authentication messages. Two alternatives are:

- Password Authentication Protocol (PAP)

- Challenge Handshake Authentication Protocol (CHAP)

To further enhance security, Cisco IOS Release 11.1 offers callback over PPP. With this LCP option, a Cisco router can act as a callback client or as a callback server.

The client makes the initial DDR call request to be called back and terminates the initial call. The callback server answers the initial call and makes the return call to the client based on its configuration statements. This option is described in RFC 1570.

Compression options increase the effective throughput on PPP connections by reducing the amount of data in the frame that must travel across the link. The protocol decompresses the frame at its destination.

Two compression protocols available in Cisco routers are Stacker and Predictor.

Error-detection mechanisms with PPP enable a process to identify fault conditions. The Quality and Magic Number options help ensure a reliable, loop-free data link.

Cisco IOS Release 11.1 and later support multilink PPP. This alternative provides load balancing over the router interfaces that PPP uses.

Packet fragmentation and sequencing, as specified in RFC 1717, splits the load for PPP and sends fragments over parallel circuits. In some cases, this "bundle" of multilink PPP pipes functions as a single logical link, improving throughput and reducing latency between peer routers. RFC 1990, *The PPP Multilink Protocol (MP)*, obsoletes RFC 1717.

PPP Session Establishment

A PPP session establishment has three phases, as shown in Figure 14–8.

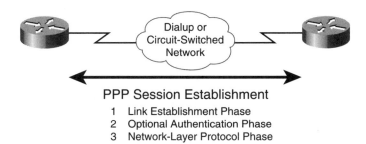

PPP Session Establishment
1   Link Establishment Phase
2   Optional Authentication Phase
3   Network-Layer Protocol Phase

Figure 14–8
PPP session establishment has three phases.

Phase 1: Link Establishment

In this phase, each PPP device sends LCP packets to configure and test the data link. LCP packets contain a Configuration Option field that allows devices to negotiate on the use of options such as the maximum receive unit, compression of certain PPP fields, and the link authentication protocol. If a Configuration Option is not included in an LCP packet, the default value for that Configuration Option is assumed.

Phase 2: Authentication (Optional)

After the link has been established and the authentication protocol decided on, the peer may be authenticated. Authentication, if used, takes place before entering the network-layer protocol phase.

PPP supports two authentication protocols: PAP and CHAP. Both of these protocols are detailed in RFC 1334, *PPP Authentication Protocols*. However, RFC 1994, *PPP Challenge Handshake Authentication Protocol*, obsoletes RFC 1334.

Phase 3: Network-Layer Protocol

In this phase, the PPP devices send NCP packets to choose and configure one or more network-layer protocols (such as IP). After each of the chosen network-layer protocols has been configured, datagrams from each network-layer protocol can be sent over the link. PPP supports several network-layer protocols including IP, IPX, AppleTalk, OSI, and so on.

Selecting a PPP Authentication Protocol

When configuring PPP authentication, you can select PAP or CHAP. In general, CHAP is the preferred protocol.

PAP

PAP provides a simple method for a remote node to establish its identity using a two-way handshake, as shown in Figure 14–9. PAP is done only upon initial link establishment.

Figure 14–9
In PAP, pass-words are sent in cleartext.

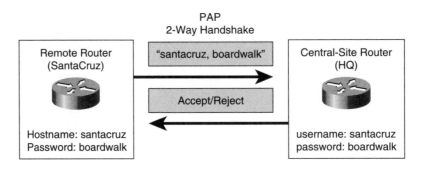

After the PPP link establishment phase is complete, a username/password pair is repeatedly sent by the remote node to the router until authentication is acknowledged or the connection is terminated.

PAP is not a strong authentication protocol. Passwords are sent across the link in clear-text, and there is no protection from playback or repeated trial-and-error attacks. (A playback attack occurs when an analyzer captures the packets and plays them back onto the network from another device.) The remote node is in control of the frequency and timing of the login attempts.

TIPS

Use PAP only if the device requiring authentication does not support CHAP.

CHAP

CHAP is used at the startup of a link, and periodically to verify the identity of the remote node using a three-way handshake.

After the PPP link establishment phase is complete, the local router sends a "challenge" message to the remote node. The remote node responds with an encrypted ID number, a secret password, and a random number. The local router checks the response value against its own calculation. If the values match, the authentication is acknowledged; otherwise, the connection is terminated immediately. Figure 14–10 summarizes this process.

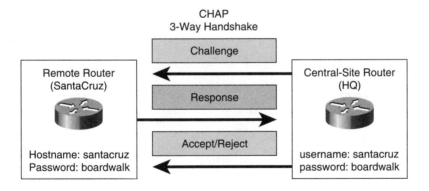

CHAP
3-Way Handshake

Challenge

Remote Router
(SantaCruz)

Response

Central-Site Router
(HQ)

Accept/Reject

Hostname: santacruz
Password: boardwalk

username: santacruz
password: boardwalk

Figure 14–10
CHAP uses a "secret" known only to authenticator and peer.

CHAP provides protection against playback attack through the use of a variable challenge value that is unique and unpredictable. The use of repeated challenges is intended to limit the time of exposure to any single attack. The local router (or a third-party authentication server such as TACACS) is in control of the frequency and timing of the challenges.

Use the **debug ppp authentication** command to display the exchange sequence as it occurs. An example of a CHAP exchange follows:

> PPP Serial1: Send CHAP challenge id=34 to remote
> PPP Serial1: CHAP challenge from P1R2
> PPP Serial1: CHAP response received from P1R2
> PPP Serial1: CHAP response id=34 received from P1R2
> PPP Serial1: Send CHAP success id=34 to remote
> PPP Serial1: Remote passed CHAP authentication.

PPP Serial1: Passed CHAP authentication with remote

Configuring PPP Authentication

The routers on each side of the WAN link must be configured for PPP authentication, as shown in Figure 14–11.

Figure 14–11
PPP authentication configuration commands.

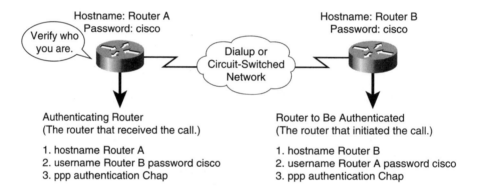

To configure PPP authentication, do the following:

1. On each router, define the username and password to expect from the remote router. The command format is:

```
Router(config)#username name password secret
```

where the parameters have the following meanings:

- *name*—This is the host name of the remote router. Note that it is case sensitive.

- *secret*—On Cisco routers, the secret password must be the same for both routers.

Add a username entry for each remote system that the local router communicates with and requires authentication from. The remote device must also have a username entry for the local router.

To enable the local router to respond to remote CHAP challenges, one username name entry must be the same as the host name, name entry that has already been assigned to your device.

2. Enter interface configuration mode for the desired interface.
3. Configure the interface for PPP encapsulation.

```
Router(config)#encapsulation ppp
```

4. Configure PPP authentication.

```
Router(config)#ppp authentication {chap ¦ chap pap ¦ pap chap
    ¦ pap}***
```

There are four different options available for PPP authentication:
- CHAP only (CHAP)
- CHAP and then PAP (CHAP PAP)
- PAP and then CHAP (PAP CHAP)
- PAP only

If both PAP and CHAP are enabled, the first method specified will be requested during link negotiation. If the peer suggests using the second method or simply refuses the first method, the second method will be tried.

The following commands can be used to simplify CHAP configuration tasks on the router:

- Using the same host name on multiple routers—When you want remote users to think they are connecting to the same router when authenticating, configure the same host name on each router.

```
Router(config-if)#ppp chap hostname hostname
```

- Use a password to authenticate to an unknown host—To limit the number of username/password entries in the router, configure a password that will be sent to hosts that want to authenticate the router.

```
Router(config-if)#ppp chap password secret
```

Verifying PPP

When PPP is configured, you can check its LCP and NCP states using the **show interfaces** command.

In Figure 14–12, for example, the administrator used this command to monitor an ISDN interface.

Figure 14–12
*The **show interfaces** command is used to verify that PPP encapsulation is configured on the interface.*

```
P1R1#show interfaces s1
Serial1 is up, line protocol is up
Hardware is HD64570
Internet address is 10.1.1.2/24
MTU 1500 bytes, BW 1544 Kbit, DLY 20000 usec, rely 255/255, load 1/255
Encapsulation PPP, loopback not set, keepalive (10 sec)
LCP Open
Open: IPCP, CDP, ATALKCP, IPXCP
Last input 00:00:04, output 00:00:00, output hang never
Last clearing of "show interface" counters never
Input quere: 0/75/0 (size/max/ drops); Total output drops: 0
Queueing strategy: weighted fair
Output queue: 0/64/0 (size/threshold/drops)
    Conversations 0/4 (active/max active)
    Reserved Conversations 0/0 (allocated/max allocated)
5 minute input rate 0 bits/sec, 0 packets/sec
5 minute output rate 0 bits/sec, 0 packets/sec
    51938 packets input, 1634908 bytes, 0 no buffer

- - more - -
```

SUMMARY

This chapter has provided an overview of WAN services and setup elements. With respect to specific WAN services that can be used with Cisco routers, this book emphasizes X.25 and Frame Relay, which are covered in more detail in the next two chapters. Remember that the routers on each end of a WAN connection must be configured with PPP. In addition, it is a good idea to configure PAP or CHAP authentication (or both) for security purposes.

Chapter Fourteen Test
Introduction to Serial Connections

Estimated Time: 15 minutes

Complete all the exercises to test your knowledge of the materials contained in this chapter. Answers are listed in Appendix A, "Chapter Test Answer Key."

Question 14.1

T   F   A WAN subscriber must know how to interface customer premises equipment to the provider service.

Question 14.2

T   F   PPP sets data-link encapsulation capable of transmitting packets from multiple protocols.

Question 14.3

For each of the following definitions, write the letter of the term that most closely matches it.

Terms:

A. Call setup service
B. Time-division multiplexing (TDM)
C. X.25 or Frame Relay service

Definitions:

____ (a) Uses a separate channel for control messages between transfer points to a called destination.
____ (b) Statistically allocates bandwidth on a single channel to multiple circuits.
____ (c) Also known as signaling.
____ (d) Uses fixed time slots to eliminate congestion.
____ (e) Packets of information share a nondedicated channel.
____ (f) Sets up and clears calls between users.
____ (g) Route is a dedicated path between sending location and receiving location.
____ (h) Uses virtual circuits to avoid call setup delays.

Question 14.4

What are the three forms of WAN services described in the chapter that you can access with Cisco routers?

A. _____

B. _____

C. _____

Question 14.5

What command do you issue to verify that your interface is configured for PPP encapsulation?

_____

Question 14.6

What command do you issue to display the CHAP exchange sequence?

_____

Configuring X.25

This chapter covers X.25 routing. It presents an overview of the X.25 protocol and explains how packets are addressed and encapsulated in X.25. It also looks at how to configure X.25 routing and verify your configuration.

X.25 OVERVIEW

X.25 is a standard that defines the connection between a terminal and a packet-switching network. X.25 offers the closest approach to worldwide data communication available. Virtually every nation uses some X.25-addressable network.

X.25 originated in the early 1970s. The networking industry commonly uses the term *X.25* to refer to the entire suite of X.25 protocols.

Engineers designed X.25 to transmit and receive data between alphanumeric "dumb" terminals through analog telephone lines. X.25 enabled dumb terminals to remotely access applications on mainframes or minicomputers.

Because modern desktop applications needed LAN-to-WAN-to-LAN data communication, engineers designed newer forms of wide-area technology: Integrated Services Digital Network (ISDN) and Frame Relay. In many situations, these newer WANs complement or extend, rather than replace, X.25.

Many different network-layer protocols can be transmitted across X.25 virtual circuits (VCs), through a process often referred to as *tunneling*. In tunneling, datagrams or other Layer 3 packets are encapsulated within the X.25 Layer 3 packets for transport across the WAN via the X.25 virtual circuit (see Figure 15–1). Each Layer 3 packet keeps the addressing that is legal for its respective protocol.

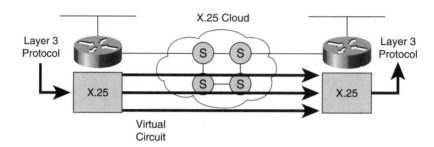

X.25 supports the following types of communication:

- IP
- AppleTalk
- Novell IPX
- Banyan VINES
- XNS
- DECnet
- ISO-CLNS
- Apollo
- Compressed TCP
- Bridging

X.25 Protocol Stack

The X.25 packet switching protocol suite compares to the lower three layers of the Open System Interconnection (OSI) model, as shown in Figure 15–2.

In general, you can think of X.25 as an overengineered data link in the internetworking world. Both X.25 at Layer 3 and Link Access Procedure Balanced (LAPB) at Layer 2 provide reliability and sliding windows. Layers 3 and 2 were designed with strong flow control and error checking to reduce the requirement for these functions external to X.25.

	OSI Reference Model			X.25 Protocol		
7	Application			•		
6	Presentation			•		
5	Session			•		
4	Transport			•		
3	Network			X.25	3	
2	Data Link			LAPB	2	
1	Physical			Physical	1	

Figure 15–2
X.25 maps to the lower layers of the OSI model.

LAPB provides a confirmed data service between two points. All data sent will be actively confirmed, and the service guarantees point-to-point ordered delivery with no drops or duplication. If this service cannot deliver, the protocol resets with a set asynchronous balanced mode (SABM) or SABM extended (SABME) that disrupts the Layer 3 service, which will in turn RESTART.

The X.25 packet level protocol (PLP) depends on the LAPB service guarantees. The network layer provides multiplexed connections over the point-to-point LAPB connection. PLP also guarantees ordered delivery with no drops or duplication. If service is disrupted, a VC will signal the possible loss of data (RESET) or will be brought down (CLEAR).

X.25 evolved in the days of analog circuits when error rates were much higher than today. For analog circuit technology at Layer 1, it is more efficient to build reliability into the network at the hardware level. With digital or fiber-optic technologies, the error rates have dropped dramatically. Newer technologies such as Frame Relay have taken advantage of drops in error rates by providing a stripped-down "unreliable" data link.

X.25 was designed in the days of alphanumeric terminals and computing on central time-sharing computers. Demands on the packet switch were lower than today. Complex applications on desktop workstations demand more bandwidth and speed. Newer technologies such as ISDN and X.25 over Frame Relay add packet-switching capability.

X.25 DTE and DCE

Each station on an X.25 attachment is either a DTE or a DCE. The X.25 DTE is typically a router or a packet assembler/disassembler (PAD). The X.25 packet-level DCE typically acts as a boundary function to the public data network (PDN) within a switch or concentrator. Figure 15–3 illustrates the relationship between the DTEs and DCEs in

Figure 15–3
DTE is usually a subscriber's router or PAD, and DCE is usually a PDN's switch or concentrator.

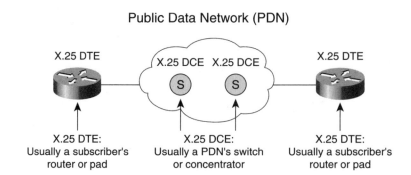

Public Data Network (PDN)

X.25 DTE   X.25 DCE   X.25 DCE   X.25 DTE

X.25 DTE:
Usually a subscriber's
router or pad

X.25 DCE:
Usually a PDN's switch
or concentrator

X.25 DTE:
Usually a subscriber's
router or pad

the PDN. The X.25 switch at the carrier site may also be called data switching equipment (DSE).

Although the terms DTE and DCE occur at all three of the layers associated with the X.25 stack, the uses shown in Figure 15–3 are independent of the physical-layer DTE/DCE.

The X.25 protocol implements virtual circuits between the X.25 DTE and X.25 DCE.

X.25 (X.121) Addressing Format

The format of X.25 addresses is defined by the ITU-T X.121 standard. Figure 15–4 shows the X.25 addressing format.

Figure 15–4
X.25 addresses follow a specific format.

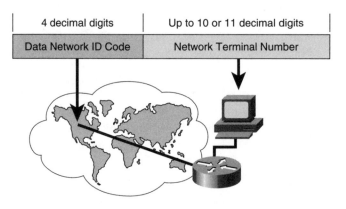

4 decimal digits	Up to 10 or 11 decimal digits
Data Network ID Code	Network Terminal Number

The first four digits specify the Data Network Identification Code (DNIC). This address field is the country code and provider number assigned by the ITU.

The remaining 10 digits specify the network terminal number (NTN); the first eight digits are assigned by the packet-switched network (PSN) provider, and the last two digits are an optimal subnumber assigned locally. These last two digits may be used to identify a particular application or device. The first eight digits along with the fourth DNIC digit form the unique address allocated to users when they come up on the X.25 network.

Private X.25 networks may assign addresses that best fit their network architecture.

Only decimal digits are legal for X.121 addresses. The router accepts an X.121 address with as few as 1 or as many as 15 digits. Some networks allow subscribers to use sub-addresses (the last two digits after the assigned base address).

For different network protocols to connect across X.25, statements are entered on the router to map the next-hop network-layer address to an X.121 address. For example, an IP network-layer address is mapped to an X.121 address to identify the next-hop host on the other side of the X.25 network.

These statements are logically equivalent to the LAN Address Resolution Protocol (ARP) that dynamically maps a network-layer address to a data-link MAC address, as shown in Figure 15–5. Maps are required for each protocol because ARP is not supported in an X.25 network.

A critical step in configuring a Cisco router for X.25 is manually mapping X.121 addresses to network-layer addresses.

Key Concept

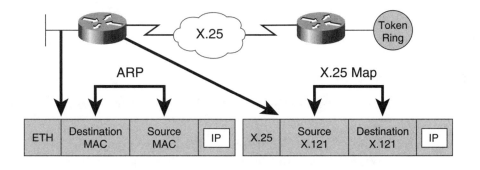

Figure 15–5
Mapping a network-layer address to an X.121 address is a manual configuration task.

X.25 Encapsulation

Movement of network-layer data through the internetwork usually involves encapsulation of datagrams inside media-specific frames, as shown in Figure 15–6. As each media frame arrives at the router and the media frame is discarded, the router analyzes the datagram and places it inside a new frame as it is forwarded.

Figure 15–6
Protocol datagrams are reliably carried inside X.25 frames.

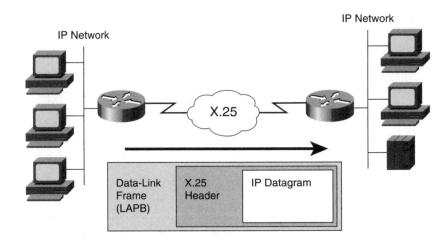

Similarly, in an X.25 environment, the LAPB frame arrives at the router, which extracts the datagram from the packet or packets. The router discards the encapsulating frame and analyzes the datagram to identify the format and next hop. Based on the route determination, the router reencapsulates the datagram in framing suitable for the outgoing media as it forwards the traffic.

X.25 Virtual Circuits

The term *virtual circuit (VC)* is used interchangeably with the terms *virtual circuit number (VCN)*, *logical channel number (LCN)*, and *virtual channel identifier (VCI)*.

A VC can be a permanent virtual circuit (PVC) or, more commonly, a switched virtual circuit (SVC). An SVC exists only for the duration of the session.

There are three phases associated with SVCs:

- Call setup

- Information transfer

- Call clear

A PVC is similar to a leased line. Both the network provider and the attached X.25 sub-scriber must provision, or make available, the virtual circuit. PVCs use no call setup or call clear that is apparent to the subscriber. Any provisioned PVCs are always present, even when no data traffic is being transferred.

As shown in Figure 15–7, VCs carry data through the X.25 cloud.

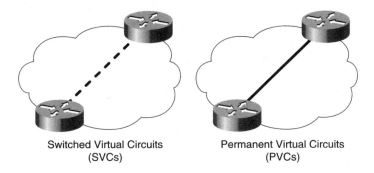

Switched Virtual Circuits
(SVCs)

Permanent Virtual Circuits
(PVCs)

Figure 15–7
*SVCs exist
only during
the call.*

The X.25 protocol offers simultaneous service to many hosts (for example, multiplex connection service). An X.25 network can support any legal configuration of SVCs and PVCs over the same physical circuit attached to the X.25 interface. However, configur-ing a large number of VCs over a serial interface may result in poor performance. X.25's original design aim assumed service for time-sharing and terminal-to-host applications, not contemporary computer-to-computer applications.

The next several sections look at how SVCs can be used to carry single or multiple pro-tocol traffic.

SVC Usage

Up to 4,095 SVCs can be configured on a single X.25 interface.

Throughput for encapsulating a specific protocol can be improved using multiple SVCs. Multiple SVCs provide a larger effective window size, especially for protocols that offer their own higher-layer resequencing. Window size refers to the amount of data that can be transferred in a single stream, without an intervening acknowledgment of receipt. The higher-layer resequencing is important because traffic can travel in different paths and may arrive out of order. In Figure 15–8, three SVCs are combined to provide greater throughput through the X.25 cloud.

A maximum of eight SVCs per protocol per destination is allowed. Combining SVCs does not benefit traditional X.25 applications such as those available from a time-sharing host.

Figure 15–8
*Combining
SVCs provides
greater
throughput.*

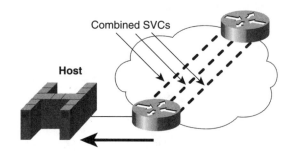

Single Protocol Virtual Circuits

The Cisco router's traditional encapsulation method enables different protocols to transport their datagrams through an X.25 cloud because the router uses separate virtual circuits, as shown in Figure 15–9.

Each protocol is specified in an individual **x25 map** command statement that references the X.121 address used to reach the destination.

Figure 15–9
*SVCs can be
configured to
handle only
one protocol if
desired.*

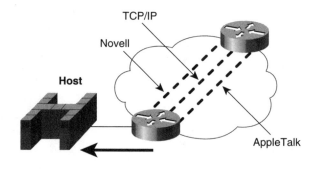

Multiprotocol Virtual Circuits

In Cisco IOS Release 10.2 and later releases, a single virtual circuit to a host can carry traffic from multiple protocols, as shown in Figure 15–10. One **x25 map** statement contains several protocol addresses mapped to a single X.121 address associated with the destination host.

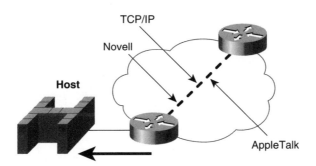

Figure 15–10
SVCs can be set up to support multiprotocol traffic.

This capability uses the method described in RFC 1356. Each of the supported protocols can map to a destination host. Because higher traffic loads are generated by routing multiple protocols over a VC, combining SVCs as described earlier in this chapter may improve throughput.

CONFIGURING X.25

When you select X.25 as a WAN protocol, you must set appropriate interface parameters, including:

- Define the X.25 encapsulation (DTE is the default).
- Assign the X.121 address (usually supplied by the PDN service provider).
- Define map statements to associate X.121 addresses with higher-level protocol addresses.

Other configuration tasks can be performed to control data throughput and to ensure compatibility with the X.25 network service provider. Commonly used parameters include the number of VCs allowed and packet size negotiation.

X.25 is a flow-controlled protocol. The default flow-control parameters must match on both sides of a link. Mismatches caused by inconsistent configurations can result in severe internetworking problems.

Both sides of an X.25 link (the DTE and the DCE) must agree on a number of parameters, including encapsulation and address mapping, as shown in Figure 15–11. They also must agree on which VC numbers to use and for what purpose.

Figure 15–11
Parameters at both ends of the link must agree.

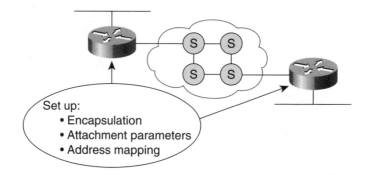

The X.25 standards mandate that the default window size of two packets and a default packet size of 128 bytes must be available. This default reflects the lowest common denominator window size. Higher packet sizes (for example, 512 or 1024 bytes) are commonly used in the United States and Europe. Packet sizes up to 4096 may be allowed by some network providers.

The following commands can be used to configure the X.25 interface parameters on a Cisco router:

- encapsulation X25
- X25 address
- X25 map

encapsulation x25 Command

Use the **encapsulation x25** command to specify the encapsulation style—**dce** or **dte**—to be used on the serial interface. An example configuration is:

```
Router(config-if)#encapsulation x25
Router(config-if)#encapsulation x25 dce
```

The router can be an X.25 DTE; typically this is the configuration when the X.25 PDN is used to transport various protocols. Or the router can be configured as an X.25 DCE, which is typical when the router acts as an X.25 switch. DTE is the default.

x25 address Command

The **x25 address** command defines the local router's X.121 address (one address per interface). The value specified must match the address designated by the X.25 PDN. The format of the command is:

```
Router(config-if)#x25 address x.121-address
```

x25 map Command

The **x25 map** command provides a static conversion of higher-level addresses to X.25 addresses. The command correlates the network-layer addresses of the peer host to the peer host's X.121 address. The format of the command is:

```
Router(config-if)#x25 map protocol address x.121-address [options]
```

where the parameters have the following meanings:

- *protocol*—Selects the protocol type. Supported protocols are **ip, xns, decnet, ipx, appletalk, vines, apollo, bridge, clns,** and **compressed tcp.**

- *address*—Specifies the protocol address (not specified for bridged or CLNS connections).

- *x.121-address*—Specifies the X.121 address. The protocol address and the X.121 addresses together must specify the complete network protocol-to-X.121 mapping.

- *options* (Optional)—Customizes the connection. One commonly used option is **broadcast.** The **broadcast** option causes the Cisco IOS software to direct any broadcasts sent through this interface to the specified X.121 address.

The following **x25 map** statement is used only to communicate with a host that understands multiple protocols over a single VC. This communication requires the multiprotocol encapsulations defined by RFC1356.

```
Router(config-if)#x25 map protocol address [protocol2 address2]* x.121-
    address [options]
```

In the preceding **x25** map command, the " * " means that a maximum of nine network protocol addresses may be associated with one host destination in a single configuration command. Bridging is not supported.

X.25 Configuration Example

In Figure 15–12, two X.25 routers are configured to connect remote company offices.

Figure 15–12
*Two X.25 rout-
ers connect
remote
offices.*

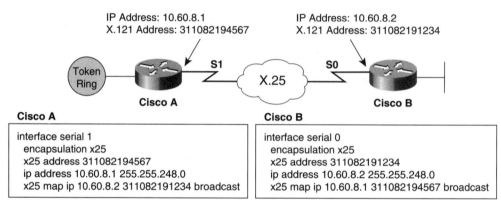

For Cisco A in Figure 15–12, the details of the configuration are as follows:

- **encapsulation x25**—Sets the encapsulation style on interface serial 1 to X.25 type.

- **x25 address 311082194567**—Establishes the X.121 address of serial 1.

- **ip address 10.60.8.1 255.255.248.0**—Specifies a Layer 3 protocol and address to associate with the X.21 address of this interface.

- **x25 map ip 10.60.8.2 311082191234 broadcast**—Maps an IP address to an X.121 address for the host (serial 0) which is at the other end of this X.25 connection. The **broadcast** option directs the interface to send routing information to neighbors over X.25.

IP routing on Cisco A forwards datagrams destined for subnet 10.60.8.0 to interface serial 1. The interface map identifies the destination to the X.25 cloud. In this typical configuration, Cisco A tries to establish an SVC to Cisco B using its X.121 source address and a destination X.121 address of 311082191234 when it sends packets to 10.60.8.2.

Upon receipt of the setup request, Cisco B identifies the remote IP address from the source X.121 address and accepts the connection. After the SVC is connected, each router uses it as a point-to-point data link for the identified destination.

The two X.25 attachments need complementary map configurations to establish the VC that will encapsulate IP datagrams.

Additional Configuration Tasks

It may be necessary to perform additional configuration steps so that the router will work correctly with the service provider network. Crucial X.25 parameters are:

- Virtual circuit range—Incoming, two-way, and outgoing
- Default packet sizes—Input and output
- Default window sizes and window modulus

These parameters must be defined, but you may not need to configure them directly because they depend on the defaults used by the service provider.

Configuring X.25 VC Ranges

Up to 4095 virtual circuits can be configured on an interface, and both ends of an X.25 connection must agree on what range of the available circuits are being used for what purpose. Table 15–1 summarizes configuration commands for virtual circuit number assignment. The complete range of virtual circuits can be allocated to PVCs, SVCs, or a combination of both, depending on your requirements. SVCs are commonly used.

Virtual Circuit Type	Range	Default	Commands
PVCs	1-4095		x25 pvc circuit
SVCs:			
Incoming only	1-4095	0	x25 lic circuit
DCE initiated	1-4095	0	x25 hic circuit
Two-way	1-4095	1	x25 ltc circuit
		1024	x25 htc circuit
Outgoing only	1-4095	0	x25 loc circuit
DTE initiated (outgoing)	1-4095	0	x25 hoc circuit

Table 15-1
Configuration commands for VC number assignment.

A low- and high-limit number must be configured to define the range, which is why two commands are needed for each category. If both limits of a range are zero, the range is unused.

The circuit numbers must be assigned so that an incoming range comes before a two-way range, both of which come before an outgoing range. Any PVCs must take a circuit number that comes before any SVC range. The following numbering scheme lists the proper order for these virtual circuit number assignment commands:

> 1 _ PVCs < (lic _ hic) < (ltc _ htc) < (loc _ hoc) _ 4095

The following decodes can be used to interpret the numbering scheme:

- lic—lowest incoming circuit number
- hic—highest incoming circuit number
- ltc—lowest two-way circuit number
- htc—highest two-way circuit number
- loc—lowest outgoing circuit number
- hoc—highest outgoing circuit number

The following example sets the virtual circuit ranges of 5 to 20 for incoming calls only (from the DCE to the DTE) and 25 to 1024 for either incoming or outgoing calls. It also specifies no virtual circuits for outgoing calls (from the DTE to the DCE).

```
x25 lic 5
x25 hic 20
x25 ltc 25
```

X.25 ignores any events on a VC number not in an assigned VC range; it considers the out-of-range VC as a protocol error. The network administrator specifies the VC ranges for an X.25 attachment. For correct operation, the X.25 DTE and DCE must have identically configured ranges. Numbers configured for any PVCs must also agree on both sides of an attachment (not necessarily end to end).

Configuring X.25 Packet Sizes

The following commands are used to configure X.25 packet sizes:

- **x25 ips**
- **x25 ops**

These commands set the default maximum input and output packet sizes, respectively. The input and output values should match unless the network supports asymmetric transmissions. The format of the command is:

```
Router(config-if)#x25 ips bytes
Router(config-if)#x25 ops bytes
```

where *bytes* refers to the maximum packet size assumed for VCs that do not negotiate a size. Supported values are: 16, 32, 64, 128, 256, 512, 1024, 2048, and 4096. The default is 128 bytes.

If the stations of an X.25 attachment conflict on the VC's maximum packet size, the VC is unlikely to work.

Packets sent across the X.25 network that exceed the specified packet size will require the router to break the packet into allowable packets with the "more bit" (M-bit) set. Packet reassembly occurs on the receiving router; this process consumes CPU cycles on both routers.

Configuring X.25 Window Parameters

X.25 uses a sliding window for flow control. Larger windows allow more packets to be in transit.

The following commands are used to configure the X.25 window size:

- x25 win
- x25 wout
- x25 modulo

Use the **x25 win** and **x25 wout** commands to set the default incoming and outgoing window sizes. The window size specifies the number of packets that can be received/sent without sending/receiving an acknowledgment. Both ends of an X.25 link must use the same default window size.

The format of the commands is:

```
Router(config-if)#x25 win packets
Router(config-if)#x25 wout packets
```

where the *packets* parameter specifies the packet window size. Possible values for window size range from one to one less than the modulus (discussed next). The default is two packets.

The **x25 modulo** command specifies the packet numbering modulus. It affects the maximum number of window sizes. For example, modulo 8 is widely used and allows virtual circuit window sizes up to 8 (using sequence numbers 0 through 7) packets. Modulo 128 is rare but allows VC window sizes up to 128 (using sequence numbers 0 through 127) packets.

The format of the command is:

```
Router(config-if)#x25 modulo modulus
```

where the parameter *modulus* can be either 8 or 128.

In Cisco IOS Release 10.2 and later, LAPB support for modulo 128 has been added, allowing greater throughput over X.25 links.

Both ends of an X.25 link must use the same modulo.

X.25 Additional Configuration Example

Figure 15–13 shows a configuration of interface serial 0. The packet sizes may not be supported by all PDNs. The window values are the maximum allowed in a modulo 8 environment.

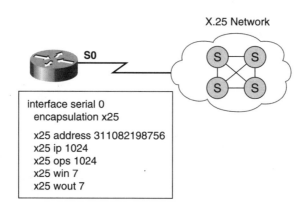

Figure 15–13
X.25 commands used to configure the router address and packet size.

```
interface serial 0
  encapsulation x25

  x25 address 311082198756
  x25 ip 1024
  x25 ops 1024
  x25 win 7
  x25 wout 7
```

An X.121 address is assigned to interface serial 0. The input and output packet and window sizes and the maximum number of virtual circuits for any protocol are also defined.

In Figure 15–13, the following commands have been specified:

- **x25 address 311082198756**—Specifies the address of the interface.

- **x25 ips/ops 1024**—Sets both input and output default packet size to 1024 to match the values defined for the network attachment. Maximum value is 4096.

- **x25 win/wout 7**—Sets both input and output window sizes to 7 to match the values defined for the network attachment.

The typical default packet size provided worldwide by PDNs is 128 bytes. In the United States and Europe, default packet sizes of 1024 are common. Other countries can also provide higher packet sizes. The Layer 3 default maximum packet size is subject to the limit that lower layers are able to support.

Setting Up the Router as a Switch

The router can be configured to switch X.25 traffic over a TCP connection, as shown in Figure 15–14. X.25-over-TCP (called XOT) is defined by RFC 1613.

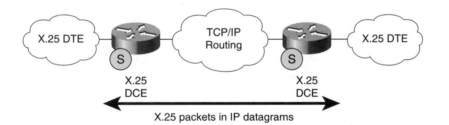

Figure 15–14
The router acts as local or remote switch.

The command to configure an XOT PVC is:

```
Router(config-if)#x25 pvc number1 tunnel address interface serial
    string pvc number2 [option]
```

The command options are **packetsize** *in out* and **windowsize** *in out*; they allow a PVC's flow control values to be defined if they differ from the interface defaults.

In this mode, the backbone comprises routers switching IP datagrams. A few X.25 devices, such as PADs, connect to each other across the routed IP backbone network.

The switching performance of IP is higher than native X.25 switching equipment. This use of a TCP/IP cloud provides customers with high-performance, concurrent switching of X.25, IP, and other protocols.

X.25 Local and XOT Switching

X.25 traffic can be routed locally between serial ports. In this case, static routing statements map X.121 addresses to serial ports. The router allows X.25 interfaces attached to different ports to make SVC connections, a capability which is called *local X.25 switching* (shown in Figure 15–15). A router configured as an XOT switch can provide significant improvement in throughput over traditional X.25 switching equipment (RFC 1613 contains information about XOT routing).

Figure 15–15
X.25 switch-
ing.

Remote X.25 switching allows X.25 interfaces attached to different routers to establish SVCs and PVCs. Remote X.25 switching is accomplished by tunneling all the X.25 call setup and data traffic between routers in a TCP connection.

The format of the command is:

```
Router(config-if)#x25 route [#position] x121-address [cud pattern]
    interface type number
```

where the parameters mean the following:

- *# position* (Optional)—A positional value that specifies the line number in the table where the entry will be placed

- *x.121-address*—Destination X.121 address pattern

- **cud** *pattern* (Optional)—Call User Data (CUD) pattern, which is a printable ASCII string

- *type-number*—The destination interface number, such as serial 0

CAUTION

The **cud** *pattern* must be the value provided by the X.25 service provider.

Monitoring X.25

Use the **show interfaces** command to display status and counter information about an interface. The output from this command also displays LAPB information. In Figure 15–16, the serial interface has its encapsulation type configured for X.25 operation.

```
Router#show interface serial0

Serial0 is up, line protocol is up
    Hardware is MK5025
    Internet address is 183.8.128.129, subnet mask is 255.255.255.128
    MYU 1500 bytes, BW 56 Kbit, DLY 20000 usec, rely 255/255, load 1/255
    Encapsulation X25, loopback not set
    LAPB DCE, state CONNECT, modulo 8, k 7, N1 12048 N2 20
            T1 3000, interface outage (partial T3) 0, T4 0
            VS 1, VR 1 Remote VR 1, Retransmissions 0
            IFRAMEs 1728559/1639143 RNRs 0/0 REJs 0/0 SABM/Es 3/2 FRMRs 0/0 DISCs 0/0
    X25 DCE, address 311012345678, state RI, modulo 8, timer 0
            Defaults: cisco encapsulation, idle 0, nvc 1
                input/output window sizes 2/2, packet sizes 128/128
            Timers: T10 60, T11 180, T12 60, T13 60, TH 0
            Channels: Incoming-only none, two-way 1-1024, Outgoing-only none
            RESTARTs 3/3 CALLs 244+235/266+262/0+0 DIAGs 0/0
    Last input 0:00:00, output 0:00:00, output hang never
    Last clearing of "show interface" counters never
    Output queue 0/40, 0 drops; input queue 2/75, 0 drops
    Five minute input rate 0 bits/sec, 3 packets/sec
    Five minute output rate 0 bits/sec, 3 packets/sec
        3370943 packets input, 113376062 bytes, 0 no buffer
        Received 1971 broadcasts, 0 runts, 0 giants
        57 input errors, 57 CRC, 0 frame, 0 overrun, 0 ignored, 0 abort
- - more - -
```

Figure 15–16
Use the show interfaces command to display status and counter information about an interface.

Additional **show** commands for X.25 include **show x25 map** and **show x25 vc**.

The **show x25** map command shows information about the configured maps (defined by the **x25 map** command), maps implicitly defined by encapsulation PVCs (defined by the **x25 pvc** command), as well as dynamic and temporary maps.

The **show x25 vc** command displays information about virtual circuits that are used for encapsulation traffic, locally switched traffic, remotely switched traffic, and CMNS switched traffic.

SUMMARY

The ubiquity of X.25 networks worldwide and their flexibility in transporting different network layer protocols over WANs make X.25 an important technology for you to understand. This chapter has defined X.25 addressing, elements, and configuration parameters, and presented examples of how to configure Cisco X.25 routers. Remember that you must statically map a network layer address to an X.121 address as a basic step in configuring X.25 on an interface. Also keep in mind that parameters must match on both ends of an X.25 connection, or the connection may work poorly or not at all.

Chapter Fifteen Test
Configuring X.25

Estimated Time: 15 minutes

Complete all the exercises to test your knowledge of the materials contained in this chapter. Answers are listed in Appendix A, "Chapter Test Answer Key."

Question 15.1

T   F   X.25 defines the lower three layers of the OSI model.

Question 15.2

T   F   LAPB is the network protocol.

Question 15.3

T   F   Tunneling of other protocols inside X.25 is supported.

Question 15.4

To configure an X.25 interface, you must:

T   F   (a) Define the interface encapsulation.

T   F   (b) Set critical parameter values for attaching to the PDN.

T   F   (c) Configure the interface X.121 address.

T   F   (d) Define any protocol to X.25 mapping.

T   F   (e) Define static IP addresses on both sides of the link.

T   F   (f) Configure one router as a master to perform setup steps.

Question 15.5

What command do you issue to specify X.25 encapsulation on your DTE interface?

_____

Question 15.6

What command do you issue to specify X.25 encapsulation on your DCE interface?

_____

16

Configuring Frame Relay

This chapter overviews Frame Relay operation and covers how to configure it for both physical interfaces and subinterfaces.

FRAME RELAY OVERVIEW

Frame Relay is a CCITT and American National Standards Institute (ANSI) standard that defines the process for sending data over a public data network. It is a next-generation protocol to X.25 and is a data-link technology (Layer 2) that is streamlined to provide high performance and efficiency. It relies on upper-layer protocols for error correction and flow control and today's dependable fiber and digital networks for reliability.

As shown in Figure 16–1, Frame Relay defines the interconnection process between your customer premises equipment (also known as DTE), such as a router, and the service provider's local access switching equipment (also known as DCE). It does not define how the data is transmitted within the service provider's Frame Relay cloud. Frame Relay is a purely Layer 2 protocol.

Frame Relay provides a means for statistically multiplexing many logical data conversations (referred to as virtual circuits) over a single physical transmission link by assigning each pair of DTEs connection identifiers. The service provider's switching equipment constructs a table mapping connection identifiers to outbound ports. When a frame is received, the switching device analyzes the connection identifier and delivers the frame to the associated outbound port. The complete path to the destination is established prior to the sending of the first frame.

Figure 16–1
Frame relay defines connection between DCEs and DTEs.

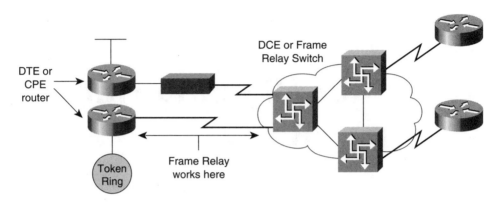

With ANSI T1.617 and ITU Q.933 (Layer 3) and Q.922 (Layer 2), Frame Relay now supports switched virtual circuits (SVCs). Cisco IOS Release 11.2 or later supports Frame Relay SVCs. You will need to determine whether your carrier supports SVCs before implementing them. Frame Relay switched virtual circuits are not covered in this book. For more information on SVCs, refer to www.cisco.com.

Frame Relay Terminology

The terminology associated with Frame Relay may be somewhat new to you, so this section briefly overviews it. Keep in mind that the terms used by Frame Relay service providers may vary from those shown here. For more information on Frame Relay, including a Frame Relay glossary, refer to the Frame Relay Forum World Wide Web page:

```
http://www.frforum.com/4000/4000index.html
```

Figure 16–2 identifies terms that are used frequently when discussing Frame Relay.

Local Access Rate

The clock speed (port speed) of the connection (local loop) to the Frame Relay cloud is the local access rate. It is the rate at which data travels into or out of the network, regardless of other settings.

Data-Link Connection Identifier (DLCI)

The DLCI is a number that identifies the logical circuit between the CPE and the Frame Relay (FR) switch. The FR switch maps the DLCIs between each pair of routers to create a PVC. DLCIs have local significance in that the identifier references the point between the local router and the Frame Relay switch to which it is connected.

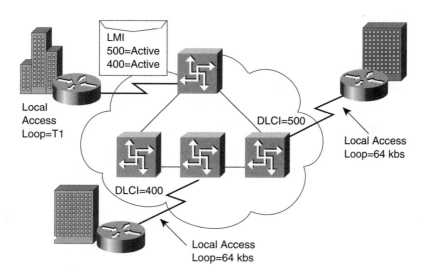

Figure 16–2
*Frame relay
elements.*

Local Management Interface (LMI)

The LMI is a signaling standard between the CPE device and the FR switch that is responsible for managing the connection and maintaining status between the devices. LMIs include support for:

- A keepalive mechanism, which verifies that data is flowing

- A multicast mechanism, which provides the network server with its local DLCI

- The multicast addressing, which gives DLCIs global rather than local significance in Frame Relay networks

- A status mechanism, which provides an ongoing status on the DLCIs known to the switch

Although the LMI is configurable, beginning in Release 11.2, the Cisco router tries to autosense which LMI type the FR switch is using by sending one or more full status requests to the FR switch. The FR switch will respond with one or more LMI types. The router configures itself with the last LMI type received. Three types of LMIs are supported:

- *cisco*—LMI type defined jointly by Cisco, StrataCom, Northern Telecom, and DEC
- *ansi*—Annex D defined by ANSI standard T1.617
- *q933a*—ITU-T Q.933 Annex A

Committed Information Rate (CIR)

CIR is the rate, in bits per second, that the Frame Relay switch agrees to use to transfer data. The rate is usually averaged over a period of time referred to as the Committed Rate Measurement Interval (Tc). (The term "MIR" for Measurement Interval Rate is used instead of CIR by at least one Frame Relay vendor.)

Oversubscription

When the sum of the CIRs on all the VCs coming in to a device exceeds the access line speed, the FR connection is said to be oversubscribed. This can occur when the access line can support the sum of CIRs purchased, but not of the CIRs plus the bursting capacities of the VCs. If oversubscription occurs, packets are dropped.

Committed Burst (Bc)

The committed burst is the maximum number of bits that the Frame Relay Network agrees to transfer during any Committed Rate Measurement Interval (Tc). The higher the Bc-to-CIR ratio is, the longer the network can handle a sustained burst.

Excess Burst

The excess burst is the maximum number of uncommitted bits that the Frame Relay switch will attempt to transfer beyond the CIR. Excess Burst is dependent on the service offerings available from your vendor but is typically limited to the port speed of the local access loop.

Forward Explicit Congestion Notification (FECN)

When a Frame Relay switch recognizes congestion in the network, it sets the FECN bit in a Frame Relay packet bound for the destination device, indicating that congestion has occurred from source to destination, as shown in Figure 16–3.

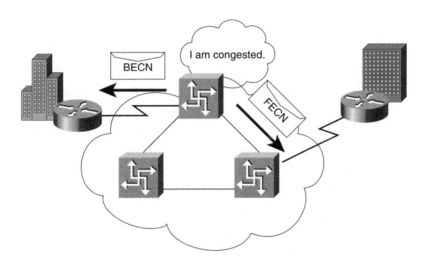

Figure 16–3
*FECN packets
indicate that
congestion
has occurred.*

Backward Explicit Congestion Notification (BECN)

When a Frame Relay switch recognizes congestion in the network, it sets the BECN bit to the source router instructing the router to reduce the rate at which it is sending packets. With Cisco IOS Release 11.2 or later, Cisco routers can respond to BECN notifications.

Discard Eligibility (DE) Indicator

When the router detects network congestion, it sets the DE bit on oversubscribed traffic. If the network experiences congestion, it will first discard those packets with the DE bit set.

Frame Relay Operation

Frame Relay is a Layer 2 protocol that describes how the DTE device communicates with and connects to a Frame Relay switch. Figure 16–4 illustrates how FR operation works.

Figure 16–4
Frame Relay operation.

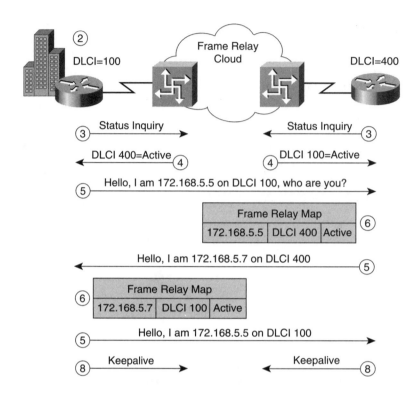

The following steps outline how the protocol operates and are keyed to Figure 16–4:

1. You order Frame Relay service from a service provider, or you create a private Frame Relay cloud.
2. Each router, either directly or through a CSU/DSU, connects to the Frame Relay switch and is assigned a DLCI.
3. When the CPE router is enabled, it sends a Status Inquiry message to the FR switch. The message notifies the switch of the router's status and asks the switch for the connection status of the other remote routers.
4. When the FR switch receives the request, it responds with a Status message that includes the DLCIs of the remote routers to which the local router can send data.
5. For each active DLCI a router can reach, the router sends an Inverse ARP request packet introducing itself and asking for each remote router to identify itself by replying with its network-layer address.

6. For each DLCI for which the router receives an Inverse ARP message, the router will create a map entry in its Frame Relay map table that includes its own DLCI and the remote router's network-layer address, as well as the state of the connection. Note that the DLCI is the router's locally configured DLCI, not the DLCI that the remote router is using. Three possible connection states appear in the Frame Relay map table:

- *Active state*—Indicates that the connection is active and that routers can exchange data

- *Inactive state*—Indicates that local connection to FR switch is working, but the remote router's connection to FR switch is not working

- *Deleted state*—Indicates that no LMI is being received from the FR switch or no service between the CPE router and FR switch is occurring

If Inverse ARP is not working, or the remote router does not support Inverse ARP, you need to configure the routes (DLCIs and IP addresses) of the remote routers. Such configurations are referred to as *static maps*; static mapping is discussed later in the "Configuring Optional Commands" section.

7. Every 60 seconds, the routers exchange Inverse ARP messages.

8. Every 10 seconds or so (this is configurable), the CPE router sends a keepalive message to the FR switch. The purpose of the keepalive message is to verify that the FR switch is still active.

The router will change the status of each DLCI, based on the response from the FR switch.

CONFIGURING FRAME RELAY

A basic Frame Relay configuration assumes that you want to configure Frame Relay on one or more physical interfaces, and that LMI and Inverse ARP are supported by the remote router(s). In this type of environment, the LMI notifies the router about the available DLCIs. Figure 16–5 illustrates a configuration for this situation.

Use the following steps to configure basic Frame Relay on a Cisco router:

1. Select the interface and enter interface configuration mode.

2. Configure a network-layer address, for example, an IP address.

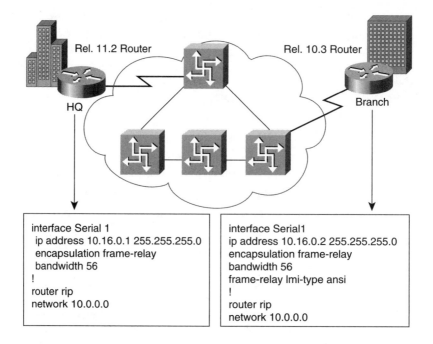

Figure 16–5
Basic Frame Relay configuration over a physical interface.

```
interface Serial 1
 ip address 10.16.0.1 255.255.255.0
 encapsulation frame-relay
 bandwidth 56
!
router rip
 network 10.0.0.0
```

```
interface Serial1
 ip address 10.16.0.2 255.255.255.0
 encapsulation frame-relay
 bandwidth 56
 frame-relay lmi-type ansi
!
router rip
 network 10.0.0.0
```

3. Select the encapsulation type used to encapsulate data traffic end-to-end. The format of the encapsulation command is:

```
Router(config-if)#encapsulation frame-relay [cisco ¦ ietf]
```

where **cisco** is the default. Use **cisco** if connecting to another Cisco router; use **ietf** if connecting to a non-Cisco router.

4. If using Cisco IOS Release 11.1 or earlier, specify the LMI-type used by the FR switch:

```
Router(config-if)#frame-relay lmi-type {ansi ¦ cisco ¦ q933i}
```

where **cisco** is the default.

With IOS Release 11.2 or later, the LMI-type is autosensed, so no configuration is needed.

5. Configure the bandwidth for the link in kilobits per second:

```
Router(config-if)#bandwidth kilobits
```

This command affects routing operation by protocols such as IGRP and EIGRP, because it is used to define the metric of the link. In addition, it allows EIGRP to determine how much data to transmit in a given amount of time. Without this command, EIGRP assumes that the bandwidth is 1.544 Mbps (T1). This command is also the basis for the statistics such as bandwidth usage.

6. If Inverse ARP was disabled on the router, re-enable it. Inverse ARP is on by default. The command form is:

```
Router(config-if)#frame-relay inverse-arp [protocol] [dlci]
```

Supported protocols include ip, ipx, appletalk, decnet, vines, and xns. The parameter *dlci* identifies the DLCI on the local interface with which you want to exchange Inverse ARP messages.

Configuring Optional Commands

Normally, Inverse ARP is used to request the next hop protocol address for a specific connection. Responses to Inverse ARP are entered in an address-to-DLCI map (Frame Relay map) table. The table is then used to route outgoing traffic. When Inverse ARP is not supported by the remote router, when configuring OSPF over Frame Relay, or when you want to control broadcast traffic when using routing, you must define the address-to-DLCI table statically. These static entries are referred to as static maps; they are implemented with the **frame-relay map** command. The full form of the command is:

```
Router(config-if)#frame-relay map protocol protocol-address dlci
   [broadcast][ietf ¦ cisco ¦ [payload-compress packet-by-packet]]
```

where the parameters and keywords have the following meanings:

- *protocol*—Defines supported protocol, bridging, or logical link control.

- *protocol-address*—Defines the network-layer address of the destination router interface.

- *dlci*—Defines the local DLCI used to connect to the remote protocol address.

- **broadcast** (Optional)—Forwards broadcasts to this address when multicast is not enabled. Use this if you want the router to forward routing updates. If not enabled, you must define static routes, and if using IPX, static SAPs.

- **ietf | cisco** (Optional)—Select the Frame Relay encapsulation type for use. Use *ietf* only if the remote router is a non-Cisco router; otherwise, use *cisco*.

- **payload-compress packet-by-packet** (Optional)—Packet-by-packet payload compression, using STAC, a Cisco proprietary compression method.

Other optional commands can be used with Frame Relay to deal with routing traffic, keepalives, and numbering.

For example, if using EIGRP, you can set what percentage of the configured bandwidth to use for EIGRP routing traffic. The default is 50 percent. Note that the syntax for this command varies, depending on whether you are setting it for EIGRP for IP, IPX, or AppleTalk. The IP syntax is as follows:

```
Router(config-if)#ip bandwidth-percent eigrp as-number percent
```

where *as-number* is the autonomous system number and *percent* is the percent of bandwidth EIGRP can use. Remember, the total bandwidth is derived from what was configured using the **bandwidth** command, or the default, which is 1.544 Mbps.

You can extend or reduce the interval at which the router interface sends keepalive (status inquiry) messages to the Frame Relay switch. The default is 10 seconds. The command form is:

```
Router(config-if)#keepalive number
```

where *number* is the interval in seconds. This value is usually two to three seconds faster (shorter interval) than the interval setting of the Frame Relay switch to ensure proper synchronization.

If an LMI-type is not used in your network or if you are doing back-to-back testing between routers, you need to specify the DLCI for each local interface by using the following command:

```
Router(config-if)#frame-relay local-dlci number
```

where *number* is the DLCI on the local interface to be used.

Review the *Cisco IOS WAN Configuration Guide* to determine if there are other optional commands that may be of interest to you.

Verifying Frame Relay Operation

After configuring Frame Relay, you can verify that the connections are active using the following **show** commands:

Command	Description
show interfaces serial	Displays information about the multicast DLCI, the DLCIs used on the Frame Relay-configured serial interface, and the LMI DLCI used for the local management interface.

Command	Description
show frame-relay pvc	Displays the status of each configured connection as well as traffic statistics. This command is also useful for viewing the number of BECN and FECN packets received by the router.
show frame-relay map	Displays the network-layer address and associated DLCI for each remote destination that the local router is connected to.
show frame-relay lmi	Displays LMI traffic statistics. For example, it shows the number of status messages exchanged between the local router and the Frame Relay switch.

SELECTING A FRAME RELAY TOPOLOGY

Frame Relay allows you to interconnect your remote sites in a variety of ways as shown in Figure 16–6. Example topologies include:

- Star topology
- Full-mesh topology
- Partial-mesh topology

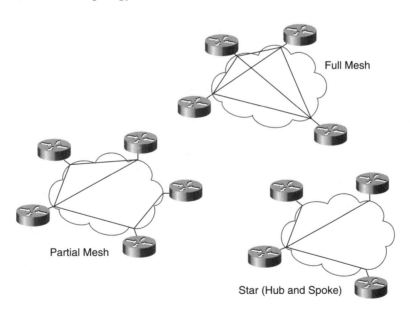

Figure 16–6
Frame Relay topologies.

Full Mesh

Partial Mesh

Star (Hub and Spoke)

Star Topology

A star topology, also known as a hub-and-spoke configuration, is the most popular Frame Relay network topology. In this topology, remote sites are connected to a central site that generally provides a service or application. This is the least expensive topology because it requires the least number of PVCs. In this scenario, the central router provides a multipoint connection because it is typically using a single interface to interconnect multiple PVCs.

Full-Mesh Topology

In a full-mesh topology, all routers have virtual circuits to all other destinations. This method, although costly, provides direct connections from each site to all other sites and allows for redundancy. For example, given three fully meshed routers, A, B, and C, if the link between router A and router B goes down, router A can still reach router B by rerouting traffic through router C. As the number of nodes in this topology increases, full-mesh topology becomes very expensive.

Partial-Mesh Topology

In a partial-mesh topology, not all sites have direct access to a central site. Depending on the traffic patterns in your network, you may want to have additional PVCs connect to remote sites that have large data traffic requirements.

REACHABILITY AND RESOURCE ISSUES FOR FRAME RELAY

In any of the possible topologies, when a single interface must be used to interconnect multiple sites, you may face reachability challenges because of the nonbroadcast multi-access (NBMA) nature of Frame Relay and its use of split horizon.

By default, a Frame Relay network provides NBMA connectivity between remote sites. Depending on the topology, routing update broadcasts received by one router may not be forwarded to all locations.

Consider the Frame Relay design shown in Figure 16–7. Split horizon reduces routing loops by not allowing a routing update received on one physical interface to be forwarded through the same physical interface. As a result, if router B in Figure 16–7 sends an update to router A, which connects multiple PVCs over a single physical interface, router A cannot send that broadcast through the same interface to the other remote routers, C and D.

To overcome this reachability problem, the router must replicate the broadcast for each active connection. That is, the distribution of broadcast traffic can be accomplished only by sending the same message to each virtual connection in order. This method requires considerable resource allocation within the router.

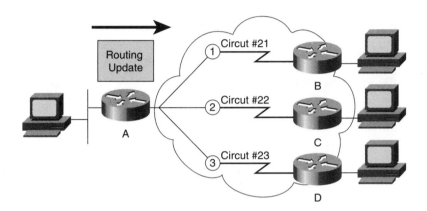

Figure 16–7
Broadcast traffic must be replicated from each active connection.

Broadcasts are not a problem if there is only a single PVC on a physical interface, because such a configuration is a point-to-point connection type.

The amount of broadcast traffic and the number of virtual circuits terminating at each router should be evaluated during the design phase of a Frame Relay network. Overhead traffic, such as routing updates, can impact the delivery of critical user data, especially when the delivery path contains low-bandwidth (56 kbps) links.

RESOLVING REACHABILITY AND RESOURCE ISSUES: SUBINTERFACES

The simplest answer to resolving the reachability issues brought on by split horizon may seem to be to turn off split horizon. Two problems exist with this solution. First, only IP allows you to disable split horizon; IPX and AppleTalk do not. (When an interface is configured with **encapsulation frame-relay**, split horizon is disabled for IP and enabled for IPX and AppleTalk, by default.)

The second problem is that disabling split horizon increases the chances of routing loops in your network.

To enable the forwarding of broadcast routing updates in a Frame Relay network, you can configure the router with logically assigned interfaces called *subinterfaces*. Subinterfaces are logical subdivisions of a physical interface, as shown in Figure 16–8. In split horizon routing environments, routing updates received on one subinterface can be sent out another subinterface. In subinterface configuration, each virtual circuit can be configured as a point-to-point connection, which allows the subinterface to act similarly to a leased line.

Figure 16–8
A single physical interface (S0) can simulate multiple logical interfaces (S0.1, S0.2, S0.3), called subinterfaces.

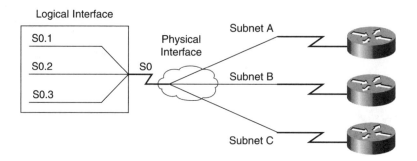

You can configure subinterfaces to support the following connection types:

- *Point-to-point*—A single subinterface is used to establish one PVC connection to another physical interface or subinterface on a remote router. In this case, the interfaces would be in the same subnet, and each interface would have a single DLCI. Each point-to-point connection is its own subnet. In this environment, broadcasts are not a problem because the routers are point-to-point and act like a leased line.

- *Multipoint*—A single subinterface is used to establish multiple PVC connections to multiple physical interfaces or subinterfaces on remote routers. In this case, all the participating interfaces would be in the same subnet, and each interface would have its own local DLCI. In this environment, because the subinterface is acting like a regular NBMA Frame Relay network, broadcast traffic is subject to the split horizon rule.

Key Concept   Instead of migrating to a routing protocol that supports turning off split horizon, subinterfaces can be used to overcome the split horizon problem.

Subinterfaces are particularly useful in a Frame Relay partial-mesh NBMA model that uses a distance vector routing protocol.

Configuring Subinterfaces

To configure subinterfaces on a physical interface, use the following steps:

1. Select the interface on which you want to create subinterfaces and enter the interface configuration mode.
2. Remove any network-layer address assigned to the physical interface. If the physical interface has an address, frames will not be received by the local subinterfaces.
3. Configure Frame Relay encapsulation, as discussed in the "Configuring Frame Relay" section.
4. Select the subinterface you want to configure. The command format is:

   ```
   Router(config-if)#interface serial number.subinterface-number
      {multipoint ¦ point-to-point}
   ```

   where the parameters and keywords have the following meanings:

   - *.subinterface-number*—Subinterface number in the range 1 to 4294967293. The interface number that precedes the period (.) must match the interface number to which this subinterface belongs.

   - **multipoint**—Select this if you want the router to forward broadcasts and routing updates that it receives. Also select this if you are routing IP and want all routers to be in the same subnet.

   - **point-to-point**—Select this if you do not want the router to forward broadcasts or routing updates and if you want each pair of point-to-point routers to have its own subnet, as shown in Figure 16–9.

   You must specify either **multipoint** or **point-to-point**; there is no default.

- Multipoint
  - Subinterfaces act as default NBMA network
  - Can save subnets because uses single subnet
  - Good for full-mesh topology

- Point-to-Point
  - Subinterfaces act as leased line
  - Each point-to-point connection requires its own subnet
  - Good for star or partial-mesh topologies

Figure 16–9
Each pair of routers has its own subnet.

5. Configure a network-layer address on the subinterface. If the subinterface is point-to-point, and you are using IP, you can use the **ip unnumbered** command:

```
Router(config-if)#ip unnumbered interface
```

If you use this command, it is recommended that the numbered interface (the one with the network-layer address) is the loopback interface. The Frame Relay link will not work if this command is pointing to an interface that is not fully operational, and a loopback interface is less likely to fail. The loopback interface is an address that represents the device itself.

6. If you configured the subinterface as point-to-point, or as multipoint with Inverse ARP enabled, you must configure the local DLCI for the subinterface to distinguish it from the physical interface. Use the following command:

```
Router(config-if)#frame-relay interface-dlci dlci-number
```

where *dlci-number* defines the local DLCI number being linked to the subinterface. This is the only way to link an LMI-derived PVC to a subinterface because LMI does not know about subinterfaces.

This command is not required for multipoint subinterfaces configured with static route maps.

Do not use this command on physical interfaces.

CAUTION

If you defined a subinterface for point-to-point communication, you cannot reassign the same subinterface number to be used for multipoint communication without first rebooting the router. Instead you can avoid using that subinterface number and use a different subinterface number.

Multipoint Subinterfaces Configuration Example

Consider the network shown in Figure 16–10.

The configuration output in Figure 16–11 shows how to configure multipoint subinterfaces. Specifically, the subinterfaces of Figure 16–10 have been configured on interface

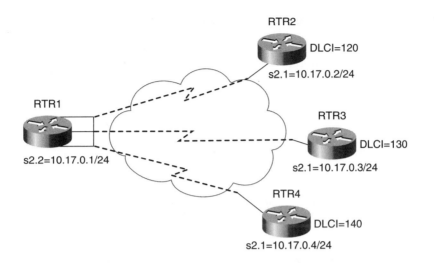

Figure 16–10
Multipoint subinterfaces.

S2.2, the central site router. With this type of configuration, the subinterface takes on the same Frame Relay characteristics as a physical interface. That is, each subinterface is NBMA and is subject to split horizon operation. The advantage, however, is that you need only a single network address.

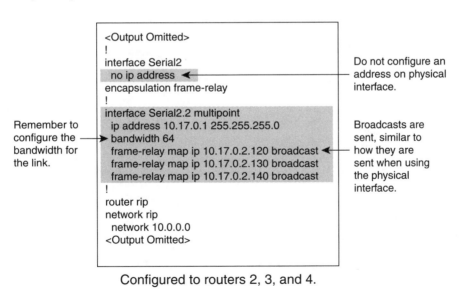

Figure 16–11
Router 1, central site, is connected to all remote sites via a single IP address with this configuration.

```
<Output Omitted>
!
interface Serial2
  no ip address   ◄
encapsulation frame-relay
!
interface Serial2.2 multipoint
  ip address 10.17.0.1 255.255.255.0
  bandwidth 64
  frame-relay map ip 10.17.0.2.120 broadcast
  frame-relay map ip 10.17.0.2.130 broadcast
  frame-relay map ip 10.17.0.2.140 broadcast
!
router rip
network rip
  network 10.0.0.0
<Output Omitted>
```

Remember to configure the bandwidth for the link.

Do not configure an address on physical interface.

Broadcasts are sent, similar to how they are sent when using the physical interface.

Configured to routers 2, 3, and 4.

Point-to-Point Subinterfaces Configuration Example

Consider the point-to-point subinterface configuration shown in Figure 16–12. Figure 16–13 shows the configuration for this Frame Relay network.

Figure 16–12
Point-to-point subinterfaces.

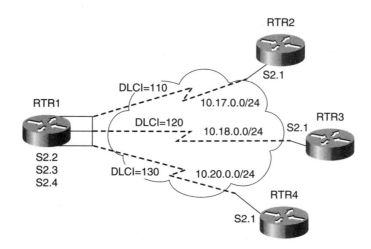

Figure 16–13
Router 1, central site, is connected to all remote sites by distinct IP addresses in this configuration.

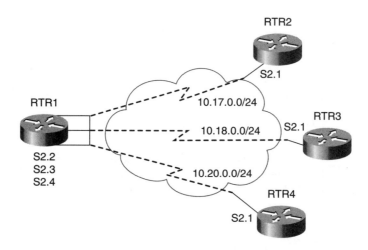

Figure 16–14 shows the configuration for Router 2 of Figure 16–12.

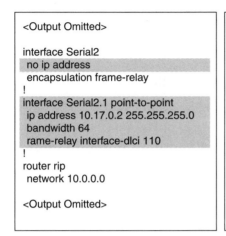

```
<Output Omitted>

interface Serial2
 no ip address
 encapsulation frame-relay
!
interface Serial2.1 point-to-point
 ip address 10.17.0.2 255.255.255.0
 bandwidth 64
 rame-relay interface-dlci 110
!
router rip
 network 10.0.0.0

<Output Omitted>
```

```
p1r2#show frame-relay map
Serial2.1 (up) : point-to-point dlci, dlci
110 (0x64, 0x1840), broadcast status defined, active

p1r2#show ip route
<Output Omitted>

   10.0.0.0/24 is subnetted, 3 subnet
R    10.18.0.0 [120/1] via 10.17.0.1, 00:00:04, Serial2.1
C    10.17.0.0 is directly connected, Serial2.1
R    10.20.0.0 [120/1] via 10.17.0.1, 00:00:04 Serial2.1
```

Figure 16–14
Router 2 at the remote site (Figure 16–12) is connected only to the central site via this configuration.

Figure 16–15 (see page 418) shows how to configure point-to-point subinterfaces using the **ip unnumbered** command.

When using the **ip unnumbered** command, remember the following:

- That the interface cannot be in the "no shut" state. That is, the interface must be in a fully operational state.

- To ensure that the most stable interface is referenced, it is recommended that you configure a loopback interface with an IP address and refer to this interface.

SUMMARY

This chapter has defined the terms and elements used in a Frame Relay network. Frame Relay allows you to configure multiple WAN connections from a single router. These connections can be physical or logical; logical connections are called subinterfaces. Sub-interfaces help overcome the potential reachability and resource usage challenges associated with Frame Relay as a result of its NBMA nature and its use of split horizon.

Figure 16–15
The ip unnum-
bered
command.

```
<Output Omitted>
!
interface Serial0
 ip address 100.4.2.1 255.255.255.0
!
ip address 172.7.2.9 255.255.255.0
 no fair-queue
 clockrate 56000
!
interface Serial2
 no ip address
 encapsulation frame-relay
!
!
interface Serial2.2 point-to-point
 ip unnumbered Serial1
 bandwidth 64
 frame-relay interface-dlci 220
!
interface Serial2.3 point-to-point
 ip unnumbered Serial0
 bandwidth 64
 frame-relay interface-dlci 230
!
<Output Omitted>
!
router rip
network 10.0.0.0
```

Chapter Sixteen Test
Configuring Frame Relay

Estimated Time: 15 minutes

Complete all the exercises to test your knowledge of the materials contained in this chapter. Answers are listed in Appendix A, "Chapter Test Answer Key."

Question 16.1

T   F   Subinterfaces enable you to set up point-to-point Frame Relay networks.

Question 16.2

T   F   Frame Relay can be used in a variety of topologies, depending on your network requirements.

Question 16.3

T   F   Frame relay routers and hubs can assist in managing traffic congestion.

Appendixes

APPENDIX A

Chapter Test
Answer Key

CHAPTER 1: "THE INTERNETWORKING MODEL"

1.1. LANs, WANs, and enterprise networks.

1.2. Layer 7: Application
Layer 6: Presentation
Layer 5: Session
Layer 4: Transport
Layer 3: Network
Layer 2: Data Link
Layer 1: Physical

1.3. [b] packets and [d] datagrams

1.4. [c] defines network addressing and determines the best path through an internetwork

1.5. [d] data, segments, packets, frames, bits

CHAPTER 2: "APPLICATIONS AND UPPER LAYERS"

2.1. Computer Applications: word processing, presentation graphics, spreadsheet, database, project planning, etc.
Network Applications: electronic mail, file transfer, remote access, client/server processes, information location, network management, etc.

Internetwork Applications: electronic data interchange, world wide web, email gateways, special-interest bulletin boards, financial transaction services, internet navigation utilities, conferencing (video, voice, data), etc.

2.2.   1F, 2C, 3E, 4D, 5A, 6B

2.3.   Host B will send back an acknowledgment indicating that data2 must be retransmitted. Host A will retransmit data2. Host B will then send an acknowledgment requesting data4. From this point, communication flow should resume normally.

CHAPTER 3: "PHYSICAL AND DATA LINK LAYERS"

3.1.   E=Ethernet; To=Token Ring; Fo=FDDI

3.2.   Ethernet II and SNAP frame types

3.3.   MAC=Media Access Control; LLC=Logical Link Control

3.4.   C: SDLC
       B: EIA/TIA-232
       E: 802.3
       F: Frame Relay
       D: Ethernet II
       G: FDDI
       A: Token Ring

CHAPTER 4: "NETWORK LAYER AND PATH DETERMINATION"

4.1. T

4.2. F

4.3. T

4.4. T

4.5. T

4.6. F

4.7. IP

4.8. Ethernet/802.3

4.9. B

4.10. A

4.11. C

CHAPTER 5: "BASIC ROUTER OPERATIONS"

5.1. RAM: show version, show running-config
NVRAM: show startup-config
Flash: show flash
Console privileged EXEC mode: enable
Interfaces: show interfaces

5.2. Type ? at the privileged EXEC mode prompt.

5.3. The **enable** command is used to enter privileged EXEC mode.

5.4. The system displays the last ten recorded commands when you enter the **show history** command.

5.5. The router quits the privileged EXEC mode and goes back to user mode.

CHAPTER 6: "CONFIGURING A ROUTER"

6.1. T

6.2. T

6.3. T

6.4. F

6.5. F

6.6. T

6.7. T

6.8. F

6.9. The **configure terminal** command is used to enter global configuration mode.

6.10. The **banner login** form of the **banner** command defines a login banner.

6.11. The **banner motd** command creates a Message of the Day banner.

6.12. You must enter a delimiting character after the **banner motd** command to indicate the end of the banner message.

6.13. The **interface serial 1** command string puts you in interface configuration mode for interface Serial1.

6.14. If you set both the enable and secret passwords, the secret password overrides the enable password and is used to enter privileged EXEC mode.

6.15. The **show controllers serial** command is used to determine if Serial1 is cabled as a DCE interface.

6.16. The **no shutdown** command is used to enable an interface.

6.17. You must be in the interface configuration mode for the specific interface you want to enable.

6.18. The **show version** command displays the current configuration register setting.

6.19. The following commands create and copy a back image in
Flash: **boot system Flash [IOS_filename]**
ROM: **boot system rom**
TFTP server: **boot system TFTP [file name][TFTP server address]**

6.20. The router must be in global configuration mode before you can issue the **boot system** command.

CHAPTER 7: "DISCOVERING AND ACCESSING OTHER CISCO ROUTERS"

7.1. T
7.2. F
7.3. T
7.4. T
7.5. F
7.6. F
7.7. F
7.8. This setting specifies the length of time between CDP updates.
7.9. On a network with a single Cisco router, or a very stable network that cannot support the extra traffic.
7.10. This parameter indicates how long the receiving device should hold a CDP packet sent from the local router.
7.11. They use multicast address 0100.0ccc.cccc and the value 2000 in the SNAP header.

CHAPTER 8: "TCP/IP OVERVIEW"

8.1. Network Interface
Internet
Transport
Application
8.2. Layer 3
8.3. Layer 4
8.4. A. ARP
B. IP
C. ICMP
D. TCP
E. TCP
F. UDP

G. UDP or IP
H. TCP
I. TCP
J. ICMP
K. UDP or IP
L. TCP
M. UDP

CHAPTER 9: "IP ADDRESS CONFIGURATION"

9.1. Class: B
Subnet: 172.16.2.0
9.2. Class: A
Subnet: 10.6.0.0
9.3. Class: A
Subnet: 10.30.36.0
9.4. Class: C
Subnet: 201.222.10.56
Broadcast: 201.222.10.63
9.5. Class: A
Subnet: 15.16.192.0
Broadcast: 15.16.199.255
9.6. Class: B
Subnet: 128.16.32.12
Broadcast: 128.16.32.15

CHAPTER 10: "IP ROUTING CONFIGURATION"

10.1. F
10.2. T
10.3. F
10.4. F
10.5. F
10.6. The networks discovered by the RIP routing protocol are preceded by the code letter R. The networks that are directly connected to the router and have been configured with the **network** command are preceded by the code letter C.
10.7. The **debug ip rip** command displays the RIP routing updates sent from and received at the router.

10.8. The **no debug ip rip** command disables the display of the RIP routing updates sent from and received at the router.

10.9. You issue the **router igrp** *autonomous-system* command to enable the IGRP routing protocol.

10.10. No. A new **router igrp** command must be issued when establishing an IGRP routing protocol for each autonomous system.

10.11. You issue the **show ip protocols** command to verify that the IGRP routing protocol is enabled.

10.12. You issue the **show ip route** command to display the current state of the IGRP routing table.

10.13. The **debug ip igrp events** command displays the IGRP routing updates events sent from the router.

10.14. IGRP uses a composite metric as its routing metric. This metric includes the following components:
(a) Bandwidth
(b) Delay
(c) Reliability
(d) Loading
(e) Maximum transmission unit

CHAPTER 11: "CONFIGURING NOVELL IPX"

11.1.

R3 Interface Name	Network Address	Encapsulation
S0	d100	hdlc
S1	c0b0	hdlc
E1	b1b0	novell-ether

11.2. Issue the **ipx routing** command to enable IPX routing on your router.

11.3. The router must be in global configuration mode before you can issue the **ipx routing** command.

11.4. You issue the **ipx network** *number* command to assign IPX network numbers on your router.

11.5. Issue the **show ipx interface** command to verify IPX address assignment on your router.

11.6. Issue the **show ipx route** command to verify entries in the routing table.

CHAPTER 12: "CONFIGURING APPLETALK"

12.1. T
12.2. T

12.3. T

12.4. F

12.5. The **appleTalk routing** command enables AppleTalk routing on the router.

12.6. Issue the **appleTalk cable-range** *cable-range* command to assign cable ranges to each interface on your router.

12.7. Issue the **appleTalk zones** *zone name* command to assign zone names to each link in your workgroup.

12.8. Issue the **show appleTalk interface** command to verify the address assignment on your router.

12.9. Issue the **show appleTalk route** command to verify entries in the routing table.

CHAPTER 13: "BASIC TRAFFIC MANAGEMENT WITH ACCESS LISTS"

13.1. You issue the **show ip interface** command to show if any access lists exist on your router.

13.2. The **ip access-group** *access-list-number* **in** command links the access list to the interface.

13.3. The router must be in interface configuration mode before you can issue this command.

13.4. You issue the **show ip interface** command to verify that a new access list exists on your router.

13.5. You issue the **show access-lists** command to display the contents of an access list.

13.6. Use the list numbers in the range of 800 to 899 when defining a standard IPX access list.

13.7. You issue the **show ipx interface** *interface* to verify that a new access list is now active on the interface.

13.8. The **show ipx access-lists** command displays the contents of all access lists.

13.9. Use the range of list numbers from 600 to 699 when you define an Apple-Talk access list.

13.10. The **show appletalk access-lists** command displays the contents of the AppleTalk access lists.

CHAPTER 14: "INTRODUCTION TO SERIAL CONNECTIONS"

14.1.   T
14.2.   T
14.3.   (a) A
           (b) B
           (c) A
           (d) B
           (e) C
           (f) A
           (g) B
           (h) C
14.4.   (A) Switched or relay services
           (B) Front end to an IBM enterprise data center computer
           (C) Connection between peer devices
14.5.   You issue the **show interfaces** *interface-number* command to verify that your interface is configured for PPP encapsulation.
14.6.   You issue the **debug ppp authentication** command to display the CHAP exchange sequence. This command is issued prior to intiating the line.

CHAPTER 15: "CONFIGURING X.25"

15.1.   T
15.2.   F
15.3.   T
15.4.   (a) T
           (b) T
           (c) T
           (d) T
           (e) F
           (f) F
15.5.   Issue the **encapsulation x25** command to specify X.25 encapsulation on your DTE interface.
15.6.   Issue the **encapsulation x25 dce** command to specify X.25 encapsulation on your DCE interface.

CHAPTER 16: "CONFIGURING FRAME RELAY"

16.1.   F
16.2.   T
16.3.   F

APPENDIX   B

Configuring DECnet

This appendix presents how to configure Cisco routers in DECnet networks. First, you'll review the DECnet protocol stack, and then you'll learn about setting global and interface parameters for DECnet. Finally, you'll see examples of various DECnet configurations and information on how to monitor a DECnet configuration.

OVERVIEW OF DECNET

DECnet is a proprietary Digital Equipment Corporation (Digital) protocol. DECnet Phase V is the current version, although DECnet Phase IV is still seen in many installations. Unless otherwise noted, the discussion in this chapter applies to Phase IV. Figure B–1 shows the Hayes DECnet architecture compared to the OSI model.

A DECnet address contains 16 bits: 6 bits for the area and 10 bits for the node address. In Figure B–2, the DECnet router is assigned address 5.14 (the area address is 5; the node address is 14). Each addressable entity is called a node; each node is assigned one address. The *area.node* address is modified and becomes a software-formatted MAC address used on all interfaces.

Traffic is localized by placing nodes in logical or physical groupings called *areas*.

DECnet Phase IV uses a distance vector protocol; path determination is based on the cost of all outgoing interfaces. Routers keep cost calculations for all hosts in their area. Due to the incorporation of logical addressing into the MAC address, no address resolution is required, as it is in IP.

	OSI Reference Model		DECnet Architecture
7	Application	7	User
			Network Mgt.
6	Presentation	6	Network Application
5	Session	5	Session Control
4	Transport	4	End-to-End Communication
3	Network	3	Routing
2	Data Link	2	Data Link
1	Physical	1	Physical

Area.Node

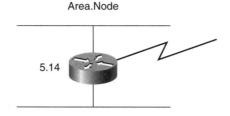

5.14

DECnet Phase IV uses the DECnet Routing Protocol (DRP). DECnet Phase V uses the standard OSI Routing Protocol (IS-IS), which is a link-state routing protocol.

Each device speaking DECnet is a node, and addresses are assigned to the node, not to individual wires or interfaces. Therefore, the entire router, as a node, is assigned one address. A 6-bit area number allows a maximum of 63 areas, and all nodes in an area must be contiguous (use an uninterrupted sequence of numbers).

Within a LAN, one router is chosen as the *designated router* (DR). The DR is always known to end nodes because of periodic DR announcements. All traffic from an end node is initially sent to the designated router for forwarding. Later, as network knowledge is learned, the end nodes use a more direct path.

Several nodes in an area are assigned intra-area router status. These nodes contain knowledge about all nodes within that area in their routing tables.

One or more nodes in an area are assigned to be an interarea router. The task of this router is to forward traffic to a specific router in another area, as shown in Figure B–3. An interarea router's routing table contains all nodes in an area plus the paths to other areas via interarea routers.

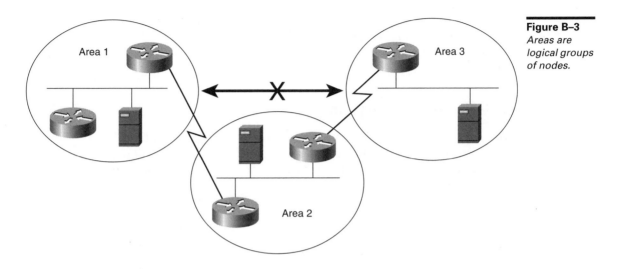

Figure B–3
Areas are logical groups of nodes.

Node Assignments

Using a 10-bit node number allows a maximum of 1023 nodes in each area. The node address is folded into the MAC address for each interface, as shown in Figure B–4.

When DECnet initializes, the modified (software-supplied) MAC address is propagated onto each interface.

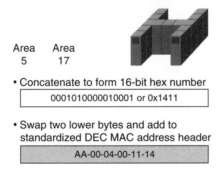

Area   Area
 5   17

• Concatenate to form 16-bit hex number

0001010000010001 or 0x1411

• Swap two lower bytes and add to standardized DEC MAC address header

AA-00-04-00-11-14

Figure B–4
Node assignments.

When a host boots, it advertises its presence. During normal operation, host reachability is advertised to local routers every 15 seconds.

End nodes (nonrouting hosts) have no knowledge of the network after they boot up. Only the designated router is known through its periodic announcements.

All traffic from the end node is forwarded to the designated router. As delivery responses from destination nodes are returned to the end node, addresses are placed in a cache. Traffic destined for nodes already in the cache will not be directed to the designated router.

Routers Pass Information

Periodic updates are sent by each router, as shown in Figure B–5. These updates contain cost information to all reachable nodes within each router's area. Each interface has an outgoing cost associated with it. Routing decisions are based on total path cost.

Figure B–5
Routers pass information.

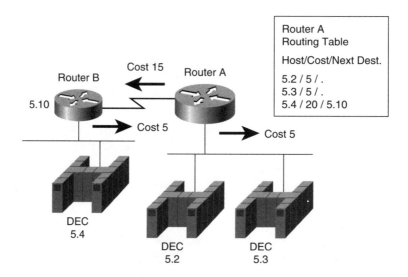

Designated Routers

As each node powers up, it advertises its presence using the Hello protocol. The Hello protocol is used by routers to advertise their presence onto the locally connected network. Periodic announcements are sent by each node to advertise its reachability.

The single DR is always known to end nodes because of periodic DR announcements, as shown in Figure B–6. All traffic from an end node is initially sent to the designated router for forwarding. Later, the end nodes use a more direct path as network knowledge is acquired.

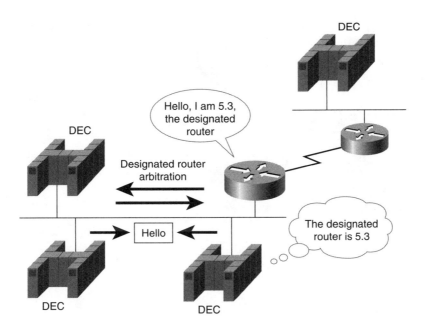

Figure B–6
The DR is known through its announcements.

Level 1 and Level 2 Routing

Routers that forward traffic within their own area are referred to as *Level 1 routers*. These routers have complete knowledge of all nodes within that area. These routers are referred to as routing-iv (a reference to Phase IV DECnet).

Routers that communicate between areas have knowledge of all nodes in their area and of the nodes that provide entry into other areas. These *Level 2 routers* are configured as area routers (as shown in Figure B–7).

DECNET CONFIGURATION COMMANDS

In this section, you learn about the global DECnet configuration parameters and the configuration commands you need to configure DECnet. This section also includes an example DECnet configuration.

DECnet Configuration Tasks

The activation of DECnet as a routing protocol requires setting global and interface parameters.

Figure B–7
Level 1 routing routes inside your area; Level 2 routing routes between areas.

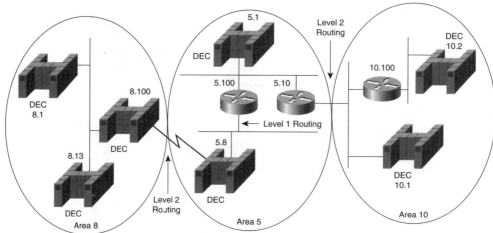

The global tasks are as follows:

- Start the DECnet routing process and assign a node address.
- Designate the router as a Level 1 or 2.

The required interface task is to assign each interface an outgoing cost. There is no default value. Figure B–8 shows an internetwork that has been configured with area addresses, node addresses, and a metric cost for crossing the WAN link.

Figure B–8
A DECnet network that is fully configured.

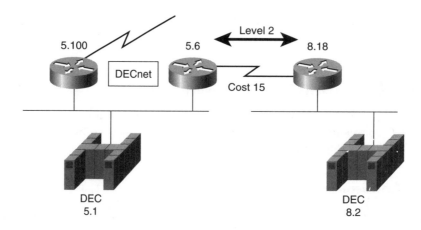

DECnet Configuration Commands

The following configuration commands are needed when routing with DECnet:

- The **decnet routing** command starts the routing process and assigns an area.node address to the entire router.

- The **decnet node-type** command establishes the routing characteristics of a router. Routers are referred to as Level 1 routers if they perform the intra-area routing task.

  Level 1 routing is specified by *routing-iv*, the default.

  Level 2 routing or interarea routing is specified by *area*.

- The **decnet cost** command enables DECnet on this interface and assigns a cost (from 1 to 63) to the interface. There are no default costs. A cost must be assigned to each interface. Suggested costs are 1 for FDDI interfaces, 4 for Ethernet interfaces, and a minimum of 10 for serial links.

  The cost assigned should be proportional to the speed of the media; the higher the bandwidth, the lower the cost associated with its use.

Refer to your Cisco documentation or go to www.cisco.com for more information on these commands.

DECnet Configuration Example

Figure B–9 shows an example of a DECnet configuration.

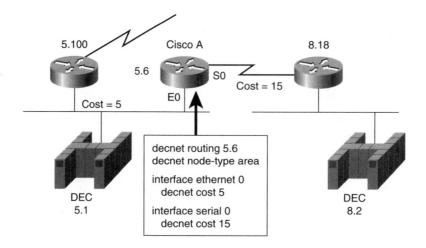

Details of the configuration in Figure B–9 are:

Command	Description
decnet routing 5.6	Enables DECnet routing and assigns the router an address of Area 5, Node 6
decnet node-type area	Defines the router as a Level 2 interarea node
decnet cost 15	Assigns an outgoing cost of 15 to the interface serial 0

The router is assigned an address of 5.6 with responsibility for connecting area 5 to other areas. Interface costs are assigned. There are no defaults.

CONFIGURING DECNET ACCESS LISTS

To configure standard and extended access lists for DECnet, the access list number an administrator chooses must be within the range of 300-399. The source address can be an entire area or an *area.node*.

The optional wildcard masks match bit-for-bit with the DECnet *area.node* address. As with other protocols, a zero in the wildcard mask indicates that the corresponding bit in the DECnet address will be checked; a one in a wildcard mask bit position means that the corresponding bit position in the DECnet address can be ignored during the access list test.

However, because the DECnet addressing uses decimal numbers, wildcard masking differs from masking used with IP addresses. For example, to mask all bits in a DECnet area address, express the "all-ones" mask as 1023 (1111111111).

If the traffic to control has its destination on another DECnet area, it must cross a Level 2 router. An access list specifying only the source address must be placed near the destination. Depending on the specific controls and masking required, this may be at the appropriate Level 2 interface.

DECnet Access List Commands

The **access-list** command is used to make an entry in a traffic filter list.

The **decnet access-group** command is used to link an access list to the selected interface.

Controlling DECnet Example

Figure B–10 shows an example of DECnet traffic control.

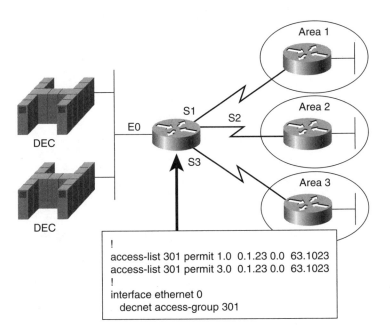

Figure B–10
Controlling DECnet example.

```
!
access-list 301 permit 1.0  0.1.23 0.0  63.1023
access-list 301 permit 3.0  0.1.23 0.0  63.1023
!
interface ethernet 0
   decnet access-group 301
```

Access list 301 is configured to allow traffic from any node in areas 1 and 3 to be forwarded out interface Ethernet 0. It implies that no other traffic will be permitted. (The end of a list contains an implicit deny all else statement.)

Monitoring DECnet

The following commands are used to monitor the progress of DECnet:

- Use the **show decnet interface** command to display status about all DECnet interfaces, including line status, timers, and access lists assigned.

- Use the **show decnet route** command to display the contents of the DECnet routing table.

- Use the **show decnet traffic** command to show different forms of traffic that have arrived at the router.

- Use the **debug decnet routing** command to display routing update messages.

For more information on configuring DECnet, refer to the Cisco documentation CD or go to www.cisco.com.

APPENDIX C

Configuring Banyan VINES

This appendix explains how to configure Banyan VINES routing on Cisco routers. First, this appendix contains a VINES overview, and then you'll learn about the configuration and monitoring commands you'll need to maintain VINES on Cisco routers.

VINES OVERVIEW

Banyan Virtual Integrated Network Service (VINES) is a proprietary protocol of Banyan Systems. The VINES protocol stack has seven layers, as does the OSI reference model. The VINES implementation differences (compared to OSI) occur at Layers 3 and 4, as shown in Figure C–1.

	OSI Reference Model			VINES Architecture
7	Application		7	Application
6	Presentation		6	Presentation
5	Session		5	Session
4	Transport		4	Interproc. Comm. Seq. Packet
3	Network		3	VINES IP
2	Data Link		2	Data Link
1	Physical		1	Physical

Figure C–1
The VINES protocol stack compared to the OSI reference model.

Layer 3 contains VINES Internet Protocol (VIP or VinesIP) as a connectionless data-gram delivery protocol. This protocol is similar to the IP in TCP/IP, and it can inter-operate in a TCP/IP environment.

Layer 4 transport protocols are the Interprocess Communications Protocol (IPC) and the Sequenced Packet Protocol (SPP). These VINES-specific protocols provide a connection-oriented, reliable transport mechanism. IPC also supports unreliable data-gram service.

Routing Update Protocol (RTP) is the network-layer protocol responsible for propagat-ing routing updates. Routing decisions are based on a *delay metric*. The administrator can assign the delay metric to each interface. If not defined statically, a delay metric based on bandwidth is assigned to the interface. The delay metric is multiplied by 200 ms (milliseconds) to represent it in a usable "time interval" format. Time interval values from different interfaces can then be compared more easily.

VINES messages are generated at 90-second intervals:

- Clients send hellos.
- Servers send hellos and updates.
- Routers send routing updates.

Cisco also supports the Sequenced Routing Update Protocol (SRTP). This more recent routing protocol uses an update-based scheme for routers and servers to communicate routing changes. SRTP is similar to the RIP routing process in its use of periodic routing updates.

A VINES address has 48 bits including a 32-bit (4-byte) network address and 16-bit (2-byte) subnet address.

Network Number

A *network number* is a unique value assigned to each server; it is the first number of the network:subnet number pairs shown in Figure C–2. Banyan servers have a hardware key that provides their addresses.

Subnet Number

The subnet number is equivalent to a host number. These values are hexadecimal format and are assigned by function: 1 is used for a server; 8001-FFFF is used for clients.

Client numbers are usually assigned incrementally beginning at 8001.

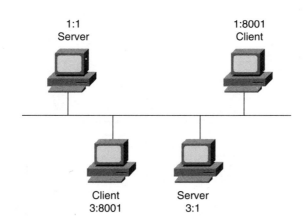

1:1
Server

1:8001
Client

Client
3:8001

Server
3:1

Figure C–2
*VINES
addressing
uses a net-
work:subnet
pair for both
servers and
clients.*

Cisco VINES Network Number

Cisco's assigned network address block is hexadecimal 300. The Cisco addresses are created from the lower 21 bits of the Ethernet or Token Ring interface Media Access Control (MAC) address. These bits are placed behind a block address of 300. The resulting value is used by the router as its network number.

For example, consider the Ethernet address 0000.0c01.58b4. The Cisco router uses the last 21 bits of the Ethernet address (0158b4) and places the Cisco block address 300 in front. The Cisco router takes the Banyan server number of 300158b4:1. The server subnet value of 1 is assigned to routers.

Host Address Assignment

A VINES client has no address on startup. A broadcast message, using the VINES Address Resolution Protocol (ARP), is sent to notify servers that a new client requires address assignment, as shown in Figure C–3.

The first server to respond to the request assigns the client address based on that server's network number and the next available subnet number.

Hello Messages

Clients send hellos at 90-second intervals. Servers send hello and update messages at 90-second intervals. Cisco routers send update messages at 90-second intervals. Routing information is included in the periodic updates.

Figure C–3
*A new client
broadcasts an
ARP request
to get an
address.*

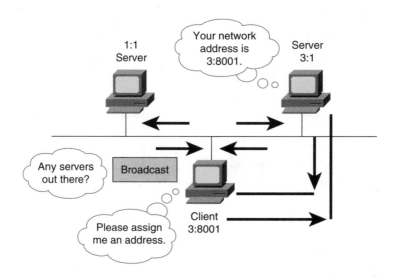

CONFIGURING VINES

Selecting VINES as a routed protocol requires configuring both global and interface parameters, as follows:

- The global task is to start the VINES routing process.

- The interface task is to assign a VINES metric on each interface. A default metric is selected if none is specified by the network administrator. The metric value is based on the bandwidth capability of each media.

- On segments where there is no VINES server, the router assigns client addresses and propagates client service broadcasts to the nearest server.

VINES Configuration Commands

The **vines routing** command starts the VINES routing process. Cisco maps to a reserved server network address based on a block of addresses assigned by Banyan. The address created contains 21 bits from the MAC address of an Ethernet or Fiber Distributed Data Interface (FDDI). Use the optional **address** field if you do not have an Ethernet or FDDI (no MAC address) present in the router.

If two routers on the same media have the same address because the addresses were not hard coded, use the optional **recompute** keyword to force random address selection.

The **vines metric** command turns on VINES processing in this interface. This configuration statement is required for each interface.

If no metric is specified, the Cisco router uses a default metric based on the bandwidth of the link, as follows:

Interface Type	Delay Metric Value
Ethernet	2
16-Mb Token Ring	2
4-Mb Token Ring	4
56-Kb Serial	45
9600 Serial	90

The **vines arp-enable** command allows the router to assign client addresses. The optional keyword **dynamic** should be used on segments that have no VINES server present. If no option is specified, the router responds to all ARP requests, even if a VINES server is present on this network.

The **vines serverless** command allows the propagation of certain broadcast packets by the router. These broadcast packets are forwarded to the nearest server. The use of the **vines serverless** command is limited to segments that do not have servers.

In Figure C–4, two Cisco routers are being configured to route VINES communications across a serial link.

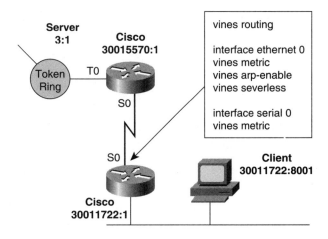

Figure C–4
A VINES configuration example, in which two Cisco routers are being configured to route VINES communications across a serial link.

- The command **vines routing** starts the VINES routing process.

- The command **vines metric** enables Ethernet 0 for VINES packet processing. A default metric will be used on the interface.

- The **vines arp-enable** command enables the router to respond to VINES ARP requests. ARP requests are issued when the client requires address assignment at startup.

- The **vines serverless** command enables the router to forward service broadcasts to the nearest VINES server.

With these commands, the VINES routing process is enabled. The router responds to address assignment requests and forwards broadcast service requests that arrive on interface Ethernet 0.

VINES Access Lists

Unlike other networking protocols, the VINES network number refers to a device: The VINES network refers to the server itself, rather than the link between devices. This reference also applies to a Cisco router configured to route VINES.

The VINES network address is 32 bits long, expressed as eight hexadecimal numbers; the subnetwork numbers refer to hosts or clients connecting to the server. The subnet number is 16 bits long expressed as a 4-bit hex number. A colon separates the VINES network and subnetwork numbers.

Subnets (nodes or hosts) use the number 1 if they are servers; or if they are clients, they use a dynamically assigned number from a hexadecimal range beginning at 8001 and ending at FFFF.

The minimal form of VINES access lists is the simple access list. For this filter, enter a source VINES address only. The router permits or denies all VINES packets using this type of access list.

Administrators using a standard VINES access list can also filter on a specific VINES protocol as well as the source and destination addresses.

With VINES, the global command for a standard access list includes the word "vines" in the command itself. It uses an access list number range of 1–100 (for example: Vines Access-List 1).

With standard access lists, the administrator also specifies a protocol by name or by number. For more details on VINES protocol keywords or ID numbers, refer to the information in the command reference document.

VINES source and destination address arguments for access list statements point to the network. The administrator uses these to control packet traffic between systems.

Once the administrator applies a VINES access list to a router interface, the router interface will permit or deny outgoing traffic matching the access list statements. Packets generated by the router itself are not subject to the access list controls.

The **vines access-list** command is used to create an entry in the traffic filter list. This list defines traffic that is "interesting" and should be either permitted or denied for forwarding. Refer to the Cisco documentation or www.cisco.com for more information on the **vines access-list** command.

MONITORING VINES OPERATION

The following commands are used to monitor the operation of VINES:

- Use the **show vines interface** command to display the status of the interface. The status information includes addresses, update timers, and presence of access lists.

- Use the **show vines route** command to display the contents of the VINES routing table, including known areas.

- Use the **show vines neighbor** command to display the contents of the neighbor table. This table contains host names, MAC addresses, encapsulation type, and interface port information.

- Use the **debug vines routing** command to display the contents of the periodic routing updates, including the incoming interface and cost to each network.

For more information on monitoring VINES operation, refer to the Cisco documentation or go to www.cisco.com.

AutoInstalling Configuration Data

The AutoInstall procedure allows a network administrator to configure a router remotely over the network. This configuration is most useful for establishing new routers in remote locations where branch office staff members have limited networking knowledge and skills.

The new router must be connected to an existing router on either a WAN or LAN link, as shown in Figure D–1. Both existing and new routers must be running Cisco IOS Release 9.1 or later for encapsulations other than Frame Relay. For Frame Relay encapsulation, both routers must be running Cisco IOS Release 10.3 or later.

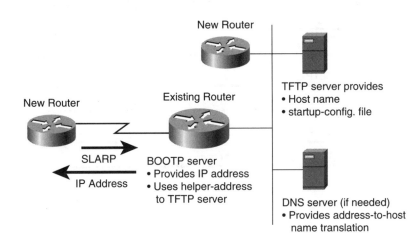

New Router

TFTP server provides
• Host name
• startup-config. file

New Router   Existing Router

SLARP

IP Address

BOOTP server
• Provides IP address
• Uses helper-address
  to TFTP server

DNS server (if needed)
• Provides address-to-host
  name translation

Figure D–1
AutoInstall enables you to configure a new router automatically and remotely.

449

The existing router acts as a Bootstrap Protocol (BOOTP) or Reverse Address Resolution Protocol (RARP) server. It must be set up to help the new router acquire its IP address. This existing router also contains a helper address for the TFTP server.

TIPS

Make sure that the new router configuration files reside on the TFTP server. Prepare new router configuration files for AutoInstall in the Cisco IOS software configuration mode. Move your new router configuration files using the **copy running-config tftp** command to store the current configuration in RAM on a network TFTP server.

This server provides a host name for the address presented by the new router. If this IP address-to-host name translation does not occur on the TFTP server, the new router uses a Domain Name System (DNS) server. The new router configuration is downloaded from a reachable TFTP server to the new router.

The AutoInstall procedure has several steps, as shown in Figure D–2.

Figure D–2
The new router acquires its IP address, host name, and configuration.

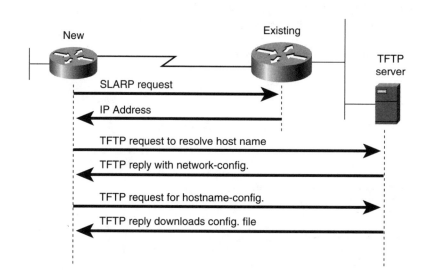

These AutoInstall steps are outlined as follows:

1. First, the new router sends a Serial Line Reverse Address Resolution Protocol (SLARP) request packet over the serial line. The existing router will reply with its IP address. If the address is the first host of the subnet, for example xx.xx.xx.1 in a class B network with subnet mask 255-255-255-0, the new router will automatically be assigned the second subnet host address for its own serial interface, for example, xx.xx.xx.2.

2. Once it has obtained an IP address, the new router requests a translation by the TFTP server to resolve this IP address into a host name. The response to this request comes in the form of a network-confg file containing the host name for the new router.

3. The new router uses its newly acquired host name to request the hostname-confg file that contains its specific configuration entries. The TFTP server downloads this file to the new router.

4. The AutoInstall process also includes several fallback requests to use if a common scenario fails to provide the proper response to the new router's requests.

   If the host name request to the TFTP server fails to provide the new router with a host name, it will fall back to another request procedure. This sends a request to the DNS server to obtain IP address-to-host name translation. Figure D–3 illustrates the fallback process.

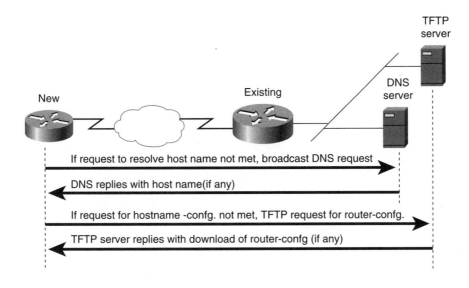

Figure D–3
If host name resolution from TFTP network-confg fails, the new router sends a request to the DNS server.

5. Later, if the new router requests a hostname-confg file, but the TFTP server cannot send the requested file, it will send a more generic configuration in a router-confg file. Then the administrator can log in to the new router and make any specific configuration changes necessary for the new router.

For AutoInstall examples, commands, and details, refer to the Cisco documentation or go to www.cisco.com.

APPENDIX E

Decimal to Hexadecimal and Binary Conversion Table

Decimal Value	Hexadecimal Value	Binary Value
0	00	0000 0000
1	01	0000 0001
2	02	0000 0010
3	03	0000 0011
4	04	0000 0100
5	05	0000 0101
6	06	0000 0110
7	07	0000 0111
8	08	0000 1000
9	09	0000 1001
10	0A	0000 1010
11	0B	0000 1011
12	0C	0000 1100
13	0D	0000 1101
14	0E	0000 1110
15	0F	0000 1111
16	10	0001 0000
17	11	0001 0001
18	12	0001 0010
19	13	0001 0011
20	14	0001 0100
21	15	0001 0101

Decimal Value	Hexadecimal Value	Binary Value
22	16	0001 0110
23	17	0001 0111
24	18	0001 1000
25	19	0001 1001
26	1A	0001 1010
27	1B	0001 1011
28	1C	0001 1100
29	1D	0001 1101
30	1E	0001 1110
31	1F	0001 1111
32	20	0010 0000
33	21	0010 0001
34	22	0010 0010
35	23	0010 0011
36	24	0010 0100
37	25	0010 0101
38	26	0010 0110
39	27	0010 0111
40	28	0010 1000
41	29	0010 1001
42	2A	0010 1010
43	2B	0010 1011
44	2C	0010 1100
45	2D	0010 1101
46	2E	0010 1110
47	2F	0010 1111
48	30	0011 0000
49	31	0011 0001
50	32	0011 0010
51	33	0011 0011
52	34	0011 0100
53	35	0011 0101
54	36	0011 0110
55	37	0011 0111
56	38	0011 1000
57	39	0011 1001
58	3A	0011 1010
59	3B	0011 1011
60	3C	0011 1100
61	3D	0011 1101

Decimal Value	Hexadecimal Value	Binary Value
62	3E	0011 1110
63	3F	0011 1111
64	40	0100 0000
65	41	0100 0001
66	42	0100 0010
67	43	0100 0011
68	44	0100 0100
69	45	0100 0101
70	46	0100 0110
71	47	0100 0111
72	48	0100 1000
73	49	0100 1001
74	4A	0100 1010
75	4B	0100 1011
76	4C	0100 1100
77	4D	0100 1101
78	4E	0100 1110
79	4F	0100 1111
80	50	0101 0000
81	51	0101 0001
82	52	0101 0010
83	53	0101 0011
84	54	0101 0100
85	55	0101 0101
86	56	0101 0110
87	57	0101 0111
88	58	0101 1000
89	59	0101 1001
90	5A	0101 1010
91	5B	0101 1011
92	5C	0101 1100
93	5D	0101 1101
94	5E	0101 1110
95	5F	0101 1111
96	60	0110 0000
97	61	0110 0001
98	62	0110 0010
99	63	0110 0011
100	64	0110 0100
101	65	0110 0101

Decimal Value	Hexadecimal Value	Binary Value
102	66	0110 0110
103	67	0110 0111
104	68	0110 1000
105	69	0110 1001
106	6A	0110 1010
107	6B	0110 1011
108	6C	0110 1100
109	6D	0110 1101
110	6E	0110 1110
111	6F	0110 1111
112	70	0111 0000
113	71	0111 0001
114	72	0111 0010
115	73	0111 0011
116	74	0111 0100
117	75	0111 0101
118	76	0111 0110
119	77	0111 0111
120	78	0111 1000
121	79	0111 1001
122	7A	0111 1010
123	7B	0111 1011
124	7C	0111 1100
125	7D	0111 1101
126	7E	0111 1110
127	7F	0111 1111
128	80	1000 0000
129	81	1000 0001
130	82	1000 0010
131	83	1000 0011
132	84	1000 0100
133	85	1000 0101
134	86	1000 0110
135	87	1000 0111
136	88	1000 1000
137	89	1000 1001
138	8A	1000 1010
139	8B	1000 1011
140	8C	1000 1100
141	8D	1000 1101

Decimal Value	Hexadecimal Value	Binary Value
142	8E	1000 1110
143	8F	1000 1111
144	90	1001 0000
145	91	1001 0001
146	92	1001 0010
147	93	1001 0011
148	94	1001 0100
149	95	1001 0101
150	96	1001 0110
151	97	1001 0111
152	98	1001 1000
153	99	1001 1001
154	9A	1001 1010
155	9B	1001 1011
156	9C	1001 1100
157	9D	1001 1101
158	9E	1001 1110
159	9F	1001 1111
160	A0	1010 0000
161	A1	1010 0001
162	A2	1010 0010
163	A3	1010 0011
164	A4	1010 0100
165	A5	1010 0101
166	A6	1010 0110
167	A7	1010 0111
168	A8	1010 1000
169	A9	1010 1001
170	AA	1010 1010
171	AB	1010 1011
172	AC	1010 1100
173	AD	1010 1101
174	AE	1010 1110
175	AF	1010 1111
176	B0	1011 0000
177	B1	1011 0001
178	B2	1011 0010
179	B3	1011 0011
180	B4	1011 0100
181	B5	1011 0101

Decimal Value	Hexadecimal Value	Binary Value
182	B6	1011 0110
183	B7	1011 0111
184	B8	1011 1000
185	B9	1011 1001
186	BA	1011 1010
187	BB	1011 1011
188	BC	1011 1100
189	BD	1011 1101
190	BE	1011 1110
191	BF	1011 1111
192	C0	1100 0000
193	C1	1100 0001
194	C2	1100 0010
195	C3	1100 0011
196	C4	1100 0100
197	C5	1100 0101
198	C6	1100 0110
199	C7	1100 0111
200	C8	1100 1000
201	C9	1100 1001
202	CA	1100 1010
203	CB	1100 1011
204	CC	1100 1100
205	CD	1100 1101
206	CE	1100 1110
207	CF	1100 1111
208	D0	1101 0000
209	D1	1101 0001
210	D2	1101 0010
211	D3	1101 0011
212	D4	1101 0100
213	D5	1101 0101
214	D6	1101 0110
215	D7	1101 0111
216	D8	1101 1000
217	D9	1101 1001
218	DA	1101 1010
219	DB	1101 1011
220	DC	1101 1100
221	DD	1101 1101

Decimal Value	Hexadecimal Value	Binary Value
222	DE	1101 1110
223	DF	1101 1111
224	E0	1110 0000
225	E1	1110 0001
226	E2	1110 0010
227	E3	1110 0011
228	E4	1110 0100
229	E5	1110 0101
230	E6	1110 0110
231	E7	1110 0111
232	E8	1110 1000
233	E9	1110 1001
234	EA	1110 1010
235	EB	1110 1011
236	EC	1110 1100
237	ED	1110 1101
238	EE	1110 1110
239	EF	1110 1111
240	F0	1111 0000
241	F1	1111 0001
242	F2	1111 0010
243	F3	1111 0011
244	F4	1111 0100
245	F5	1111 0101
246	F6	1111 0110
247	F7	1111 0111
248	F8	1111 1000
249	F9	1111 1001
250	FA	1111 1010
251	FB	1111 1011
252	FC	1111 1100
253	FD	1111 1101
254	FE	1111 1110
255	FF	1111 1111

Password Recovery

This appendix explains several password recovery techniques for Cisco routers and Catalyst switches. You can perform password recovery on most of the platforms without changing hardware jumpers, but all platforms require the router to be reloaded. Password recovery can be done only from the console port physically attached to the router.

There are three ways to restore access to a router when the password is lost. You can view the password, change the password, or erase the configuration and start over as if the box were new.

Each procedure follows these basic steps:

1. Configure the router to boot up without reading the configuration memory (NVRAM). This is sometimes called the "test system mode."
2. Reboot the system.
3. Access enable mode (which can be done without a password if you are in test system mode).
4. View or change the password, or erase the configuration.
5. Reconfigure the router to boot up and read the NVRAM as it normally does.
6. Reboot the system.

CAUTION

Some password recovery requires a terminal to issue a BREAK signal; you must be familiar with how your terminal or PC terminal emulator issues this signal. For example, in Pro-Comm, the keys Alt-B will by default generate the BREAK signal, and in Windows Terminal you press Break or Ctrl-Break. Windows Terminal also allows you to define a function key as BREAK. From the terminal window, select Function Keys and define one as BREAK by filling in the characters ^$B (Shift 6, Shift 4, and Capital B).

The following sections contain detailed instructions for specific Cisco routers and Catalyst switches. Locate your product at the beginning of each section to determine which technique to use.

TECHNIQUE #1

Relevant devices are: all Cisco AGS, Cisco 2000 Series, Cisco 2500 Series, Cisco 3000 Series, 680x0-Based Cisco 4000 Series, Cisco 7000 Series Running Cisco IOS 10.0 or Later in ROMs, IGS Series Running Cisco IOS 9.1 or Later in ROMs.

This technique can be used on the Cisco 7000 and Cisco 7010 only if the router has Cisco IOS 10.0 ROMs installed on the RP card. It may be booting Flash Cisco IOS 10.0 software, but it needs the actual ROMs on the processor card as well.

1. Attach a terminal or PC with terminal emulation to the console port of the router.
2. Type **show version** and record the setting of the configuration register. It is usually 0x2102 or 0x102.
3. Power the router down, and then up.
4. Press the Break key on the terminal within 60 seconds of the power up. You will see the > prompt with no router name. If you don't, the terminal is not sending the correct Break signal. In that case, check the terminal or terminal emulation setup.
5. Type **o/r 0x42** at the > prompt to boot from Flash or **o/r 0x41** to boot from the boot ROMs. (Note that this is the letter "o," not the numeral zero.) If you have Flash and it is intact, 0x42 is the best setting. Use 0x41 only if the Flash is erased or not installed.

CAUTION

If you use 0x41, you can only view or erase the configuration. You cannot change the password.

---

6. Type **i** at the > prompt. The router will reboot but will ignore its saved configuration.
7. Answer no to all the setup questions.
8. Type **enable** at the Router> prompt. You'll be in enable mode and see the Router# prompt.
9. Choose one of these three options:
   - To view the password, type **show config.**

   - To change the password (in case it is encrypted, for example), do the following:

        a. Type **config mem** to copy the NVRAM into memory.

        b. Type **wr term.**

           If you have **enable secret xxxx**, then:

           Type **config term** and make the changes.

           Type **enable secret <password>.**

           Press **Ctrl-Z.**

           If you do not, then:

           Type **enable password <password>.**

           Press **Ctrl-Z.**

        c. Type **write mem** to commit the changes.

   - To erase the config, type **write erase.**
10. Type **config term** at the prompt.
11. Type **config-register 0x2102,** or whatever value you recorded in step 2.
12. Press **Ctrl-Z** to quit from the editor.
13. Type **reload** at the prompt. You do not need to write memory.

TECHNIQUE #2

Relevant devices are: Cisco 1003, Cisco 4500, IDT Orion-Based Cisco 3600, or Motorola 860-Based Cisco 2600.

1. Attach a terminal or PC with terminal emulation to the console port of the router.
2. Type **show version** and record the setting of the configuration register. It is usually 0x2102 or 0x102.
3. Power the router down, and then up.
4. Press the Break key on the terminal within 60 seconds of the power up.

    You will see the rommon> prompt. If you don't, the terminal is not sending the correct Break signal. In that case, check the terminal or terminal emulation setup.
5. Type **confreg** at the rommon> prompt.
6. Answer **y** to the "Do you wish to change configuration[y/n]?" prompt.
7. Answer **n** to all the questions that appear until you reach the "ignore system config info[y/n]?" prompt. Answer **y**.
8. Answer **n** to the remaining questions until you reach the "change boot characteristics[y/n]?" prompt. Answer **y**.
9. At the "enter to boot:" prompt, type **2** followed by a carriage return.

    If Flash is erased, type **1**. If all Flash is erased, the 4500 must be returned to Cisco for service.

CAUTION

If you use "1," you can only view or erase the configuration. You cannot change the password.

---

10. A configuration summary is printed. Answer no to the "Do you wish to change configuration[y/n]?" prompt.
11. Type **reset** at the rommon> prompt, or power-cycle your 4500 or 7500.
12. Once it boots up, answer no to all the Setup questions.
13. Type **enable** at the Router> prompt. You'll be in enable mode and see the Router# prompt.

14. Choose one of these three options:
- To view the password, type **show config**.
- To change the password (in case it is encrypted, for example):

   a. Type **config mem** to copy the NVRAM into memory.

   b. Type **wr term**.

   If you have **enable secret xxxx**, then:

   Type **config term** and make the changes.

   Type **enable secret <password>**.

   Press **Ctrl-Z**.

   If you do not, then:

   Type **enable password <&ltpassword>**.

   Press **Ctrl-Z**.

   c. Type **write mem** to commit the changes.
- To erase the config, type **write erase**.

15. Type **config term** at the prompt.
16. Type **config-register 0x2102** or whatever value you recorded in step 2.
17. Press **Ctrl-Z** to quit from the editor.
18. Type **reload** at the prompt. You do not need to write memory.

Password recovery procedures for Cisco Catalyst switches and older Cisco routers and communication servers can be found at www.cisco.com/warp/customer/701/22.htm.

Frame Relay Examples and Configurations

This appendix includes AppleTalk and IPX Frame Relay configuration examples, and a Frame Relay switching example.

APPLETALK OVER FRAME RELAY EXAMPLE

In the example shown in Figure G–1, the two routers communicate with each other using AppleTalk over the Frame Relay network.

On router A, the following commands configure AppleTalk over Frame Relay:

Command	Description
encapsulation frame-relay	Enables Frame Relay
appletalk cable-range 18-18 18.47	Enables an extended AppleTalk network and sets the cable range and node address
appletalk zone eng	Configures the zone name for the connected AppleTalk network
frame-relay map appletalk 18.65 23 broadcast	Maps the remote 18.65 AppleTalk address to DLCI number 23

467

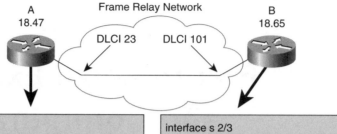

Figure G–1
*Using Apple-
Talk over the
Frame Relay
network.*

```
interface s 0
   ip address 172.21.48.24 255.255.255.0
   encapsulation frame-relay
   appletalk cable-range 18-18 18.47
   appletalk zone eng
   frame-rely map appletalk 18.65 23 broadcast
```

```
interface s 2/3
   ip address 172.21.48.31 255.255.255.0
   encapsulation frame-relay
   appletalk cable-range 18-18 18.65
   appletalk zone eng
   frame-rely map appletalk 18.47 101 broadcast
```

On router B, the following commands are used:

Command	Description
encapsulation frame-relay	Enables Frame Relay
appletalk cable-range 18-18 18.65	Enables an extended AppleTalk network and sets the cable range and node address
appletalk zone eng	Configures the zone name for the connected AppleTalk network
frame-relay map appletalk 18.47 101 broadcast	Maps the remote 18.47 AppleTalk address to DLCI number 101

In the example shown in Figure G–1, Frame Relay is configured on the main interface.

CONFIGURING IPX OVER FRAME RELAY EXAMPLE

In the example shown in Figure G–2, router A has two IPX networks corresponding to Frame Relay subinterfaces. Router B has a single statically mapped Frame Relay-based IPX network.

On router A, the following commands are used to configure IPX over Frame Relay:

Command	Description
Interface S0	Defines the interface
Encapsulation frame-relay	Enables Frame Relay
interface s 0.1 point-to-point	Configures subinterface S0.1 as a point-to-point interface
ipx network 1	Configures IPX network 1 on the S0.1 subinterface
frame-relay interface-dlci 23	Configures network 1 using DLCI 23
interface s 0.2 point-to-point	Configures subinterface S0.2 as a point-to-point interface
ipx network 2	Configures network 2 on subinterface S0.2
frame-relay interface-dlci 27	Configures Inverse ARP on network 2 using DLCI 27

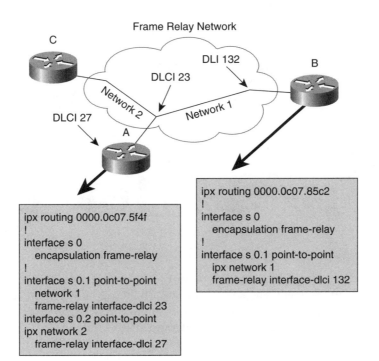

Figure G–2
Router A connects to routers B and C through a Frame Relay cloud.

On router B, the following commands are used:

Command	Description
interface s 0.1 point-to-point	Configures subinterface S0.1 as a point-to-point interface for Frame Relay
frame-relay interface-dlci 132	Configures network 1 using DLCI 132

FRAME RELAY SWITCHING EXAMPLE

Local Frame Relay switching enables the Cisco router to switch Frame Relay frames between interfaces based on the DLCI number in the frame header. A router interface performing local PVC switching is usually configured as a Frame Relay DCE (switch), as shown in the top portion of Figure G–3.

Figure G–3
Switching can be configured for local or remote operations.

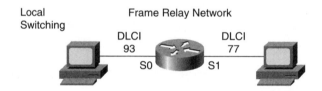

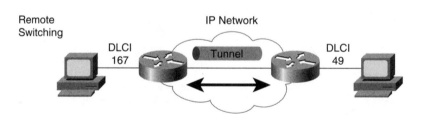

Remote Frame Relay switching enables the router to encapsulate Frame Relay frames in IP datagrams and tunnel them across an IP backbone, as shown in the lower portion of Figure G–3. The Cisco generic routing encapsulation (GRE) tunnel protocol is used for remote Frame Relay switching. The router is usually configured as a Frame Relay DCE.

The configuration in Figure G–4 shows how to define Frame Relay switching for an IP connection.

The router is configured as a remote Frame Relay switch. Traffic arriving on S0 using DLCI 167 will be switched to output interface S1 and DLCI 43 will be used in the

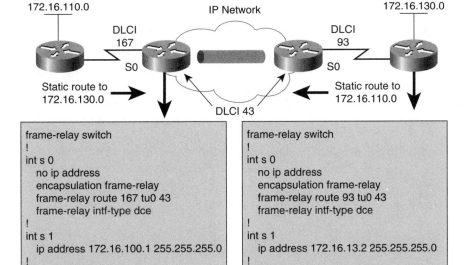

Figure G–4
*A remote
Frame Relay
switch
configuration.*

source identifier. The traffic will be carried through the IP network using a GRE tunnel having a next-hop destination of 172.16.100.1. The tunnel uses the same DLCI number.

On the left side of Figure G–4, the parameters of the **frame-relay route** command have the following meanings:

- **167**—The DLCI of the arriving (source) traffic to be switched

- **tu0**—The outgoing interface to use

- **43**—The outgoing DLCI to be used when forwarding the traffic

The other critical commands are as follows:

Command	Description
frame-relay intf-type dce	Establishes interface S0 as the DCE. In this back-to-back Frame Relay connection, one interface must act as the DCE.
tunnel source serial 1	Defines that software-only tunnel interface 0 will use physical interface serial 1 as the entry into the tunnel.
tunnel destination 172.16.13.2	Defines that the tunnel will deliver traffic to IP address 172.16.13.2 as the tunnel destination.

H

Glossary

A

AARP—AppleTalk Address Resolution Protocol. Protocol in the AppleTalk protocol stack that maps a data-link address to a network address.

AARP probe packets—Packets transmitted by AARP that determine if a randomly selected node ID is being used by another node in a nonextended AppleTalk network. If the node ID is not being used, the sending node uses that node ID. If the node ID is being used, the sending node chooses a different ID and sends more AARP probe packets.

ABM—Asynchronous Balanced Mode. An HDLC (and derivative protocol) communication mode supporting peer-oriented, point-to-point communications between two stations, where either station can initiate transmission.

access list—List kept by routers to control access through or to the router for a number of services (for example, to prevent packets with a certain IP address from leaving a particular interface on the router).

access method—1. Generally, the way in which network devices access the network medium. 2. Software within an SNA processor that controls the flow of information through a network.

ACK—See *acknowledgment*.

acknowledgment—Notification sent from one network device to another to acknowledge that some event (for example, receipt of a message) occurred. Sometimes abbreviated ACK. Compare to *NAK*.

active monitor—Device responsible for performing maintenance functions on a Token Ring. A network node is selected to be the active monitor if it has the highest MAC address on the ring. The active monitor is responsible for such ring maintenance tasks as ensuring that tokens are not lost and that frames do not circulate indefinitely.

adapter—See *NIC*.

address—Data structure or logical convention used to identify a unique entity, such as a particular process or network device.

address mapping—Technique that allows different protocols to interoperate by translating addresses from one format to another. For example, when routing IP over X.25, the IP addresses must be mapped to the X.25 addresses so that the IP packets can be transmitted by the X.25 network.

address mask—Bit combination used to describe which portion of an address refers to the network or subnet and which part refers to the host. Sometimes referred to simply as *mask*.

address resolution—Generally, a method for resolving differences between computer addressing schemes. Address resolution usually specifies a method for mapping network layer (Layer 3) addresses to data link layer (Layer 2) addresses.

Address Resolution Protocol—See *ARP*.

adjacency—Relationship formed between selected neighboring routers and end nodes for the purpose of exchanging routing information. Adjacency is based on the use of a common media segment.

Advanced Research Projects Agency—See *ARPA*.

advertising—Router process in which routing or service updates are sent so that other routers on the network can maintain lists of usable routes.

AEP—AppleTalk Echo Protocol. Used to test connectivity between two AppleTalk nodes. One node sends a packet to another node and receives a duplicate, or echo, of that packet.

AFP—AppleTalk Filing Protocol. Presentation-layer protocol that allows users to share data files and application programs that reside on a file server. AFP supports AppleShare and Mac OS File Sharing.

agent—1. Generally, software that processes queries and returns replies on behalf of an application. 2. In NMSs, process that resides in all managed devices and reports the values of specified variables to management stations.

algorithm—Well-defined rule or process for arriving at a solution to a problem. In networking, algorithms are commonly used to determine the best route for traffic from a particular source to a particular destination.

ANSI—American National Standards Institute. Voluntary organization composed of corporate, government, and other members that coordinates standards-related activities, approves U.S. national standards, and develops positions for the United States in international standards organizations. ANSI helps develop international and U.S. standards relating to, among other things, communications and networking. ANSI is a member of the IEC and the ISO.

AppleTalk—Series of communications protocols designed by Apple Computer consisting of two phases. Phase 1, the earlier version, supports a single physical network that can have only one network number and be in one zone. Phase 2 supports multiple logical networks on a single physical network and allows networks to be in more than one zone. See also *zone*.

application—Program that performs a function directly for a user. FTP and Telnet clients are examples of network applications.

application layer—Layer 7 of the OSI reference model. This layer provides services to application processes (such as e-mail, file transfer, and terminal emulation) that are outside the OSI model. The application layer identifies and establishes the availability of intended communication partners (and the resources required to connect with them), synchronizes cooperating applications, and establishes agreement on procedures for error recovery and control of data integrity. Corresponds roughly with the transaction services layer in the SNA model. See also *data link layer*, *network layer*, *physical layer*, *presentation layer*, *session layer*, and *transport layer*.

area—Logical set of network segments (CLNS-, DECnet-, or OSPF-based) and their attached devices. Areas are usually connected to other areas via routers, making up a single autonomous system.

ARP—Address Resolution Protocol. Internet protocol used to map an IP address to a MAC address. Defined in RFC 826. Compare with *RARP*.

ARPA—Advanced Research Projects Agency. Research and development organization that is part of DoD. ARPA is responsible for numerous technological advances in communications and networking. ARPA evolved into DARPA, and then back into ARPA again (in 1994).

ARPANET—Advanced Research Projects Agency Network. Landmark packet-switching network established in 1969. ARPANET was developed in the 1970s by BBN and funded by ARPA (and later DARPA). It eventually evolved into the Internet. The term *ARPA-NET* was officially retired in 1990.

ASCII—American Standard Code for Information Interchange. 8-bit code for character representation (7 bits plus parity).

Asynchronous Balanced Mode—See *ABM*.

Asynchronous Transfer Mode—See *ATM*.

asynchronous transmission—Term describing digital signals that are transmitted without precise clocking. Such signals generally have different frequencies and phase relationships. Asynchronous transmissions usually encapsulate individual characters in control bits (called start and stop bits) that designate the beginning and end of each character. Compare with *synchronous transmission*.

ATM—Asynchronous Transfer Mode. International standard for cell relay in which multiple service types (such as voice, video, or data) are conveyed in fixed-length (53-byte) cells. Fixed-length cells allow cell processing to occur in hardware, thereby reducing transit delays. ATM is designed to take advantage of high-speed transmission media such as E3, SONET, and T3.

ATM Forum—International organization jointly founded in 1991 by Cisco Systems, NET/ADAPTIVE, Northern Telecom, and Sprint that develops and promotes standards-based implementation agreements for ATM technology. The ATM Forum expands on official standards developed by ANSI and ITU-T, and develops implementation agreements in advance of official standards.

ATP—AppleTalk Transaction Protocol. Transport-level protocol that provides a loss-free transaction service between sockets. The service allows exchanges between two socket clients in which one client requests the other to perform a particular task and to report the results. ATP binds the request and response together to ensure the reliable exchange of request-response pairs.

AURP—AppleTalk Update-Based Routing Protocol. Method of encapsulating Apple-Talk traffic in the header of a foreign protocol, allowing the connection of two or more discontiguous AppleTalk internetworks through a foreign network (such as TCP/IP) to form an AppleTalk WAN. This connection is called an AURP tunnel. In addition to its encapsulation function, AURP maintains routing tables for the entire AppleTalk WAN by exchanging routing information between exterior routers.

authentication—In security, the verification of the identity of a person or process.

B

backbone—Part of a network that acts as the primary path for traffic that is most often sourced from, and destined for, other networks.

bandwidth—Difference between the highest and lowest frequencies available for network signals. The term is also used to describe the rated throughput capacity of a given network medium or protocol.

bandwidth reservation—Process of assigning bandwidth to users and applications served by a network. Involves assigning priority to different flows of traffic based on how critical and delay-sensitive they are. This makes the best use of available bandwidth, and if the network becomes congested, lower-priority traffic can be dropped. Sometimes called *bandwidth allocation*.

Banyan VINES—See *VINES*.

Basic Rate Interface—See *BRI*.

B channel—bearer channel. In ISDN, a full-duplex, 64-kbps channel used to send user data. Compare to *D channel*, *E channel*, and *H channel*.

binary—Numbering system characterized by ones and zeros (1 = on; 0 = off).

BOOTP—Bootstrap Protocol. Protocol used by a network node to determine the IP address of its Ethernet interfaces to affect network booting.

Bootstrap Protocol—See *BOOTP*.

BRI—Basic Rate Interface. ISDN interface composed of two B channels and one D channel for circuit-switched communication of voice, video, and data. Compare with *PRI*.

bridge—Device that connects and passes packets between two network segments that use the same communications protocol. Bridges operate at the data link layer (Layer 2) of the OSI reference model. In general, a bridge will filter, forward, or flood an incoming frame based on the MAC address of that frame.

broadcast—Data packet that will be sent to all nodes on a network. Broadcasts are identified by a broadcast address. Compare with *multicast* and *unicast*. See also *broadcast address*.

broadcast address—Special address reserved for sending a message to all stations. Generally, a broadcast address is a MAC destination address of all ones. Compare with *multicast address* and *unicast address*. See also *broadcast*.

broadcast domain—Set of all devices that will receive broadcast frames originating from any device within the set. Broadcast domains are typically bounded by routers (or in a switched network, by virtual LANs) because routers do not forward broadcast frames.

bus topology—Linear LAN architecture in which transmissions from network stations propagate the length of the medium and are received by all other stations. Compare with *ring topology, star topology,* and *tree topology.*

C

cable range—Range of network numbers that is valid for use by nodes on an extended AppleTalk network. The cable range value can be a single network number or a contiguous sequence of several network numbers. Node addresses are assigned based on the cable range value.

caching—Form of replication in which information learned during a previous transaction is used to process later transactions.

call setup time—Time required to establish a switched call between DTE devices.

carrier—Electromagnetic wave or alternating current of a single frequency, suitable for modulation by another, data-bearing signal.

CCITT—Consultative Committee for International Telegraph and Telephone. International organization responsible for the development of communications standards. Now called the ITU-T. See *ITU-T.*

CDDI—Copper Distributed Data Interface. Implementation of FDDI protocols over STP and UTP cabling. CDDI transmits over relatively short distances (about 100 meters), providing data rates of 100 Mbps using a dual-ring architecture to provide redundancy. Based on the ANSI Twisted-Pair Physical Medium Dependent (TPPMD) standard. Compare with *FDDI.*

Challenge Handshake Authentication Protocol—See *CHAP.*

CHAP—Challenge Handshake Authentication Protocol. Security feature supported on lines using PPP encapsulation that prevents unauthorized access. CHAP does not itself prevent unauthorized access; it merely identifies the remote end. The router or access server then determines if that user is allowed access. Compare to *PAP.*

circuit—Communications path between two or more points.

circuit group—Grouping of associated serial lines that link two bridges. If one of the serial links in a circuit group is in the spanning tree for a network, any of the serial links in the circuit group can be used for load balancing. This load-balancing strategy avoids data ordering problems by assigning each destination address to a particular serial link.

client—Node or software program (front-end device) that requests services from a server.

client/server computing—Term used to describe distributed computing (processing) network systems in which transaction responsibilities are divided into two parts: client (front end) and server (back end). Both terms (*client* and *server*) can be applied to software programs or actual computing devices. Also called *distributed computing (processing)*. Compare with *peer-to-peer computing*.

client/server model—Common way to describe network services and the model user processes (programs) of those services. Examples include the nameserver/nameresolver paradigm of the DNS and fileserver/file-client relationships such as NFS and diskless hosts.

CO—central office. Local telephone company office to which all local loops in a given area connect and in which circuit switching of subscriber lines occurs.

coding—Electrical techniques used to convey binary signals.

common carrier—Licensed, private utility company that supplies communication services to the public at regulated prices.

congestion—Traffic in excess of network capacity.

congestion avoidance—Mechanism by which an ATM network controls traffic entering the network to minimize delays. To use resources most efficiently, lower-priority traffic is discarded at the edge of the network if conditions indicate that it cannot be delivered.

connectionless—Term used to describe data transfer without the existence of a virtual circuit. Compare with *connection-oriented*. See also *virtual circuit*.

connection-oriented—Term used to describe data transfer that requires the establishment of a virtual circuit. See also *connectionless* and *virtual circuit*.

console—DTE through which commands are entered into a host.

convergence—Speed and ability of a group of internetworking devices running a specific routing protocol to agree on the topology of an internetwork after a change in that topology.

count to infinity—Problem that can occur in routing algorithms that are slow to converge, in which routers continuously increment the hop count to particular networks. Typically, some arbitrary hop-count limit is imposed to prevent this problem.

CSMA/CD—carrier sense multiple access with collision detection. Media-access mechanism wherein devices ready to transmit data first check the channel for a carrier. If no carrier is sensed for a specific period of time, a device can transmit. If two devices transmit at once, a collision occurs and is detected by all colliding devices. This collision subsequently delays retransmissions from those devices for some random length of time. CSMA/CD access is used by Ethernet and IEEE 802.3.

CSU—channel service unit. Digital interface device that connects end-user equipment to the local digital telephone loop. Often referred to together with DSU, as CSU/DSU.

D

DARPA—Defense Advanced Research Projects Agency. U.S. government agency that funded research for and experimentation with the Internet. Evolved from ARPA, and then, in 1994, back to ARPA. See also *ARPA*.

DAS—1. dual attachment station. Device attached to both the primary and the secondary FDDI rings. Dual attachment provides redundancy for the FDDI ring: If the primary ring fails, the station can wrap the primary ring to the secondary ring, isolating the failure and retaining ring integrity. Also called a Class A station. Compare with *SAS*. 2. dynamically assigned socket. Socket that is dynamically assigned by DDP upon request by a client. In an AppleTalk network, the sockets numbered 128 to 254 are allocated as DASs.

data flow control layer—Layer 5 of the SNA architectural model. This layer determines and manages interactions between session partners, particularly data flow. Corresponds to the session layer of the OSI model. See also *data-link control layer, path control layer, physical control layer, presentation services layer, transaction services layer,* and *transmission control layer.*

datagram—Logical grouping of information sent as a network layer unit over a transmission medium without prior establishment of a virtual circuit. IP datagrams are the primary information units in the Internet. The terms *cell, frame, message, packet,* and *segment* are also used to describe logical information groupings at various layers of the OSI reference model and in various technology circles.

data-link control layer—Layer 2 in the SNA architectural model. Responsible for the transmission of data over a particular physical link. Corresponds roughly to the data link layer of the OSI model. See also *data flow control layer, path control layer, physical control layer, presentation services layer, transaction services layer,* and *transmission control layer.*

data link layer—Layer 2 of the OSI reference model. Provides transit of data across a physical link. The data link layer is concerned with physical addressing, network topology, line discipline, error notification, ordered delivery of frames, and flow control. The IEEE divided this layer into two sublayers: the MAC sublayer and the LLC sublayer. Sometimes simply called *link layer.* Roughly corresponds to the data-link control layer of the SNA model.

DCE—1. data communications equipment (EIA expansion) 2. data circuit-terminating equipment (ITU-T expansion). Devices and connections of a communications network that comprise the network end of the user-to-network interface. The DCE provides a physical connection to the network, forwards traffic, and provides a clocking signal used to synchronize data transmission between DCE and DTE devices. Modems and interface cards are examples of DCE. Compare with *DTE.*

D channel—1. delta channel. Full-duplex, 16-kbps (BRI) or 64-kbps (PRI) ISDN channel. Compare with B channel, E channel, and H channel. 2. In SNA, a device that connects a processor and main storage with peripherals.

DDP—Datagram Delivery Protocol. AppleTalk network layer protocol responsible for the socket-to-socket delivery of datagrams over an AppleTalk internetwork.

DDR—dial-on-demand routing. Technique whereby a router can automatically initiate and close a circuit-switched session as transmitting stations demand. The router spoofs keepalives so that end stations treat the session as active. DDR permits routing over ISDN or telephone lines sometimes using an external ISDN terminal adaptor or modem.

DECnet—Group of communications products (including a protocol suite) developed and supported by Digital Equipment Corporation. DECnet/OSI (also called *DECnet Phase V*) is the most recent iteration and supports both OSI protocols and proprietary Digital protocols. Phase IV Prime supports inherent MAC addresses that allow DECnet nodes to coexist with systems running other protocols that have MAC address restrictions.

DECnet Routing Protocol (DRP)—Proprietary routing scheme introduced by Digital Equipment Corporation in DECnet Phase III. In DECnet Phase V, DECnet completed its transition to OSI routing protocols (ES-IS and IS-IS).

default route—Routing table entry that is used to direct frames for which a next hop is not explicitly listed in the routing table.

demarc—Demarcation point between carrier equipment and CPE.

demultiplexing—The separating of multiple input streams that have been multiplexed into a common physical signal back into multiple output streams. See also *multiplexing*.

designated router—OSPF router that generates LSAs for a multiaccess network and has other special responsibilities in running OSPF. Each multiaccess OSPF network that has at least two attached routers has a designated router that is elected by the OSPF Hello protocol. The designated router enables a reduction in the number of adjacencies required on a multiaccess network, which in turn reduces the amount of routing protocol traffic and the size of the topological database.

destination address—Address of a network device that is receiving data. See also *source address*.

destination service access point—See *DSAP*.

DHCP—Dynamic Host Configuration Protocol. Provides a mechanism for allocating IP addresses dynamically so that addresses automatically can be reused when hosts no longer need them.

dial-on-demand routing—See *DDR*.

dial-up line—Communications circuit that is established by a switched-circuit connection using the telephone company network.

distance vector routing algorithm—Class of routing algorithms that iterate on the number of hops in a route to find a shortest-path spanning tree. Distance vector routing algorithms call for each router to send its entire routing table in each update, but only to its neighbors. Distance vector routing algorithms can be prone to routing loops, but are computationally simpler than link state routing algorithms. Also called Bellman-Ford routing algorithm. See also *link state routing algorithm*.

DNS—Domain Name System. System used in the Internet for translating names of network nodes into addresses.

DoD—Department of Defense. U.S. government organization that is responsible for national defense. The DoD has frequently funded communication protocol development.

DSAP—destination service access point. SAP of the network node designated in the Destination field of a packet. Compare with *SSAP*. See also *SAP* (service access point).

DTE—data terminal equipment. Device at the user end of a user-network interface that serves as a data source, destination, or both. DTE connects to a data network through a DCE device (for example, a modem) and typically uses clocking signals generated by the DCE. DTE includes such devices as computers, routers, and multiplexers. Compare with *DCE*.

dual attachment station—See *DAS*.

dual counter-rotating rings—Network topology in which two signal paths, whose directions are opposite each other, exist in a token-passing network. FDDI and CDDI are based on this concept.

dual-homed station—Device attached to multiple FDDI concentrators to provide redundancy.

dual homing—Network topology in which a device is connected to the network by way of two independent access points (points of attachment). One access point is the primary connection, and the other is a standby connection that is activated in the event of a failure of the primary connection.

dynamic routing—Routing that adjusts automatically to network topology or traffic changes. Also called *adaptive routing*. Requires that a routing protocol be run between routers.

E

E1—Wide-area digital transmission scheme used predominantly in Europe that carries data at a rate of 2.048 Mbps. E1 lines can be leased for private use from common carriers. Compare with *T1*.

E3—Wide-area digital transmission scheme used predominantly in Europe that carries data at a rate of 34.368 Mbps. E3 lines can be leased for private use from common carriers. Compare with *T3*.

E channel—echo channel. 64-kbps ISDN circuit-switching control channel. The E channel was defined in the 1984 ITU-T ISDN specification, but was dropped in the 1988 specification. Compare with *B channel, D channel,* and *H channel*.

EEPROM—electrically erasable programmable read-only memory. EPROM that can be erased using electrical signals applied to specific pins.

EIA—Electronic Industries Association. Group that specifies electrical transmission standards. The EIA and TIA have developed numerous well-known communications standards, including EIA/TIA-232 and EIA/TIA-449.

encapsulation—Wrapping of data in a particular protocol header. For example, upper-layer data is wrapped in a specific Ethernet header before network transit. Also, when bridging dissimilar networks, the entire frame from one network can simply be placed in the header used by the data link layer protocol of the other network. See also *tunneling*.

EPROM—erasable programmable read-only memory. Nonvolatile memory chips that are programmed after they are manufactured and, if necessary, can be erased by some means and reprogrammed. Compare with *EEPROM* and *PROM*.

Ethernet—Baseband LAN specification invented by Xerox Corporation and developed jointly by Xerox, Intel, and Digital Equipment Corporation. Ethernet networks use CSMA/CD and run over a variety of cable types at 10, 100, and 1000 Mbps. Ethernet is similar to the IEEE 802.3 series of standards.

excess rate—Traffic in excess of the insured rate for a given connection. Specifically, the excess rate equals the maximum rate minus the insured rate. Excess traffic is delivered only if network resources are available and can be discarded during periods of congestion. Compare with *insured rate* and *maximum rate*.

F

FDDI—Fiber Distributed Data Interface. LAN standard, defined by ANSI X3T9.5, specifying a 100-Mbps token-passing network using fiber-optic cable, with transmission distances of up to 2 km. FDDI uses a dual-ring architecture to provide redundancy. Compare with *CDDI* and *FDDI II*.

FDDI II—ANSI standard that enhances FDDI. FDDI II provides isochronous transmission for connectionless data circuits and connection-oriented voice and video circuits. Compare with *FDDI*.

Fiber Distributed Data Interface—See *FDDI*.

fiber-optic cable—Physical medium capable of conducting modulated light transmission. Compared with other transmission media, fiber-optic cable is more expensive but is not susceptible to electromagnetic interference. Sometimes called *optical fiber*.

File Transfer Protocol—See *FTP*.

filter—Generally, a process or device that screens network traffic for certain characteristics, such as source address, destination address, or protocol, and determines whether to forward or discard that traffic based on the established criteria.

firewall—A device that controls who may access a private network and is itself immune to penetration.

Flash memory—Nonvolatile storage that can be electrically erased and reprogrammed so that software images can be stored, booted, and rewritten as necessary. Flash memory was developed by Intel and is licensed to other semiconductor companies.

flash update—Routing update sent asynchronously in response to a change in the network topology. Compare with *routing update*.

flat addressing—Scheme of addressing that does not use a logical hierarchy to determine location.

flow—Stream of data traveling between two endpoints across a network (for example, from one LAN station to another). Multiple flows can be transmitted on a single circuit.

flow control—Technique for ensuring that a transmitting entity does not overwhelm a receiving entity with data. When the buffers on the receiving device are full, a message is sent to the sending device to suspend the transmission until the data in the buffers has been processed. In IBM networks, this technique is called *pacing*.

forwarding—Process of sending a frame toward its ultimate destination by way of an internetworking device.

fragment—Piece of a larger packet that has been broken down to smaller units. In Ethernet networks, also sometimes referred to as a frame less than the legal limit of 64 bytes.

fragmentation—Process of breaking a packet into smaller units when transmitting over a network medium that cannot support the original size of the packet.

frame—Logical grouping of information sent as a data link layer unit over a transmission medium. Often refers to the header and trailer, used for synchronization and error control, that surround the user data contained in the unit. The terms *cell*, *datagram*, *message*, *packet*, and *segment* are also used to describe logical information groupings at various layers of the OSI reference model and in various technology circles.

frame forwarding—Mechanism by which frame-based traffic, such as HDLC and SDLC, traverses an ATM network.

Frame Relay—Industry-standard, switched data link layer protocol that handles multiple virtual circuits using a form of HDLC encapsulation between connected devices. Frame Relay is more efficient than X.25, the protocol for which it is generally considered a replacement. See also *X.25*.

FTP—File Transfer Protocol. Application protocol, part of the TCP/IP protocol stack, used for transferring files between network nodes. FTP is defined in RFC 959.

full duplex—Capability for simultaneous data transmission between a sending station and a receiving station. Compare with *half duplex* and *simplex*.

full mesh—Term describing a network in which devices are organized in a mesh topology, with each network node having either a physical circuit or a virtual circuit connecting it to every other network node. A full mesh provides a great deal of redundancy, but because it can be prohibitively expensive to implement, it is usually reserved for network backbones. See also *mesh* and *partial mesh*.

G

gateway—In the IP community, an older term referring to a routing device. Today, the term *router* is used to describe nodes that perform this function, and *gateway* refers to a special-purpose device that performs an application layer conversion of information from one protocol stack to another. Compare with *router*.

Gb—gigabit. Approximately 1,000,000,000 bits.

Get Nearest Server—See *GNS*.

gigabit—Abbreviated Gb.

GNS—Get Nearest Server. Request packet sent by a client on an IPX network to locate the nearest active server of a particular type. An IPX network client issues a GNS request to solicit either a direct response from a connected server or a response from a router that tells it where on the internetwork the service can be located. GNS is part of the IPX SAP. See also *IPX* and *SAP* (Service Advertisement Protocol).

H

half duplex—Capability for data transmission in only one direction at a time between a sending station and a receiving station. Compare with *full duplex* and *simplex*.

handshake—Sequence of messages exchanged between two or more network devices to ensure transmission synchronization before sending user data.

hardware address—See *MAC address*.

H channel—high-speed channel. Full-duplex ISDN primary rate channel operating at 384 Kbps. Compare with *B channel, D channel,* and *E channel*.

HDLC—High-Level Data Link Control. Bit-oriented synchronous data link layer protocol developed by ISO. HDLC specifies a data encapsulation method on synchronous serial links using frame characters and checksums.

header—Control information placed before data when encapsulating that data for network transmission. Compare with *trailer*.

hello packet—Multicast packet that is used by routers using certain routing protocols for neighbor discovery and recovery. Hello packets also indicate that a client is still operating and network-ready.

holddown—State into which a route is placed so that routers will neither advertise the route nor accept advertisements about the route for a specific length of time (the holddown period). Holddown is used to flush bad information about a route from all routers in the network. A route is typically placed in holddown when a link in that route fails.

hop—Passage of a data packet from one network node, typically a router, to another. See also *hop count*.

hop count—Routing metric used to measure the distance between a source and a destination. RIP uses hop count as its sole metric. See also *hop* and *RIP*.

host—Computer system on a network. Similar to *node*, except that host usually implies a computer system, whereas *node* generally applies to any networked system, including access servers and routers. See also *node*.

host address—See *host number*.

host number—Part of an IP address that designates which node on the subnetwork is being addressed. Also called a *host address*.

HTML—Hypertext Markup Language. Simple hypertext document formatting language that uses tags to indicate how a given part of a document should be interpreted by a viewing application, such as a Web browser.

HTTP—Hypertext Transfer Protocol. The protocol used by Web browsers and Web servers to transfer files, such as text and graphics files.

hub—1. Generally, a term used to describe a device that serves as the center of a star-topology network and connects end stations. Operates at Layer 1 of the OSI model. 2. In Ethernet and IEEE 802.3, an Ethernet multiport repeater, sometimes called a *concentrator*.

hybrid network—Internetwork made up of more than one type of network technology, including LANs and WANs.

Hypertext Transfer Protocol—See *HTTP*.

I

IANA—Internet Assigned Numbers Authority. Organization operated under the auspices of the ISOC as a part of the IAB. IANA delegates authority for IP address-space allocation and domain-name assignment to the InterNIC and other organizations. IANA also maintains a database of assigned protocol identifiers used in the TCP/IP stack, including autonomous system numbers.

ICMP—Internet Control Message Protocol. Network layer Internet protocol that reports errors and provides other information relevant to IP packet processing. Documented in RFC 792.

IEEE—Institute of Electrical and Electronics Engineers. Professional organization whose activities include the development of communications and network standards. IEEE LAN standards are the predominant LAN standards today.

IEEE 802.2—IEEE LAN protocol that specifies an implementation of the LLC sublayer of the *data link layer*. IEEE 802.2 handles errors, framing, flow control, and the network layer (Layer 3) service interface. Used in IEEE 802.3 and IEEE 802.5 LANs. See also *IEEE 802.3* and *IEEE 802.5*.

IEEE 802.3—IEEE LAN protocol that specifies an implementation of the physical layer and the MAC sublayer of the data link layer. IEEE 802.3 uses CSMA/CD access at a variety of speeds over a variety of physical media. Extensions to the IEEE 802.3 standard specify implementations for Fast Ethernet. Physical variations of the original IEEE 802.3 specification include 10Base2, 10Base5, 10BaseF, 10BaseT, and 10Broad36. Physical variations for Fast Ethernet include 100BaseTX and 100BaseFX.

IEEE 802.5—IEEE LAN protocol that specifies an implementation of the physical layer and MAC sublayer of the *data link layer*. IEEE 802.5 uses token passing access at 4 or 16 Mbps over STP or UTP cabling and is functionally and operationally equivalent to IBM Token Ring. See also *Token Ring*.

IETF—Internet Engineering Task Force. Task force consisting of over 80 working groups responsible for developing Internet standards. The IETF operates under the auspices of ISOC.

IGP—Interior Gateway Protocol. Internet protocol used to exchange routing information within an autonomous system. Examples of common Internet IGPs include IGRP, OSPF, and RIP.

Institute of Electrical and Electronics Engineers—See *IEEE*.

insured rate—The long-term data throughput, in bits or cells per second, that an ATM network commits to support under normal network conditions. The insured rate is 100 percent allocated; the entire amount is deducted from the total trunk bandwidth along the path of the circuit. Compare with *excess rate* and *maximum rate*.

Integrated Services Digital Network—See *ISDN*.

interface—1. Connection between two systems or devices. 2. In routing terminology, a network connection on the router. 3. In telephony, a shared boundary defined by common physical interconnection characteristics, signal characteristics, and meanings of interchanged signals. 4. Boundary between adjacent layers of the OSI model.

Internet—Largest global internetwork, connecting tens of thousands of networks worldwide and having a "culture" that focuses on research and standardization based on real-life use. Many leading-edge network technologies come from the Internet community. The Internet evolved in part from ARPANET. At one time called the DARPA Internet, not to be confused with the general term *internet*.

Internet protocol—Any protocol that is part of the TCP/IP protocol stack. See *IP*. See also *TCP/IP*.

internetwork—Collection of networks interconnected by routers and other devices that functions (generally) as a single network

internetworking—General term used to refer to the industry devoted to connecting networks together. The term can refer to products, procedures, and technologies.

Internetwork Packet Exchange—See *IPX*.

InterNIC—Organization that serves the Internet community by supplying user assistance, documentation, training, registration service for Internet domain names, network addresses, and other services. Formerly called *NIC*.

interoperability—Capability of computing equipment manufactured by different vendors to communicate with one another successfully over a network.

IP—Internet Protocol. Network layer protocol in the TCP/IP stack offering a connectionless internetwork service. IP provides features for addressing, type-of-service specification, fragmentation and reassembly, and security. Defined in RFC 791. IPv4 (Internet Protocol version 4) is a connectionless, best-effort packet switching protocol. See also *IPv6*.

IP address—32-bit address assigned to hosts using TCP/IP. An IP address belongs to one of five classes (A, B, C, D, or E) and is written as 4 octets separated by periods (dotted decimal format). Each address consists of a network number, an optional subnetwork number, and a host number. The network and subnetwork numbers together are used for routing, while the host number is used to address an individual host within the network or subnetwork. A subnet mask is used to extract network and subnetwork information from the IP address. CIDR provides a new way of representing IP addresses and subnet masks. Also called an *Internet address*.

IP datagram—Fundamental unit of information passed across the Internet. Contains source and destination addresses along with data and a number of fields that define such things as the length of the datagram, the header checksum, and flags to indicate if the datagram can be (or was) fragmented.

IPv6—IP version 6. Replacement for the current version of IP (version 4). IPv6 includes support for flow ID in the packet header, which can be used to identify flows. Formerly called IPng (IP next generation).

IPX—Internetwork Packet Exchange. NetWare network layer (Layer 3) protocol used for transferring data from servers to workstations. IPX is similar to IP and XNS.

IPXWAN—IPX wide-area network. Protocol that negotiates end-to-end options for new links. When a link comes up, the first IPX packets sent across are IPXWAN packets negotiating the options for the link. When the IPXWAN options are successfully determined, normal IPX transmission begins. Defined by RFC 1362.

ISDN—Integrated Services Digital Network. Communication protocol, offered by telephone companies, that permits telephone networks to carry data, voice, and other source traffic.

ITU-T—International Telecommunication Union Telecommunication Standardization Sector (ITU-T) (formerly the Committee for Internatiional Telegraph and Telephone ([CCITT]). An international organization that develops communication standards. See also *CCITT*.

K

Kb—kilobit. Approximately 1,000 bits.

kBps—kilobytes per second.

kbps—kilobits per second.

keepalive interval—Period of time between each keepalive message sent by a network device.

kilobit—Abbreviated Kb.

kilobits per second—Abbreviated kbps.

kilobyte—Abbreviated KB.

kilobytes per second—Abbreviated kBps.

L

LAN—local-area network. High-speed, low-error data network covering a relatively small geographic area (up to a few thousand meters). LANs connect workstations, peripherals, terminals, and other devices in a single building or other geographically limited area. LAN standards specify cabling and signaling at the physical and data link layers of the OSI model. Ethernet, FDDI, and Token Ring are widely used LAN technologies. Compare with *MAN* and *WAN*.

LAPB—Link Access Procedure Balanced. Data link layer protocol in the X.25 protocol stack. LAPB is a bit-oriented protocol derived from HDLC. See also *HDLC* and *X.25*.

LAPD—Link Access Procedure on the D channel. ISDN data link layer protocol for the D channel. LAPD was derived from the LAPB protocol and is designed primarily to satisfy the signaling requirements of ISDN basic access. Defined by ITU-T Recommendations Q.920 and Q.921.

LAT—local-area transport. A network virtual terminal protocol developed by Digital Equipment Corporation.

leased line—Transmission line reserved by a communications carrier for the private use of a customer. A leased line is a type of dedicated line.

link—Network communications channel consisting of a circuit or transmission path and all related equipment between a sender and a receiver. Most often used to refer to a WAN connection. Sometimes referred to as a line or a transmission link.

Link Access Procedure Balanced—See *LAPB*.

Link Access Procedure on the D channel—See *LAPD*.

link layer—See *data link layer*.

link-layer address—See *MAC address*.

link-state routing algorithm—Routing algorithm in which each router broadcasts or multicasts information regarding the cost of reaching each of its neighbors to all nodes in the internetwork. Link-state algorithms create a consistent view of the network and are therefore not prone to routing loops, but they achieve this at the cost of relatively greater computational difficulty and more widespread traffic (compared with distance vector routing algorithms). Compare with *distance vector routing algorithm.*

LLC—logical link control. Higher of the two data link layer sublayers defined by the IEEE. The LLC sublayer handles error control, flow control, framing, and MAC-sublayer addressing. The most prevalent LLC protocol is IEEE 802.2, which includes both connectionless and connection-oriented variants.

load balancing—In routing, the capability of a router to distribute traffic over all its network ports that are the same distance from the destination address. Good load-balancing algorithms use both line speed and reliability information. Load balancing increases the use of network segments, thus increasing effective network bandwidth.

local-area network—See *LAN.*

local loop—Line from the premises of a telephone subscriber to the telephone company CO.

local traffic filtering—Process by which a bridge filters out (drops) frames whose source and destination MAC addresses are located on the same interface on the bridge, thus preventing unnecessary traffic from being forwarded across the bridge. Defined in the IEEE 802.1 standard.

loop—Route where packets never reach their destination but simply cycle repeatedly through a constant series of network nodes.

loopback test—Test in which signals are sent and then directed back toward their source from some point along the communications path. Loopback tests are often used to test network interface usability.

M

MAC—Media Access Control. Lower of the two sublayers of the *data link layer* defined by the IEEE. The MAC sublayer handles access to shared media, such as whether token passing or contention will be used. See also *data link layer* and *LLC.*

MAC address—Standardized *data link layer* address that is required for every device that connects to a LAN. Other devices in the network use these addresses to locate specific devices in the network and to create and update routing tables and data structures. MAC addresses are 6 bytes long and are controlled by the IEEE. Also known as a *hardware address*, *MAC-layer address*, or *physical address*. Compare with *network address.*

MAC address learning—Service that characterizes a learning switch in which the source MAC address of each received packet is stored so that future packets destined for that address can be forwarded only to the switch interface on which that address is located. Packets destined for unrecognized broadcast or multicast addresses are forwarded out every switch interface except the originating one. This scheme helps minimize traffic on the attached LANs. MAC address learning is defined in the IEEE 802.1 standard.

MAC-layer address—See *MAC address*.

MAN—metropolitan-area network. Network that spans a metropolitan area. Generally, a MAN spans a larger geographic area than a LAN, but a smaller geographic area than a WAN. Compare with *LAN* and *WAN*.

Management Information Base—See *MIB*.

mask—See *address mask* and *subnet mask*.

MAU—media attachment unit. Device used in Ethernet and IEEE 802.3 networks that provides the interface between the AUI port of a station and the common medium of the Ethernet. The MAU, which can be built into a station or can be a separate device, performs physical layer functions including the conversion of digital data from the Ethernet interface, collision detection, and injection of bits onto the network. Sometimes referred to as a media access unit, also abbreviated MAU, or as a transceiver. In Token Ring, a MAU is known as a multistation access unit and is usually abbreviated MSAU to avoid confusion.

maximum rate—Maximum total data throughput allowed on a given virtual circuit, equal to the sum of the insured and uninsured traffic from the traffic source. The uninsured data might be dropped if the network becomes congested. The maximum rate, which cannot exceed the media rate, represents the highest data throughput the virtual circuit will ever deliver, measured in bits or cells per second. Compare with *excess rate* and *insured rate*.

MB—megabyte. Approximately 1,000,000 bytes.

Mb—megabit. Approximately 1,000,000 bits.

MBS—maximum burst size. In an ATM signaling message, burst tolerance is conveyed through the MBS, which is coded as a number of cells. The burst tolerance together with the SCR and the GCRA determine the MBS that can be transmitted at the peak rate and still be in conformance with the GCRA.

Mbps—megabits per second.

media—Plural of medium. Various physical environments through which transmission signals pass. Common network media include twisted-pair, coaxial, and fiber-optic cable, and the atmosphere (through which microwave, laser, and infrared transmission occurs). Sometimes called *physical media*.

Media Access Control—See *MAC*.

media access unit—See *MAU*.

megabit—Abbreviated Mb. Approximately 1,000,000 bits.

megabits per second—Abbreviated Mbps.

megabyte—Abbreviated MB. Approximately 1,000,000 bytes.

mesh—Network topology in which devices are organized in a manageable, segmented manner with many, often redundant, interconnections strategically placed between network nodes. See also *full mesh* and *partial mesh*.

message—Application layer (Layer 7) logical grouping of information, often composed of a number of lower-layer logical groupings such as packets. The terms *datagram*, *frame*, *packet*, and *segment* are also used to describe logical information groupings at various layers of the OSI reference model and in various technology circles.

metric—See *routing metric*.

MIB—Management Information Base. Database of network management information that is used and maintained by a network management protocol such as SNMP. The value of a MIB object can be changed or retrieved using SNMP commands, usually through a GUI network management system. MIB objects are organized in a tree structure that includes public (standard) and private (proprietary) branches.

MSAU—multistation access unit. Wiring concentrator to which all end stations in a Token Ring network connect. The MSAU provides an interface between these devices and the Token Ring interface of a router. Sometimes abbreviated MAU.

MTU—maximum transmission unit. Maximum packet size, in bytes, that a particular interface can handle.

multicast—Single packets copied by the network and sent to a specific subset of network addresses. These addresses are specified in the Destination Address Field. Compare with *broadcast* and *unicast*.

multicast address—Single address that refers to multiple network devices. Synonymous with group address. Compare with *broadcast address* and *unicast address*. See also *multicast*.

multiplexing—Scheme that allows multiple logical signals to be transmitted simultaneously across a single physical channel. Compare with *demultiplexing*.

multistation access unit—See *MSAU*.

multivendor network—Network using equipment from more than one vendor. Multivendor networks pose many more compatibility problems than single-vendor networks. Compare with *single-vendor network*.

N

NAK—negative acknowledgment. Response sent from a receiving device to a sending device indicating that the information received contained errors. Compare to *acknowledgment*.

name resolution—Generally, the process of associating a name with a network address.

name server—Server connected to a network that resolves network names into network addresses.

NAT—Network Address Translation. Mechanism for reducing the need for globally unique IP addresses. NAT allows an organization with addresses that are not globally unique to connect to the Internet by translating those addresses into globally routable address space. Also known as *Network Address Translator*.

NAUN—nearest active upstream neighbor. In Token Ring or IEEE 802.5 networks, the closest upstream network device from any given device that is still active.

NBP—Name Binding Protocol. AppleTalk transport-level protocol that translates a character string name into the DDP address of the corresponding socket client. NBP enables AppleTalk protocols to understand user-defined zones and device names by providing and maintaining translation tables that map names to their corresponding socket addresses.

neighboring routers—In OSPF, two routers that have interfaces to a common network. On multiaccess networks, neighbors are dynamically discovered by the OSPF Hello protocol.

NetBEUI—NetBIOS Extended User Interface. Enhanced version of the NetBIOS protocol used by network operating systems such as LAN Manager, LAN Server, Windows for Workgroups, and Windows NT. NetBEUI formalizes the transport frame and adds additional functions. NetBEUI implements the OSI LLC2 protocol.

NetBIOS—Network Basic Input/Output System. API used by applications on an IBM LAN to request services from lower-level network processes. These services might include session establishment and termination, and information transfer.

NetWare—Popular distributed NOS developed by Novell. Provides transparent remote file access and numerous other distributed network services.

NetWare Link Services Protocol—See *NLSP.*

NetWare Loadable Module—See *NLM.*

network—Collection of computers, printers, routers, switches, and other devices that are able to communicate with each other over some transmission medium.

network address—Network layer address referring to a logical, rather than a physical, network device. Also called a *protocol address*. Compare with *MAC address.*

Network Address Translation—See *NAT.*

network administrator—Person responsible for the operation, maintenance, and management of a network.

network analyzer—Hardware or software device offering various network troubleshooting features, including protocol-specific packet decodes, specific preprogrammed troubleshooting tests, packet filtering, and packet transmission.

Network Basic Input/Output System—See *NetBIOS.*

network byte order—Internet-standard ordering of the bytes corresponding to numeric values.

network interface—Boundary between a carrier network and a privately owned installation.

network interface card—See *NIC.*

network layer—Layer 3 of the OSI reference model. This layer provides connectivity and path selection between two end systems. The network layer is the layer at which routing occurs. Corresponds roughly with the path control layer of the SNA model. See also *application layer, data link layer, physical layer, presentation layer, session layer,* and *transport layer.*

network management—Generic term used to describe systems or actions that help maintain, characterize, or troubleshoot a network.

network number—Part of an IP address that specifies the network to which the host belongs.

NFS—Network File System. As commonly used, a distributed file system protocol suite developed by Sun Microsystems that allows remote file access across a network. In actuality, NFS is simply one protocol in the suite. NFS protocols include NFS, RPC, XDR, and others. These protocols are part of a larger architecture that Sun refers to as ONC.

NIC—1. network interface card. Board that provides network communication capabilities to and from a computer system. Also called an adapter. 2. Network Information Center. Organization whose functions have been assumed by the InterNIC. See *InterNIC*.

NLM—NetWare Loadable Module. Individual program that can be loaded into memory and function as part of the NetWare NOS.

NLSP—NetWare Link Services Protocol. Link-state routing protocol based on IS-IS.

node—1. Endpoint of a network connection or a junction common to two or more lines in a network. Nodes can be processors, controllers, or workstations. Nodes, which vary in routing and other functional capabilities, can be interconnected by links and serve as control points in the network. *Node* is sometimes used generically to refer to any entity that can access a network and is frequently used interchangeably with *device*. 2. In SNA, the basic component of a network and the point at which one or more functional units connect channels or data circuits.

nonextended network—AppleTalk Phase 2 network that supports addressing of up to 253 nodes and only 1 zone.

nonseed router—In AppleTalk, a router that must first obtain, and then verify, its configuration with a seed router before it can begin operation. See also *seed router*.

non-stub area—Resource-intensive OSPF area that carries a default route, static routes, intra-area routes, interarea routes, and external routes. Non-stub areas are the only OSPF areas that can have virtual links configured across them, and are the only areas that can contain an ASBR. Compare with *stub area*.

NOS—network operating system. Generic term used to refer to what are really distributed file systems. Examples of NOSs include LAN Manager, NetWare, NFS, VINES, and Windows NT.

Novell IPX—See *IPX*.

NTP—Network Time Protocol. Protocol built on top of TCP that ensures accurate local time-keeping with reference to radio and atomic clocks located on the Internet. This protocol is capable of synchronizing distributed clocks within milliseconds over long time periods.

NVRAM—nonvolatile RAM. RAM that retains its contents when a unit is powered off.

O

octet—8 bits. In networking, the term *octet* is often used (rather than *byte*) because some machine architectures employ bytes that are not 8 bits long.

ODI—Open Data-Link Interface. Novell specification providing a standardized interface for NICs (network interface cards) that allows multiple protocols to use a single NIC.

Open Shortest Path First—See *OSPF*.

Open System Interconnection—See *OSI*.

Open System Interconnection reference model—See *OSI reference model*.

OSI—Open System Interconnection. International standardization program created by ISO and ITU-T to develop standards for data networking that facilitate multivendor equipment interoperability.

OSI Presentation Address—Address used to locate an OSI Application entity. It consists of an OSI Network Address and up to three selectors, one each for use by the transport, session, and presentation entities.

OSI reference model—Open System Interconnection reference model. Network architectural model developed by ISO and ITU-T. The model consists of seven layers, each of which specifies particular network functions such as addressing, flow control, error control, encapsulation, and reliable message transfer. The lowest layer (the physical layer) is closest to the media technology. The lower two layers are implemented in hardware and software, while the upper five layers are implemented only in software. The highest layer (the application layer) is closest to the user. The OSI reference model is used universally as a method for teaching and understanding network functionality. Similar in some respects to SNA. See *application layer, data link layer, network layer, physical layer, presentation layer, session layer*, and *transport layer*.

OSPF—Open Shortest Path First. Link-state, hierarchical IGP routing algorithm proposed as a successor to RIP in the Internet community. OSPF features include least-cost routing, multipath routing, and load balancing. OSPF was derived from an early version of the IS-IS protocol.

OUI—Organizational Unique Identifier. 3 octets assigned by the IEEE in a block of 48-bit LAN addresses.

P

packet—Logical grouping of information that includes a header containing control information and (usually) user data. Packets are most often used to refer to network layer units of data. The terms *datagram, frame, message*, and *segment* are also used to describe logical information groupings at various layers of the OSI reference model and in various technology circles.

packet internet groper—See *ping*.

PAP—Password Authentication Protocol. Authentication protocol that allows PPP peers to authenticate one another. The remote router attempting to connect to the local router is required to send an authentication request. Unlike CHAP, PAP passes the password and host name or username in the clear (unencrypted). PAP does not itself prevent unauthorized access but merely identifies the remote end. The router or access server then determines if that user is allowed access. PAP is supported only on PPP lines. Compare with *CHAP*.

parallel transmission—Method of data transmission in which the bits of a data character are transmitted simultaneously over a number of channels. Compare with *serial transmission*.

partial mesh—Network in which devices are organized in a mesh topology, with some network nodes organized in a full mesh, but with others that are only connected to one or two other nodes in the network. A partial mesh does not provide the level of redundancy of a full mesh topology but is less expensive to implement. Partial mesh topologies are generally used in the peripheral networks that connect to a fully meshed backbone.

Password Authentication Protocol—See *PAP*.

path control layer—Layer 3 in the SNA architectural model. This layer performs sequencing services related to proper data reassembly. The path control layer is also responsible for routing. Corresponds roughly with the network layer of the OSI model. See also *data flow control layer, data-link control layer, physical control layer, presentation services layer, transaction services layer,* and *transmission control layer*.

payload—Portion of a cell, frame, or packet that contains upper-layer information (data).

peer-to-peer computing—Peer-to-peer computing calls for each network device to run both client and server portions of an application. Also describes communication between implementations of the same OSI reference model layer in two different network devices. Compare with *client-server computing*.

permanent virtual circuit—See *PVC*.

PHY—1. physical sublayer. One of two sublayers of the FDDI physical layer. 2. physical layer. In ATM, the physical layer provides for the transmission of cells over a physical medium that connects two ATM devices. The PHY is comprised of two sublayers: PMD and TC.

physical address—See *MAC address*.

physical control layer—Layer 1 in the SNA architectural model. This layer is responsible for the physical specifications for the physical links between end systems. Corresponds to the physical layer of the OSI model. See also *data flow control layer, data-link control layer, path control layer, presentation services layer, transaction services layer,* and *transmission control layer.*

physical layer—Layer 1 of the OSI reference model. The physical layer defines the electrical, mechanical, procedural, and functional specifications for activating, maintaining, and deactivating the physical link between end systems. Corresponds with the physical control layer in the SNA model. See also *application layer, data link layer, network layer, presentation layer, session layer,* and *transport layer.*

ping—packet internet groper. ICMP echo message and its reply. Often used in IP networks to test the reachability of a network device.

PLP—packet level protocol. Network layer protocol in the X.25 protocol stack. Sometimes called *X.25 Level 3* and *X.25 Protocol.* See also *X.25.*

point-to-multipoint connection—One of two fundamental connection types. In ATM, a point-to-multipoint connection is a unidirectional connection in which a single source end-system (known as a root node) connects to multiple destination end-systems (known as leaves). Compare with *point-to-point connection.*

point-to-point connection—One of two fundamental connection types. In ATM, a point-to-point connection can be a unidirectional or bidirectional connection between two ATM end-systems. Compare with *point-to-multipoint connection.*

Point-to-Point Protocol—See *PPP.*

poison reverse updates—Routing updates that explicitly indicate that a network or subnet is unreachable, rather than implying that a network is unreachable by not including it in updates. Poison reverse updates are sent to defeat large routing loops.

port—1. Interface on an internetworking device (such as a router). 2. In IP terminology, an upper-layer process that receives information from lower layers. Ports are numbered, and many are associated with a specific process. For example, SMTP is associated with port 25. A port number of this type is called a well-known address. 3. To rewrite software or microcode so that it will run on a different hardware platform or in a different software environment than that for which it was originally designed.

POST—power-on self test. Set of hardware diagnostics that runs on a hardware device when that device is powered up.

PPP—Point-to-Point Protocol. Successor to SLIP that provides router-to-router and host-to-network connections over synchronous and asynchronous circuits. Whereas SLIP was designed to work with IP, PPP was designed to work with several network layer protocols, such as IP, IPX, and ARA. PPP also has built-in security mechanisms, such as CHAP and PAP. PPP relies on two protocols: LCP and NCP.

presentation layer—Layer 6 of the OSI reference model. This layer ensures that information sent by the application layer of one system will be readable by the application layer of another. The presentation layer is also concerned with the data structures used by programs and therefore negotiates data transfer syntax for the application layer. Corresponds roughly with the presentation services layer of the SNA model. See also *application layer*, *data link layer*, *network layer*, *physical layer*, *session layer*, and *transport layer*.

presentation services layer—Layer 6 of the SNA architectural model. This layer provides network resource management, session presentation services, and some application management. Corresponds roughly with the presentation layer of the OSI model.

PRI—Primary Rate Interface. ISDN interface to primary rate access. Primary rate access consists of a single 64-Kbps D channel plus 23 (T1) or 30 (E1) B channels for voice or data. Compare to *BRI*.

priority queuing—Routing feature in which frames in an interface output queue are prioritized based on various characteristics such as protocol, packet size, and interface type.

PROM—programmable read-only memory. ROM that can be programmed using special equipment. PROMs can be programmed only once. Compare with *EPROM*.

protocol—Formal description of a set of rules and conventions that govern how devices on a network exchange information.

protocol address—See *network address*.

protocol stack—Set of related communications protocols that operate together and, as a group, address communication at some or all of the seven layers of the OSI reference model. Not every protocol stack covers each layer of the model, and often a single protocol in the stack will address a number of layers at once. TCP/IP is a typical protocol stack.

proxy—Entity that, in the interest of efficiency, essentially stands in for another entity.

proxy Address Resolution Protocol—See *proxy ARP*.

proxy ARP—proxy Address Resolution Protocol. Variation of the ARP protocol in which an intermediate device (for example, a router) sends an ARP response on behalf of an end node to the requesting host. Proxy ARP can lessen bandwidth use on slow-speed WAN links.

PVC—permanent virtual circuit. Virtual circuit that is permanently established. PVCs save bandwidth associated with circuit establishment and tear down in situations where certain virtual circuits must exist all the time. In ATM terminology, called a *permanent virtual connection*. Compare with *SVC*.

Q

QoS—quality of service. Measure of performance for a transmission system that reflects its transmission quality and service availability.

queue—1. Generally, an ordered list of elements waiting to be processed. 2. In routing, a backlog of packets waiting to be forwarded over a router interface.

queuing delay—Amount of time that data must wait before it can be transmitted onto a statistically multiplexed physical circuit.

R

RAM—random-access memory. Volatile memory that can be read and written by a microprocessor.

random-access memory—See *RAM*.

RARP—Reverse Address Resolution Protocol. Protocol in the TCP/IP stack that provides a method for finding IP addresses based on MAC addresses. Compare with *ARP*.

reassembly—The putting back together of an IP datagram at the destination after it has been fragmented either at the source or at an intermediate node.

redirect—Part of the ICMP and ES-IS protocols that allows a router to tell a host that using another router would be more effective.

redundancy—1. In internetworking, the duplication of devices, services, or connections so that, in the event of a failure, the redundant devices, services, or connections can perform the work of those that failed. 2. In telephony, the portion of the total information contained in a message that can be eliminated without loss of essential information or meaning.

Request For Comments—See *RFC*.

RFC—Request For Comments. Document series used as the primary means for communicating information about the Internet. Some RFCs are designated by the IAB as Internet standards. Most RFCs document protocol specifications such as Telnet and FTP, but some are humorous or historical. RFCs are available online from numerous sources.

ring—Connection of two or more stations in a logically circular topology. Information is passed sequentially between active stations. Token Ring, FDDI, and CDDI are based on this topology.

ring topology—Network topology that consists of a series of repeaters connected to one another by unidirectional transmission links to form a single closed loop. Each station on the network connects to the network at a repeater. While logically a ring, ring topologies are most often organized in a closed-loop star. Compare with *bus topology*, *star topology*, and *tree topology*.

RIP—Routing Information Protocol. IGP supplied with UNIX BSD systems. The most common IGP in the Internet. RIP uses hop count as a routing metric.

RMON—remote monitoring. MIB agent specification described in RFC 1271 that defines functions for the remote monitoring of networked devices. The RMON specification provides numerous monitoring, problem detection, and reporting capabilities.

ROM—read-only memory. Nonvolatile memory that can be read, but not written, by the microprocessor.

routed protocol—Protocol that can be routed by a router. A router must be able to interpret the logical internetwork as specified by that routed protocol. Examples of routed protocols include AppleTalk, DECnet, and IP.

route map—Method of controlling the redistribution of routes between routing domains.

route summarization—Consolidation of advertised network numbers in OSPF and IS-IS. In OSPF, this causes a single summary route to be advertised to other areas by an area border router.

router—Network layer device that uses one or more metrics to determine the optimal path along which network traffic should be forwarded. Routers forward packets from one network to another based on network layer information contained in routing updates. Occasionally called a *gateway* (although this definition of *gateway* is becoming increasingly outdated).

routing—Process of finding a path to a destination host. Routing is very complex in large networks because of the many potential intermediate destinations a packet might traverse before reaching its destination host.

routing metric—Method by which a routing algorithm determines that one route is better than another. This information is stored in routing tables and sent in routing updates. Metrics include bandwidth, communication cost, delay, hop count, load, MTU, path cost, and reliability. Sometimes referred to simply as a *metric*.

routing protocol—Protocol that accomplishes routing through the implementation of a specific routing algorithm. Examples of routing protocols include IGRP, OSPF, and RIP.

routing table—Table stored in a router or some other internetworking device that keeps track of routes to particular network destinations and, in some cases, metrics associated with those routes.

Routing Table Maintenance Protocol—See *RTMP*.

routing update—Message sent from a router to indicate network reachability and associated cost information. Routing updates are typically sent at regular intervals and after a change in network topology. Compare with *flash update*.

RPF—Reverse Path Forwarding. Multicasting technique in which a multicast datagram is forwarded out of all but the receiving interface if the receiving interface is the one used to forward unicast datagrams to the source of the multicast datagram.

RSVP—Resource Reservation Protocol. Protocol that supports the reservation of resources across an IP network. Applications running on IP end systems can use RSVP to indicate to other nodes the nature (bandwidth, jitter, maximum burst, and so forth) of the packet streams they want to receive. RSVP depends on IPv6. Also known as *Resource Reservation Setup Protocol*.

RTMP—Routing Table Maintenance Protocol. Apple Computer's proprietary routing protocol. RTMP establishes and maintains the routing information that is required to route datagrams from any source socket to any destination socket in an AppleTalk network. Using RTMP, routers dynamically maintain routing tables to reflect changes in topology. RTMP was derived from RIP.

RTP—1. Routing Table Protocol. VINES routing protocol based on RIP. Distributes network topology information and aids VINES servers in finding neighboring clients, servers, and routers. Uses delay as a routing metric. 2. Rapid Transport Protocol. Provides pacing and error recovery for APPN data as it crosses the APPN network. With RTP, error recovery and flow control are done end-to-end rather than at every node. RTP prevents congestion rather than reacts to it. 3. Real-Time Transport Protocol. One of the IPv6 protocols. RTP is designed to provide end-to-end network transport functions for applications transmitting real-time data, such as audio, video, or simulation data, over multicast or unicast network services. RTP provides services such as payload type identification, sequence numbering, timestamping, and delivery monitoring to real-time applications.

S

SAP—1. service access point. Field defined by the IEEE 802.2 specification that identifies the upper layer process and is part of an address specification. Thus, the destination plus the DSAP define the recipient of a packet. The same applies to the SSAP. 2. Service Advertising Protocol. IPX protocol that provides a means of informing network clients, via routers and servers, of available network resources and services.

SAS—single attachment station. Device attached only to the primary ring of an FDDI ring. Also known as a Class B station. Compare with *DAS*. See also *FDDI*.

SDLC—Synchronous Data Link Control. SNA data link layer communications protocol. SDLC is a bit-oriented, full-duplex serial protocol that has spawned numerous similar protocols, including HDLC and LAPB.

secondary station—In bit-synchronous data link layer protocols such as HDLC, a station that responds to commands from a primary station. Sometimes referred to simply as a *secondary*.

seed router—Router in an AppleTalk network that has the network number or cable range built in to its port descriptor. The seed router defines the network number or cable range for other routers in that network segment and responds to configuration queries from nonseed routers on its connected AppleTalk network, allowing those routers to confirm or modify their configurations accordingly. Each AppleTalk network must have at least one seed router.

segment—1. Section of a network that is bounded by bridges, routers, or switches. 2. In a LAN using a bus topology, a *segment* is a continuous electrical circuit that is often connected to other such segments with repeaters. 3. Term used in the TCP specification to describe a single transport layer unit of information. The terms *datagram*, *frame*, *message*, and *packet* are also used to describe logical information groupings at various layers of the OSI reference model and in various technology circles.

Sequenced Packet Exchange—See *SPX*.

serial transmission—Method of data transmission in which the bits of a data character are transmitted sequentially over a single channel. Compare with *parallel transmission*.

server—Node or software program that provides services to clients.

service access point—See *SAP*.

Service Advertising Protocol—See *SAP*.

session—1. Related set of connection-oriented communications transactions between two or more network devices. 2. In SNA, a logical connection enabling two NAUs to communicate.

session layer—Layer 5 of the OSI reference model. This layer establishes, manages, and terminates sessions between applications and manages data exchange between presentation layer entities. Corresponds to the data flow control layer of the SNA model.

shortest-path routing—Routing that minimizes distance or path cost through application of an algorithm.

simplex—Capability for transmission in only one direction between a sending station and a receiving station. Broadcast television is an example of a simplex technology. Compare with full duplex and half duplex.

single-vendor network—Network using equipment from only one vendor. Single-vendor networks rarely suffer compatibility problems. See also *multivendor network*.

sliding window flow control—Method of flow control in which a receiver gives transmitter permission to transmit data until a window is full. When the window is full, the transmitter must stop transmitting until the receiver advertises a larger window. TCP, other transport protocols, and several data link layer protocols use this method of flow control.

SLIP—Serial Line Internet Protocol. Standard protocol for point-to-point serial connections using a variation of TCP/IP. Predecessor of PPP.

SMI—Structure of Management Information. Document (RFC 1155) specifying rules used to define managed objects in the MIB.

SNA—Systems Network Architecture. Large, complex, feature-rich network architecture developed in the 1970s by IBM. Similar in some respects to the OSI reference model, but with a number of differences. SNA is essentially composed of seven layers. See *data flow control layer*, *data-link control layer*, *path control layer*, *physical control layer*, *presentation services layer*, *transaction services layer*, and *transmission control layer*.

SNMP—Simple Network Management Protocol. Network management protocol used almost exclusively in TCP/IP networks. SNMP provides a means to monitor and control network devices, and to manage configurations, statistics collection, performance, and security.

socket—1. Software structure operating as a communications end point within a network device (similar to a port). 2. Addressable entity within a node connected to an AppleTalk network; sockets are owned by software processes known as *socket clients*. AppleTalk sockets are divided into two groups: SASs, which are reserved for clients such as AppleTalk core protocols, and DASs, which are assigned dynamically by DDP upon request from clients in the node. An AppleTalk socket is similar in concept to a TCP/IP port.

socket number—8-bit number that identifies a socket. A maximum of 254 different socket numbers can be assigned in an AppleTalk node.

source address—Address of a network device that is sending data.

spanning tree—Loop-free subset of a Layer 2 (switched) network topology.

spanning-tree algorithm—Algorithm used by the Spanning-Tree Protocol to create a spanning tree. Sometimes abbreviated as STA.

Spanning-Tree Protocol—Bridge protocol that uses the spanning-tree algorithm, enabling a learning switch to dynamically work around loops in a switched network topology by creating a spanning tree. Switches exchange BPDU messages with other bridges to detect loops, and then remove the loops by shutting down selected switch interfaces. If the primary link fails, a standby link is activated. Refers to both the IEEE 802.1 Spanning-Tree Protocol standard and the earlier Digital Equipment Corporation Spanning-Tree Protocol upon which it is based. The IEEE version supports switch domains and allows the switch to construct a loop-free topology across an extended LAN. The IEEE version is generally preferred over the Digital version. Sometimes abbreviated as STP.

SPF—shortest path first algorithm. Routing algorithm that iterates on length of path to determine a shortest-path spanning tree. Commonly used in link-state routing algorithms. Sometimes called *Dijkstra's algorithm*.

split-horizon updates—Routing technique in which information about routes is prevented from exiting the router interface through which that information was received. Split-horizon updates are useful in preventing routing loops.

spoofing—1. Scheme used by routers to cause a host to treat an interface as if it were up and supporting a session. The router spoofs replies to keepalive messages from the host in order to convince that host that the session still exists. Spoofing is useful in routing environments such as DDR, in which a circuit-switched link is taken down when there is no traffic to be sent across it in order to save toll charges. 2. The act of a packet illegally claiming to be from an address from which it was not actually sent. Spoofing is designed to foil network security mechanisms such as filters and access lists.

SPX—Sequenced Packet Exchange. Reliable, connection-oriented protocol that supplements the datagram service provided by network layer (Layer 3) protocols. Novell derived this commonly used NetWare transport protocol from the SPP of the XNS protocol suite.

SQE—signal quality error. In Ethernet, transmission sent by a transceiver back to the controller to let the controller know whether the collision circuitry is functional. Also called *heartbeat*.

SSAP—source service access point. The SAP of the network node designated in the Source field of a packet. Compare to *DSAP*. See also *SAP*.

standard—Set of rules or procedures that are either widely used or officially specified.

star topology—LAN topology in which end points on a network are connected to a common central switch by point-to-point links. A ring topology that is organized as a star implements a unidirectional closed-loop star, instead of point-to-point links. Compare with *bus topology*, *ring topology*, and *tree topology*.

static route—Route that is explicitly configured and entered into the routing table, by default. Static routes take precedence over routes chosen by dynamic routing protocols.

stub area—OSPF area that carries a default route, intra-area routes, and interarea routes, but does not carry external routes. Virtual links cannot be configured across a stub area, and they cannot contain an ASBR. Compare to *non-stub area*.

subnet—See *subnetwork*.

subnet address—Portion of an IP address that is specified as the subnetwork by the subnet mask.

subnet mask—32-bit address mask used in IP to indicate the bits of an IP address that are being used for the subnet address. Sometimes referred to simply as *mask*.

subnetwork—1. In IP networks, a network sharing a particular subnet address. Subnetworks are networks arbitrarily segmented by a network administrator in order to provide a multilevel, hierarchical routing structure while shielding the subnetwork from the addressing complexity of attached networks. Sometimes called a *subnet*. 2. In OSI networks, a collection of ESs and ISs under the control of a single administrative domain and using a single network access protocol.

SVC—switched virtual circuit. Virtual circuit that is dynamically established on demand and is torn down when transmission is complete. SVCs are used in situations where data transmission is sporadic. Called a *switched virtual connection* in ATM terminology. Compare with *PVC*.

synchronous transmission—Term describing digital signals that are transmitted with precise clocking. Such signals have the same frequency, with individual characters encapsulated in control bits (called start bits and stop bits) that designate the beginning and end of each character. Compare with *asynchronous transmission*.

T

T1—Digital WAN carrier facility. T1 transmits DS-1-formatted data at 1.544 Mbps through the telephone-switching network, using AMI or B8ZS coding. Compare with *E1*.

T3—Digital WAN carrier facility. T3 transmits DS-3-formatted data at 44.736 Mbps through the telephone switching network. Compare with *E3*.

TACACS—Terminal Access Controller Access Control System. Authentication protocol, developed by the DDN community, that provides remote access authentication and related services, such as event logging. User passwords are administered in a central database rather than in individual routers, providing an easily scalable network security solution.

TCP—Transmission Control Protocol. Connection-oriented transport layer protocol that provides reliable full-duplex data transmission. TCP is part of the TCP/IP protocol stack.

TCP/IP—Transmission Control Protocol/Internet Protocol. Common name for the suite of protocols developed by the U.S. DoD in the 1970s to support the construction of worldwide internetworks. TCP and IP are the two best-known protocols in the suite.

Telnet—Standard terminal emulation protocol in the TCP/IP protocol stack. Telnet is used for remote terminal connection, enabling users to log in to remote systems and use resources as if they were connected to a local system. Telnet is defined in RFC 854.

throughput—Rate of information arriving at, and possibly passing through, a particular point in a network system.

timeout—Event that occurs when one network device expects to hear from another network device within a specified period of time but does not. The resulting timeout usually results in a retransmission of information or the dissolving of the session between the two devices.

Time To Live—See TTL.

token—Frame that contains control information. Possession of the token allows a network device to transmit data onto the network.

token bus—LAN architecture using token passing access over a bus topology. This LAN architecture is the basis for the IEEE 802.4 LAN specification.

token passing—Access method by which network devices access the physical medium in an orderly fashion based on possession of a small frame called a token. Contrast with *circuit switching* and *contention*.

Token Ring—Token-passing LAN developed and supported by IBM. Token Ring runs at 4 or 16 Mbps over a ring topology. Similar to IEEE 802.5.

TokenTalk—Apple Computer's data-link product that allows an AppleTalk network to be connected by Token Ring cables.

topology—Physical arrangement of network nodes and media within an enterprise networking structure.

traceroute—Program available on many systems that traces the path a packet takes to a destination. It is mostly used to debug routing problems between hosts. There is also a traceroute protocol defined in RFC 1393.

traffic management—Techniques for avoiding congestion and shaping and policing traffic. Allows links to operate at high levels of utilization by scaling back lower-priority, delay-tolerant traffic at the edge of the network when congestion begins to occur.

trailer—Control information appended to data when encapsulating the data for network transmission. Compare with *header*.

transaction services layer—Layer 7 in the SNA architectural model. Represents user application functions, such as spreadsheets, word-processing, or electronic mail, by which users interact with the network. Corresponds roughly with the application layer of the OSI reference model. See also *data flow control layer, data-link control layer, path control layer, physical control layer, presentation services layer,* and *transmission control layer.*

transmission control layer—Layer 4 in the SNA architectural model. This layer is responsible for establishing, maintaining, and terminating SNA sessions, sequencing data messages, and controlling session level flow. Corresponds to the transport layer of the OSI model. See also *data flow control layer, data-link control layer, path control layer, physical control layer, presentation services layer,* and *transaction services layer.*

Transmission Control Protocol—See *TCP.*

transport layer—Layer 4 of the OSI reference model. This layer is responsible for reliable network communication between end nodes. The transport layer provides mechanisms for the establishment, maintenance, and termination of virtual circuits, transport fault detection and recovery, and information flow control. Corresponds to the transmission control layer of the SNA model. See also *application layer, data link layer, network layer, physical layer, presentation layer,* and *session layer.*

trap—Message sent by an SNMP agent to an NMS, console, or terminal to indicate the occurrence of a significant event, such as a specifically defined condition or a threshold that was reached.

tree topology—LAN topology similar to a bus topology, except that tree networks can contain branches with multiple nodes. Transmissions from a station propagate the length of the medium and are received by all other stations. Compare with *bus topology, ring topology,* and *star topology.*

TTL—Time To Live. Field in an IP header that indicates how long a packet is considered valid.

tunneling—Architecture that is designed to provide the services necessary to implement any standard point-to-point encapsulation scheme.

U

UDP—User Datagram Protocol. Connectionless transport layer protocol in the TCP/IP protocol stack. UDP is a simple protocol that exchanges datagrams without acknowledgments or guaranteed delivery, requiring that error processing and retransmission be handled by other protocols. UDP is defined in RFC 768.

unicast—Message sent to a single network destination. Compare with *broadcast* and *multicast.*

unicast address—Address specifying a single network device. Compare with *broadcast address* and *multicast address.*

URL—universal resource locator. Standardized addressing scheme for accessing hypertext documents and other services using a browser.

V

VINES—Virtual Integrated Network Service. NOS developed and marketed by Banyan Systems.

virtual circuit—Logical circuit created to ensure reliable communication between two network devices. A virtual circuit is defined by a VPI/VCI pair and can be either permanent (PVC) or switched (SVC). Virtual circuits are used in Frame Relay and X.25. In ATM, a virtual circuit is called a virtual channel. Sometimes abbreviated *VC.*

W

WAN—wide-area network. Data communications network that serves users across a broad geographic area and often uses transmission devices provided by common carriers. Frame Relay, SMDS, and X.25 are examples of WANs. Compare with *LAN* and *MAN.*

watchdog packet—Used to ensure that a client is still connected to a NetWare server. If the server has not received a packet from a client for a certain period of time, it sends that client a series of watchdog packets. If the station fails to respond to a predefined number of watchdog packets, the server concludes that the station is no longer connected and clears the connection for that station.

watchdog spoofing—Subset of spoofing that refers specifically to a router acting especially for a NetWare client by sending watchdog packets to a NetWare server to keep the session between client and server active. Useful when the client and server are separated by a DDR WAN link.

watchdog timer—1. Hardware or software mechanism that is used to trigger an event or an escape from a process unless the timer is periodically reset. 2. In NetWare, a timer that indicates the maximum period of time that a server will wait for a client to respond to a watchdog packet. If the timer expires, the server sends another watchdog packet (up to a set maximum).

X

X.25—ITU-T standard that defines how connections between DTE and DCE are maintained for remote terminal access and computer communications in PDNs. X.25 specifies LAPB, a data link layer protocol, and PLP, a network layer protocol. Frame Relay has to some degree superseded X.25.

Z

ZIP—Zone Information Protocol. AppleTalk session layer protocol that maps network numbers to zone names. ZIP is used by NBP to determine which networks contain nodes that belong to a zone.

zone—In AppleTalk, a logical group of network devices.

zone multicast address—Data-link-dependent multicast address at which a node receives the NBP broadcasts directed to its zone.

Index